Lecture Notes in Computer Science 3104

Commenced Publication in 1973
Founding and Former Series Editors:
Gerhard Goos, Juris Hartmanis, and Jan van Leeuwen

Editorial Board

T0218651

Springer

Berlin
Heidelberg
New York
Hong Kong
London
Milan
Paris
Tokyo

Rastislav Královič   Ondrej Sýkora (Eds.)

Structural Information and Communication Complexity

11th International Colloquium, SIROCCO 2004
Smolenice Castle, Slowakia, June 21-23, 2004
Proceedings

Springer

Volume Editors

Rastislav Královič
Comenius University, Department of Computer Science
84248 Bratislava, Slowakia
E-mail: kralovic@dcs.fmph.uniba.sk

Ondrej Sýkora
Loughborough University, Department of Computer Science
Loughborough, Leicestershire LE11 3TU, UK
E-mail: o.sykora@lboro.ac.uk

Library of Congress Control Number: 2004107846

CR Subject Classification (1998): F.2, C.2, G.2, E.1

ISSN 0302-9743
ISBN 3-540-22230-8 Springer-Verlag Berlin Heidelberg New York

Springer-Verlag is a part of Springer Science+Business Media

springeronline.com

© Springer-Verlag Berlin Heidelberg 2004
Printed in Germany

Typesetting: Camera-ready by author, data conversion by Olgun Computergrafik
Printed on acid-free paper   SPIN: 11012894   06/3142   5 4 3 2 1 0

Preface

The Colloquium on Structural Information and Communication Complexity (SIROCCO) is an annual meeting focused on the relationship between computing and communication. Over its 11 years of existence, SIROCCO has gained a considerable respect and has become an acknowledged forum bringing together specialists interested in the fundamental principles underlying all computing through communication.

SIROCCO 2004 was the 11th in this series, held in Smolenice Castle, June 21–23, 2004. Previous SIROCCO colloquia took place in Ottawa (1994), Olympia (1995), Siena (1996), Ascona (1997), Amalfi (1998), Lacanau-Océan (1999), L'Aquila (2000), Val de Nuria (2001), Andros (2002), and Umeå(2003). The colloquium in 2004 was special in the respect that, for the first time, the proceedings were published in the Lecture Notes in Computer Science series of Springer–Verlag.

SIROCCO has always encouraged high-quality research focused on the study of those factors which are significant for the computability and the communication complexity of problems, and on the interplay between structure, knowledge, and complexity. It covers topics as distributed computing, mobile computing, optical computing, parallel computing, communication complexity, information dissemination, routing protocols, distributed data-structures, models of communication, network topologies, high-speed interconnection networks, wireless networks, sense of direction, structural properties, and topological awareness. The 56 contributions submitted to this year's SIROCCO were subject to a thorough refereeing process and 26 high quality submissions were selected for publication. We thank the Program Committee members for their profound and careful work. Our gratitude extends to the numerous subreferees for their valuable refereeing. We also acknowledge the effort of all authors who submitted their contributions.

We thank the invited speakers at this colloquium, Paul Spirakis (Patras) and Shmuel Zaks (Haifa) for accepting our invitation to share their insights on new developments in their areas of interest. Paul Spirakis delivered a talk about "*Algorithmic aspects in congestion games*" and Shmuel Zaks presented "*Results and research directions in ATM and optical networks*".

We would like to express our sincere gratitude to the conference chair David Peleg (Rehovot) for his enthusiasm and invaluable consultations, and to the organizing team chaired by Dana Pardubská and Imrich Vrťo.

Our deepest respect belongs to our late friend and colleague Peter Ružička who started the preparation of SIROCCO 2004.

June 2004

Rastislav Královič
Ondrej Sýkora

Organization

Conference Chair:	David Peleg (Rehovot)
Program Chairs:	Rastislav Královič (Bratislava)
	Ondrej Sýkora (Loughborough)
Local Arrangements:	Dana Pardubská (Comenius University)
	Imrich Vrťo (Slovak Academy of Sciences)

Program Committee

Bernadette Charron-Bost (Palaiseau)
Bogdan Chlebus (Denver),
Francesc Comellas (Barcelona),
Krzysztof Diks (Warsaw),
Stefan Dobrev (Ottawa),
Robert Elssässer (Paderborn),
Michele Flammini (L'Aquila),
Cyril Gavoille (Bordeaux),
Evangelos Kranakis (Carleton),

Luděk Kučera (Praha),
Jan van Leeuwen (Utrecht),
Marios Mavronicolas (Nicosia),
Linda Pagli (Pisa),
Heiko Schröder (Melbourne),
Iain Stewart (Durham),
Savio Tse (Hongkong),
Peter Widmayer (Zurich),
Shmuel Zaks (Haifa)

Referees

Luzi Anderegg
Vittorio Bilò
Maurizio Bonuccelli
Hajo Broersma
Valentina Ciriani
Robert Dabrowski
O. Delmas
Charles Delorme
Joerg Derungs
Y. Dourisboure
Eran Edirisinghe
Guillaume Fertin
Faith Fich
Pierre Fraigniaud
Martin Gairing
Leszek Gąsieniec
Michael Gatto
E. Godard
N. Hanusse
Jan van den Heuvel
Michael Hoffman
Roy S. C. Ho
Ralf Klasing

Lukasz Kowalik
Richard Královič
Slawomir Lasota
Fessant Fabrice Le
Ulf Lorenz
Thomas Lücking
David Manlove
Bernard Mans
Ján Maňuch
Giovanna Melideo
Marcin Mucha
Alfredo Navarra
Marc Nunkesser
Dana Pardubská
Paolo Penna
Iain Phillips
Nadia Pisanti
Giuseppe Prencipe
Geppino Pucci
André Raspaud
Stefan Rührup
N. Saheb
Piotr Sankowski

Mordechai Shalom
Stefan Schamberger
Riccardo Silvestri
Ladislav Stacho
Martin Stanek
Gabor Szabo
Gerard Tel
S. Tixeuil
Pavel Tvrdík
Pavol Ďuriš
L. Viennot
Klaus Volbert
Imrich Vrťo
Martin Škoviera
Tomasz Walen
Rui Wang
Mirjam Wattenhofer
Birgitta Weber
Jennifer Welch
Andreas Woclaw
David Wood
Qin Xin
A. Zemmari

Table of Contents

Traffic Grooming
in a Passive Star WDM Network

Eric Angel, Evripidis Bampis, and Fanny Pascual

LaMI – Université d'Évry Val d'Essonne,
CNRS UMR 8042 – 523 Place des Terrasses,
91000 Évry, France
{angel,bampis,fpascual}@lami.univ-evry.fr

Abstract. We consider the traffic grooming problem in passive WDM star networks. Traffic grooming is concerned with the development of techniques for combining low speed traffic components onto high speed channels in order to minimize network cost. We first prove that the traffic grooming problem in star networks is NP-hard for a more restricted case than the one considered in [2]. Then, we propose a polynomial time algorithm for the special case where there are two wavelengths per fiber using matching techniques. Furthermore, we propose two reductions of our problem to two combinatorial optimization problems, the *constrained multiset multicover problem* [3], and the *demand matching problem* [4] allowing us to obtain a polynomial time H_{2C} (resp. $2 + \frac{4}{5}$) approximation algorithm for the minimization (resp. maximization) version of the problem, where C is the capacity of each wavelength.

Keywords: star, traffic grooming, WDM network, approximation, algorithm.

1 Introduction

Recently, in order to utilize bandwidth more effectively, new models appeared allowing several independent traffic streams to share the bandwidth of a lightpath. It is in general impossible to setup lightpaths between every pair of edge routers and thus it is natural to consider that traffic is electronically switched (groomed) onto new lightpaths toward the destination node. The introduction of electronic switching increases the degree of connectivity among the edge routers while at the same time it may reduce significantly wavelength requirements for a given traffic demand. The drawback of this approach is that the introduction of expensive active components (optical transceivers and electronic switches) may increase the cost of the network. These observations motivated R. Dutta and G.N. Rouskas [2] to study the traffic grooming problem that we consider in this paper in order to find a tradeoff between the cost of the network and its performance.

   We focus on star networks composed by a set of transmitters, a set of receivers and a hub, and the goal is to minimize the *total amount of electronic switching*. This cost function measures exactly the amount of electronic switching inside

R. Královič and O. Sýkora (Eds.): SIROCCO 2004, LNCS 3104, pp. 1–12, 2004.

the network (but it only indirectly captures the transceiver cost). Our interest to star networks besides their simplicity, which allows us to provide the first approximation algorithms with performance guarantee for this variant of the traffic grooming problem, is also motivated by their use in the interconnection of LANs or MANs with a wide area backbone.

1.1   Problem Definition

We consider a network in the form of a star with $N + 1$ nodes. There is a single *hub* node which is connected to every other node by a physical link. All the nodes, except the hub, are divided into two groups V_1 and V_2: the nodes in V_1 are the transmitters and the nodes in V_2 are the receivers. The hub is numbered 0 and the N other nodes are numbered from 1 to N in some arbitrary order. Each physical link consists of a fiber, and each fiber can carry W wavelengths. Each wavelength has a capacity C, expressed in units of some arbitrary rate. We represent a traffic pattern by a demand matrix $T = [t_{ij}]$, where integer t_{ij} denotes the number of traffic streams (each unit demand) from node $i \in V_1$ to node $j \in V_2$. We do not allow the traffic demands to be greater than the capacity of a lightpath, i.e. for all $(i, j), 0 \le t_{ij} \le C$.

The hub has both optical and electronic switching capabilities: it let some lightpaths pass through transparently, while it may terminate some others. Traffic on terminated lightpaths is electronically switched (groomed) onto a new lightpath towards the destination node. A traffic demand (or request) t_{ij} must have its own wavelength from i to the hub and from the hub to j to be optically routed, whereas it can share a wavelength with some other traffic demands if it is electronically switched. The goal we consider in this paper is to *minimize the total amount of electronic switching at the hub*. This problem is often called *the traffic grooming problem*.

R. Dutta and G.N. Rouskas considered in [2] the traffic grooming problem in several network topologies, including a star network. However there are differences between their problem and ours: in [2], each node of the network, including the hub, can be a transmitter and a receiver, and traffic demands between two nodes are allowed to be greater than the capacity of a wavelength (i.e. it is possible that $t_{ij} > C$ for some i, j). To distinguish the two problems, we will call their problem the traffic grooming problem in an active star, and our problem *the traffic grooming problem in a passive star* (see Section 4 for an integer linear programming formulation of the problem). Once we know which traffic demands are optically routed, the wavelength assignment problem is easy in the case of a passive star network.

There are in fact two versions of the traffic grooming problem: either we want to minimize the total amount of electronic switching at the hub (this is the *minimization* version), or we want to maximize the total amount of traffic which is optically routed (this is the *maximization* version). These two versions are equivalent (i.e. an optimal solution for one is also an optimal solution for the other one) because the optimal solution of the maximization problem is equal to the sum of all the traffic demands, minus the optimal solution of the minimization problem.

Our results are as follows. First, we show in Section 2 that the traffic grooming problem in a passive star is NP-Complete, in both the minimization and the maximization versions of the problem. Then we show in Section 3 that these problems are polynomially solvable if there are only two wavelengths per fiber ($W = 2$): we give an algorithm which gives an optimal solution. In Section 4, we show that we cannot deduce a constant approximation guarantee of the maximization (resp. minimization) version from a constant approximation guarantee of the minimization (resp. maximization) version of the problem, and we give two approximation algorithms. The first one concerns the minimization version: we transform our problem in a constrained multiset multicover problem [3], and we get an approximation guarantee of H_{2C}. The second approximation algorithm concerns the maximization version: we transform our problem in a demand matching problem in a bipartite graph [4], and we obtain an approximation guarantee of $(2 + \frac{4}{5})$. We conclude the paper in Section 5.

2 Hardness Results

Let us show in this section that the decision version of the grooming problem in a passive star is NP-Complete.

In order to do this proof, we were inspired by the proof of R. Dutta and G.N. Rouskas in [2]: in this paper they showed that their traffic grooming problem is NP-complete. They reduced the decision version of the Knapsack problem to their problem. We do the same reduction, replacing traffic demands t_{ij} greater than C by several traffic demands of the same weight from i, or to j. They also used traffic demands to the hub (or from the hub). We replace these traffic demands by traffic demands to some new nodes (or from some new nodes) and we force these traffic demands to be switched electronically at the hub.

We reduce the decision version of the Knapsack problem [1] to our grooming problem: let $Q \in Z^+$, is there a solution of our grooming problem in which the amount of optically routed traffic is greater or equal to Q? An instance of the Knapsack problem is given by a finite set U of cardinality n, for each element $u_i \in U$ a weight $w_i \in Z^+$, and a value $v_i \in Z^+, \forall i \in \{1, 2, ..., n\}$, a target weight $B \in Z^+$, and a target value $K \in Z^+$. The problem asks whether there exists a binary vector $X = \{x_1, x_2, ..., x_n\}$ such that $\sum_{i=1}^{n} x_i w_i \leq B$, and $\sum_{i=1}^{n} x_i v_i \geq K$. Given such an instance, we construct a star network using the following transformation: $W = n$, $C = max_i(w_i + v_i) + 1$, $Q = K + \sum_{i=1}^{n}(C - w_i - v_i)$ and the traffic matrix is represented on the Figure 1. In this figure the nodes are the nodes of the star network (the hub is not represented), and the links represent the traffic demands. Traffic demands equal to 0 are not represented, and the value on the link from a node a to a node b is $t_{a,b}$. Nodes from $n + 1$ to $n + 10$ represent each one a node of the network, but nodes $i_\alpha, j_\alpha, k_\alpha, l_\alpha, m_\alpha, p_\alpha$ and q_α represent each one several nodes:

For the nodes i_α, α ranges from 1 to n (i.e. i_α represents the nodes $i_1, i_2, ..., i_n$);
for the nodes j_α, α ranges from 1 to $\lfloor \frac{(n-2)C}{C-1} \rfloor$;
for the nodes k_α, α ranges from 1 to $\lfloor \frac{\sum_{k=1}^{n}(w_k - B)}{C-1} \rfloor$;
for the nodes l_α, α ranges from 1 to $\lfloor \frac{nC - (C-1)}{C-1} \rfloor$;
for the nodes m_α, α ranges from 1 to $\lfloor \frac{nC - ((\sum_{k=1}^{n}(w_k - B)) \bmod (C-1))}{C-1} \rfloor$;
for the nodes p_α, α ranges from 1 to $\lfloor \frac{n}{C-1} \rfloor$;
for the nodes q_α, α ranges from 1 to $\lfloor \frac{nC - n((n-2)C \bmod (C-1))}{C-1} \rfloor$;
and for the nodes r_α, α ranges from 1 to $\lfloor \frac{nC - \sum_{k=1}^{n} w_k}{C-1} \rfloor$.

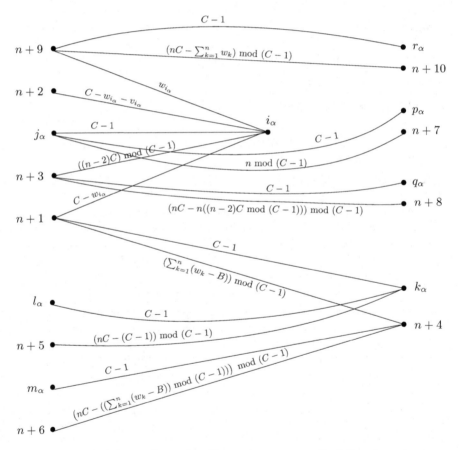

Fig. 1. Illustration of the traffic matrix. Transmitters are on the left and receivers on the right

Lemma 1. *Let a be a transmitter and b a receiver. It is not possible to have a lightpath from a to b, if $(a, b) \neq (n+1, i_\alpha)$ or $(a, b) \neq (n+2, i_\alpha)$.*

Proof. Let us show that each traffic demand different from 0 between each couple (a, b) of nodes in $V_1 \times V_2$ $((a, b) \neq (n+1, i_\alpha)$ and $(n+2, i_\alpha))$, cannot be optically routed. In order to show that it is not possible to route $t_{a,b}$ optically, we will see that either the sum of the traffic streams from a, or the sum of the traffic streams to b, is equal to nC, and that $t_{a,b}$ is smaller than C.

- $\forall x \in V_2, t_{n+9,x}$ cannot be optically routed. Indeed:

$$\sum_{x \in V_2} t_{n+9,x} = \sum_\beta t_{n+9,r_\beta} + t_{n+9,n+10} + \sum_\beta t_{n+9,i_\beta}$$
$$= \lfloor \tfrac{nC - \sum_{k=1}^n w_k}{C-1} \rfloor (C-1) + (nC - \sum_{k=1}^n w_k) \bmod (C-1)$$
$$+ \sum_{k=1}^n w_k$$
$$= nC$$

and $t_{n+9,r_\beta} < C$, $t_{n+9,n+10} < C$, $t_{n+9,i_\alpha} < C$.

- $\forall x \in V_2, t_{j_\alpha,x}$ cannot be optically routed. Indeed:

$$\sum_{x \in V_2} t_{j_\alpha,x} = \sum_\beta t_{j_\alpha,i_\beta} + \sum_\beta t_{j_\alpha,p_\beta} + t_{j_\alpha,n+7}$$
$$= n(C-1) + \lfloor \tfrac{n}{C-1} \rfloor (C-1) + n \bmod (C-1)$$
$$= nC$$

and $t_{j_\alpha,i_\beta} < C$, $t_{j_\alpha,p_\beta} < C$, $t_{j_\alpha,n+7} < C$.

- $\forall x \in V_2, t_{n+3,x}$ cannot be optically routed. Indeed:

$$\sum_{x \in V_2} t_{n+3,x} = \sum_\beta t_{n+3,i_\beta} + \sum_\beta t_{n+3,q_\beta} + t_{n+3,n+8}$$
$$= n((n-2)C \bmod (C-1)) + \lfloor \tfrac{nC - n((n-2)C \bmod (C-1))}{C-1} \rfloor (C-1)$$
$$+ (nC - n((n-2)C \bmod (C-1))) \bmod (C-1)$$
$$= nC$$

and $t_{n+3,i_\alpha} < C$, $t_{n+3,q_\alpha} < C$, $t_{n+3,n+8} < C$.

- $\forall x \in V_1, t_{x,k_\alpha}$ cannot be optically routed. Indeed:

$$\sum_{x \in V_1} t_{x,k_\alpha} = t_{n+1,k_\alpha} + \sum_\beta t_{l_\beta,k_\alpha} + t_{n+5,k_\alpha}$$
$$= (C-1) + \lfloor \tfrac{nC - (C-1)}{C-1} \rfloor (C-1) + (nC - (C-1)) \bmod (C-1)$$
$$= nC$$

and $t_{n+1,k_\alpha} < C$, $t_{l_\beta,k_\alpha} < C$, $t_{n+5,k_\alpha} < C$.

- $\forall x \in V_1, t_{x,n+4}$ cannot be optically routed. Indeed:

$$\sum_{x \in V_1} t_{x,n+4} = \sum_\beta t_{m_\beta,n+4} + t_{n+6,n+4} + t_{n+1,n+4}$$
$$= \tfrac{nC - ((\sum_{k=1}^n (w_k - B)) \bmod (C-1))}{C-1} (C-1) + (\lfloor nC - ((\sum_{k=1}^n (w_k -$$

$$B)) \bmod (C-1)) \rfloor) \bmod (C-1) + (\sum_{k=1}^n (w_k - B)) \bmod (C-1)$$
$$= nC$$

and $t_{n+1,n+4} < C$, $t_{m_\beta,n+4} < C$, $t_{n+6,n+4} < C$. □

Lemma 2. *Let $\alpha \in \{1, ..., n\}$. Traffic demands t_{n+1,i_α} and t_{n+2,i_α} cannot be optically routed simultaneously.*

Proof. The node i_α receives from the hub a total traffic equal to: $t_{n+9,i_\alpha} + \sum_\beta t_{j_\beta,i_\alpha} + t_{n+3,i_\alpha} = w_{i_\alpha} + \lfloor \tfrac{(n-2)C}{C-1} \rfloor (C-1) + ((n-2)C) \bmod (C-1) = (n-2)C + w_{i_\alpha} > (n-2)C$

Since $W = n$, there is at most one wavelength left to have a lightpath to the node i_α. □

Lemma 3. *Let $\alpha \in \{1, ..., n\}$. It is possible to have a lightpath from $n+1$ to i_α, or from $n+2$ to i_α.*

Proof. — Let us show that it is possible to have a lightpath from $n+1$ to i_α:
$\sum_{x \in V_1, x \neq n+1} t_{x,i_\alpha} = (n-2)C + w_{i_\alpha} + (C - w_{i_\alpha} - v_{i_\alpha}) = (n-1)C - v_{i_\alpha} \leq (n-1)C$
and, since $\exists i_\alpha \in \{1, ..., n\}, B \geq w_{i_\alpha}$ (otherwise the instance of the Knapsack problem would be trivial) and $C = max_{i_\alpha}(v_{i_\alpha} + w_{i_\alpha}) + 1$, we have $\sum_{x \in V_2, x \neq i_\alpha} t_{n+1,x} = \sum_{k=1}^{n}(w_k - B) \leq (n-1)C$. So there is enough bandwidth to have a lightpath from $n+1$ to i_α.
  — Let us show that it is possible to have a lightpath from $n+2$ to i_α:
$\sum_{x \in V_1, x \neq n+2} t_{x,i_\alpha} = (n-2)C + w_{i_\alpha} + (C - w_{i_\alpha}) = (n-1)C \leq (n-1)C$
and $\sum_{x \in V_2, x \neq i_\alpha} t_{n+2,x} = 0 \leq (n-1)C$. So there is enough bandwidth to have a lightpath from $n+2$ to i_α.   □

Since t_{n+1,i_α} and t_{n+2,i_α} ($\alpha \in \{1, ..., n\}$) are the only traffic demands which can be optically routed (Lemma 1) and since, for each $\alpha \in \{1, ..., n\}$ it is possible to have a lightpath from $n+1$ to i_α, or from $n+2$ to i_α, (Lemma 3) but not both (Lemma 2), we need only to consider solutions in which there is a lightpath from exactly one of the nodes $n+1$, $n+2$, to each node i_α to determine the satisfiability of the instance.

Let X denote a candidate solution of the Knapsack instance. Consider the solution of the grooming problem in which X (respectively, $\overline{X}$) represents the indicator vector of the lightpaths formed from node $n+1$ (resp., $n+2$). Nodes i_α are numbered from 1 to n: let $\alpha \in \{1, ..., n\}$, we have $i_\alpha = \alpha$. Applying the transformation to the satisfiability criteria of Knapsack, we obtain:

$\sum_{i=1}^{n} x_i w_i \leq B$
$\Leftrightarrow \sum_{i=1}^{n} x_i(C - t_{n+1,i}) \leq \sum_{i=1}^{n}(C - t_{n+1,i}) - (t_{n+1,n+4} + \sum_\beta t_{n+1,k_\beta})$
$\Leftrightarrow \sum_{i=1}^{n}(\overline{x_i} t_{n+1,i}) + (t_{n+1,n+4} + \sum_\beta t_{n+1,k_\beta}) \leq (n - \sum_{i=1}^{n} x_i)C$

This inequality means that the amount of electronically routed traffic demands (the left hand side of the inequality) has to be smaller than or equal to the capacity of a link, C, times the number of links available (i.e. n minus the number of traffic demands which are electronically routed).

$\sum_{i=1}^{n} x_i v_i \geq K$
$\Leftrightarrow \sum_{i=1}^{n} x_i(t_{n+1,i} - t_{n+2,i}) \geq Q - \sum_{i=1}^{n} t_{n+2,i}$
$\Leftrightarrow \sum_{i=1}^{n}(x_i t_{n+1,i} + \overline{x_i} t_{n+2,i}) \geq Q$

This inequality means that the total amount of optically routed traffic has to be greater than or equal to Q.

Therefore, a vector X either satisfies both the Knapsack and the grooming instance, or fails to satisfy both. Hence, the grooming instance is satisfiable if and only if the Knapsack instance is.

Theorem 1. *The decision versions of the minimization and the maximization versions of the grooming problem in a passive star are NP-Complete.*

Proof. We already proved that the decision version of the minimization version of the grooming problem in a passive star is NP-Complete. Since we can easily switch from a version to the other one ($OPT_{max} = (\sum_{i \in V_1, j \in V_2} t_{ij}) - OPT_{min}$), the decision version of the maximization version of the grooming problem in a passive star is also NP-Complete.   □

3 Polynomial Time Algorithm for $W = 2$

Let us show that the grooming problem in a passive star is polynomially solvable when the number of wavelengths on each fiber, W, is equal to 2. We will give a polynomial time algorithm which gives an optimal solution of this problem.

First of all, let us remark that in each row (or column) of T where there is at least three values different from 0, we can at most route one traffic demand optically, because each lightpath which is not electronically switched at the hub needs a wavelength for him only, and we have only two wavelengths per fiber. On the contrary, when there is in a row (or column) only two values different from 0, it may be possible to route optically both. We transform the matrix T in a matrix T' in which it is possible to route optically at most one traffic demand for each row, and one traffic demand for each column: if a row of T has two and only two values t_{ij} and $t_{ij'}$ different from 0, we create two rows $i1$ and $i2$ in T' such that $t'_{i1,j} = t_{ij}$, $t'_{i2,j} = t_{ij'}$ and the other values in these rows are 0. Similarly, if a column of T has two and only two values t_{ij} and $t_{i'j}$ different from 0, we create two columns $j1$ and $j2$ in T' such that $t'_{i,j1} = t_{ij}$, $t'_{i,j2} = t_{i'j}$ and the other values in these columns are 0. In this way, there is at most one request per row and one request per column which can be optically routed.

Let us now transform our traffic matrix T' in another matrix $M = [m_{ij}]$, in which we will look for a maximum weighted matching. In order to do that, we will apply the following rule: if a traffic demand t'_{ij} cannot be optically routed (i.e. $\sum_{k \neq j} t'_{ik} > C$ or $\sum_{k \neq i} t'_{kj} > C$) then $m_{ij} = -\infty$. Otherwise, $m_{ij} = t'_{ij}$. Since there is in M at most one request per row and one request per column which can be optically routed, and since a traffic demand m_{ij} is different from $-\infty$ if and only if it is allowed to be optically routed, the result of a maximum weighted matching in the bipartite graph whose adjacent matrix is M, is an optimal result for the traffic grooming problem whose traffic matrix is T.

Theorem 2. *The minimization and the maximization versions of the traffic grooming problem in a passive star are polynomially solvable if the number of wavelengths per fiber is equal to two.*

Proof. The optimal solution of the minimization version and the optimal solution of the maximization version are the same, and the algorithm above solves polynomially these problems.   □

4 Approximation Algorithms

Theorem 3. *It is not possible to deduce a constant approximation guarantee for the maximization (resp. minimization) version of the traffic grooming prob-*

lem in a passive star network from a constant approximation algorithm for the minimization (resp. maximization) version.

Proof. Let OPT_{max} be the cost of an optimal solution of the maximization version (i.e. the maximum amount of traffic which can be optically routed) and OPT_{min} the cost of an optimal solution of the minimization version (i.e. the minimum amount of traffic which is electronically switched at the hub in a feasible solution). Let S be a solution of the traffic grooming problem in a passive star. Let denote $c_{max}(S)$ the cost of the maximization version (i.e. $c_{max}(S)$ is the amount of traffic which is optically routed in the solution S) and $c_{min}(S)$ the cost of the minimization version (i.e. $c_{min}(S)$ is the amount of traffic which is electronically switched at the hub in the solution S).

Let ε_1 and ε_2 be two real numbers such that $0 \le \varepsilon_1 < 1$ and $0 \le \varepsilon_2$. Let us show that it is not possible to deduce a $(1 - \varepsilon_1)$-approximation guarantee for the maximization version from a $(1 + \varepsilon_2)$-approximation algorithm for the minimization version: Consider an instance I of the problem, where we can at most route only one traffic stream optically ($OPT_{max} = 1$ and $OPT_{min} = (\sum_{i \in V_1, j \in V_2} t_{ij}) - 1$). Consider a solution S of I where all the traffic demands are electronically switched at the hub ($c_{max}(S) = 0$ and $c_{min}(S) = \sum_{i \in V_1, j \in V_2} t_{ij}$). If $\sum_{i \in V_1, j \in V_2} t_{ij}$ is large enough, we have $\frac{c_{min}(S)}{OPT_{min}} = \frac{\sum_{i \in V_1, j \in V_2} t_{ij}}{(\sum_{i \in V_1, j \in V_2} t_{ij}) - 1} \le 1 + \varepsilon_2$ but there is no $\varepsilon_1 < 1$ such that $\frac{c_{max}(S)}{OPT_{max}} = \frac{0}{1}$ is greater than or equal to $1 - \varepsilon_1$.

Let us show that it is possible to have an instance such that $\sum_{i \in V_1, j \in V_2} t_{ij}$ is as big as we wish and where we can at most route one traffic stream optically: consider the instance where we have $2W$ transmitters, 2 receivers, and the traffic matrix $[t_{ij}]$ is such that $t_{1,1} = 1$; $\forall i \in \{2, 3, ..., 2W\}$, $t_{i,1} = 0$; and $\forall i \in \{1, ..., 2W\}$, $t_{i,2} = \frac{C}{2}$. The only traffic demand which can be optically routed is $t_{1,1}$ because the second receiver receives WC traffic streams, and each traffic demand is different from C.

Similarly, let us show that it is not possible to deduce a $(1+\varepsilon_2)$-approximation guarantee for the minimization version from a $(1 - \varepsilon_1)$-approximation algorithm for the maximization version: Consider an instance I of the problem, where all the traffic demands can be optically routed ($OPT_{max} = \sum_{i \in V_1, j \in V_2} t_{ij}$ and $OPT_{min} = 0$). It is trivial that such an instance exists. Consider a solution S of I where all the traffic streams, except two, are optically routed ($c_{max}(S) = (\sum_{i \in V_1, j \in V_2} t_{ij}) - 2$ and $c_{min}(S) = 2$). If $\sum_{i \in V_1, j \in V_2} t_{ij}$ is large enough, we have $\frac{c_{max}(S)}{OPT_{max}} = \frac{(\sum_{i \in V_1, j \in V_2} t_{ij}) - 2}{\sum_{i \in V_1, j \in V_2} t_{ij}} \ge 1 - \varepsilon_1$ but there is no $\varepsilon_2 \ge 0$ such that $\frac{c_{min}(S)}{OPT_{min}} = \frac{2}{0}$ is smaller than or equal to $1 + \varepsilon_2$. □

4.1   Approximation Algorithm for the Minimization Version

Let us give an integer programming formulation of the minimization version of the traffic grooming problem in a passive star. Let denote $x_{ij} \in \{0, 1\}$ a variable which indicates whether the traffic demand t_{ij} is optically routed ($x_{ij} = 0$) or electronically switched at the hub ($x_{ij} = 1$). The objective is to:

$$\text{Minimize} \sum_{i \in V_1, j \in V_2} x_{ij} t_{ij} \tag{1}$$

We have two types of constraints:

constraints on the frequencies:

$$\forall i \in V_1, \sum_{j \in V_2} x_{ij} \geq |V_2| - W \tag{2}$$

$$\forall j \in V_2, \sum_{i \in V_1} x_{ij} \geq |V_1| - W \tag{3}$$

constraints on the traffic:

$$\forall i \in V_1, \sum_{j \in V_2} (1 - x_{ij})(C - t_{ij}) \leq WC - \sum_{j \in V_2} t_{ij} \tag{4}$$

$$\forall j \in V_2, \sum_{i \in V_1} (1 - x_{ij})(C - t_{ij}) \leq WC - \sum_{i \in V_1} t_{ij} \tag{5}$$

Inequalities (2) (resp.(3)) mean that at most W traffic demands per transmitter (resp. receiver) can be optically routed, because we need one wavelength for each traffic demand optically routed. Inequalities (4) and (5) mean that the unused space ($C - t_{ij}$) left when t_{ij} is optically routed, has to be smaller than the free space available (i.e. the total amount of bandwidth minus the total amount of traffic demands). It is also equivalent to say that the amount of electronically routed traffic demands has to be smaller than the capacity of a link, C, times the number of links available (i.e. W minus the number of traffic demands which are optically routed).

Constrains (4) and (5) are equivalent to:

$$\forall i \in V_1, \sum_{j \in V_2} x_{ij}(C - t_{ij}) \geq C(|V_2| - W) \tag{6}$$

$$\forall j \in V_2, \sum_{i \in V_1} x_{ij}(C - t_{ij}) \geq C(|V_1| - W) \tag{7}$$

Note that the constraints on the traffic imply the constraints on the frequencies: if we divide by C the constraints on the frequencies we have:

$$\forall i \in V_1, \sum_{j \in V_2} x_{ij}(1 - \frac{t_{ij}}{C}) \geq |V_2| - W \tag{8}$$

$$\forall j \in V_2, \sum_{i \in V_1} x_{ij}(1 - \frac{t_{ij}}{C}) \geq |V_1| - W \tag{9}$$

Since $t_{ij} \leq C$, these last inequalities imply the constraints on the frequencies (2) and (3).

So we can formulate our problem in the following way:

$$\text{Minimize} \sum_{i \in V_1, j \in V_2} x_{ij} t_{ij} \tag{10}$$

subject to:

$$\forall i \in V_1, \sum_{j \in V_2} x_{ij}(C - t_{ij}) \geq C(|V_2| - W) \tag{11}$$

$$\forall j \in V_2, \sum_{i \in V_1} x_{ij}(C - t_{ij}) \geq C(|V_1| - W) \tag{12}$$

$$\forall i \in V_1, j \in V_2, x_{ij} \in \{0, 1\} \tag{13}$$

Theorem 4. *There exists a polynomial time H_{2C}-approximation algorithm for the minimization version of the traffic grooming problem in a passive star.*

Proof. We will transform the minimization version of the traffic grooming problem in a passive star into a constrained multiset multicover problem. Given a universal set $\mathcal{U}$, a collection of subsets of $\mathcal{U}$, $T = \{S_1, S_2, ..., S_k\}$, and a cost function $c : T \to Q^+$, the *set cover* problem asks for a minimum cost sub-collection $\mathcal{C} \in T$ that covers all the elements of $\mathcal{U}$ (i.e. $\bigcup_{S \in \mathcal{C}} S = \mathcal{U}$). The *multiset multicover* is a natural generalization of the set cover problem: in this problem each element e occurs in a multiset S with arbitrary multiplicity denoted $m(S, e)$, and each element e has an integer coverage requirement r_e, which specifies how many times e has to be covered. In the *constrained multiset multicover* problem, each subset $S \in T$ is chosen at most once. Thus the integer program is: Minimize $\sum_S c(S) x_S$ subject to $\sum m(S, e) x_S \geq r_e$ and $x_S \in \{0, 1\}$.

Let us show the transformation of the traffic grooming problem into the constrained multiset multicover problem. This transformation comes directly from the integer programming formulation of the problem: for each request from the transmitter $i \in V_1$, denoted by e_i, to the receiver $j \in V_2$, denoted by r_j, we create the subset S_{ij} which contains $(C - t_{ij})$ times e_i and $(C - t_{ij})$ times r_j. The cost of this subset is $c(S_{ij}) = t_{ij}$. The covering requirement of each element $e \in V_1$ is $r_e = C(|V_2| - W)$, and the covering requirement of each element $e \in V_2$ is $r_e = C(|V_1| - W)$.

S. Rajagopalan and V. Vazirani give in [3] a greedy approximation algorithm for the constrained multiset multicover problem. This algorithm consists in iteratively picking the most cost-effective set from T and removing this set from T. The cost-effectiveness of a set S is the average cost at which it covers new elements, i.e. the cost of S divided by the number of its elements which are not yet covered. They proved that this algorithm has an approximation guarantee of H_k, the k-th harmonic number (i.e. $H_k = 1 + \frac{1}{2} + ... + \frac{1}{k}$), where k is the size of the largest multiset in the given instance. In our case, the size of a multiset

S_{ij} is $2C - 2t_{ij}$, which is smaller than $2C$. So we obtain a solution of the traffic grooming problem which has an approximation guarantee of $H_{2C} \leq log(2C) + 1$.

$\square$

4.2   Approximation Algorithm for the Maximization Version

The maximization version of the traffic grooming problem in a passive star is the following one:

$$\text{Maximize} \sum_{i \in V_1, j \in V_2} y_{ij} t_{ij} \tag{14}$$

subject to:

$$\forall i \in V_1, \sum_{j \in V_2} y_{ij}(C - t_{ij}) \leq CW - \sum_{j \in V_2} t_{ij} \tag{15}$$

$$\forall j \in V_2, \sum_{i \in V_1} y_{ij}(C - t_{ij}) \leq CW - \sum_{i \in V_1} t_{ij} \tag{16}$$

$$\forall i \in V_1, j \in V_2, y_{ij} \in \{0, 1\} \tag{17}$$

Here y_{ij} indicates whether t_{ij} is optically routed ($y_{ij} = 1$) or electronically switched at the hub ($y_{ij} = 0$). This integer programming formulation of the problem is obtained by replacing x_{ij} in the integer programming formulation of the minimization version by $1 - y_{ij}$.

Theorem 5. *There exists a polynomial time $(2 + \frac{4}{5})$-approximation algorithm for the maximization version of the traffic grooming problem in a passive star.*

Proof. Let us now transform this problem into a *demand matching problem* [4]. The demand matching problem is the following one: take a graph $G = (V, E)$ and let each node $v \in V$ have an integral *capacity*, denoted by b_v. Let each edge $e = (u, v) \in E$ have an integral *demand*, denoted by d_e. In addition, associated with each edge $e \in E$ is a *profit*, denoted by p_e. A *demand matching* is a subset $M \subseteq E$ such that $\sum_{e \in \delta(v) \cap M} d_e \leq b_v$ for each node v. Here $\delta(v)$ denotes the set of edges of G incident to v. The demand matching problem is to find a demand matching which maximizes $\sum_{e \in M} p_e$. Thus the integer program is: Maximize $\sum_{e \in E} \frac{p_e}{d_e} x_e$ subject to: $\forall v \in V, \sum_{e \in \delta(v)} x_e \leq b_v$ and $\forall e \in E, x_e \in \{0, d_e\}$.

B. Shepherd and A. Vetta showed in [4] that a randomized algorithm provides a factor $(2 + \frac{4}{5})$-approximation guarantee for the demand matching problem in bipartite graphs.

Let us show the transformation of the traffic grooming problem in a passive star into the demand matching problem: $\forall e \in (V_1 \times V_2)$, $x_e = y_e(C - t_e)$, $p_e = t_e$, $d_e = C - t_e$. $\forall v \in V_1$, $b_v = CW - \sum_{j \in V_2} t_{vj}$ and $\forall v \in V_2$, $b_v = CW - \sum_{i \in V_1} t_{iv}$.

Since there is a factor $(2 + \frac{4}{5})$-approximation algorithm for the demand matching problem, there is also a factor $(2 + \frac{4}{5})$-approximation algorithm for the traffic grooming problem in a passive star.

$\square$

5   Conclusion

We showed in this paper that the traffic grooming problem in a passive star is NP-Complete, in both the minimization and the maximization versions of the problem. We showed that these problems are polynomially solvable if there are two wavelengths per fiber: we gave an algorithm which gives an optimal solution. We showed that we cannot deduce a constant approximation guarantee of the maximization (resp. minimization) version from a constant approximation guarantee of the minimization (resp. maximization) version of the problem. We gave two approximation algorithms and we obtained an approximation guarantee of H_{2C} for the minimization version and an approximation guarantee of $(2 + \frac{4}{5})$ for the maximization version. Since the solutions returned by these algorithms are all solutions of the maximization version as well as solutions of the minimization version of the problem, it would be interesting to program these algorithms and compare their results.

References

1. M.R. Garey, D.S. Johnson, Computers and Intractability. H.Freeman and Co., New York, 1979.
2. R. Dutta, G.N. Rouskas, On Optimal Traffic Grooming in WDM Rings. IEEE Journal on selected areas in communications, Vol. 20, No. 1, January 2002.
3. S. Rajagopalan, V.V. Vazirani, Primal-dual RNC approximation algorithms for set cover and covering integer programs. SIAM Journal on Computing, 28:526-541, 1999.
4. B. Shepherd, A. Vetta, The demand matching problem. IPCO 2002, published in Lecture Notes in Computer Science Vol. 2337, pages 457-474, 2002.

The Price of Anarchy in All-Optical Networks

Vittorio Bilò and Luca Moscardelli

Dipartimento di Informatica
Università di L'Aquila
Via Vetoio, loc. Coppito 67010 L'Aquila
{bilo,moscardelli}@di.univaq.it

Abstract. In this paper we consider all-optical networks in which a service provider has to satisfy a given set of communication requests. Each request is charged a cost depending on its wavelength and on the wavelengths of the other requests met along its path in the network. Under the assumption that each request is issued by a selfish agent, we seek for payment strategies which can guarantee the existence of a pure Nash equilibrium, that is an assignment of paths to the requests so that no request can lower its cost by choosing a different path in the network. For such strategies, we bound the loss of performance of the network (price of anarchy) by comparing the number of wavelengths used by the worst pure Nash equilibrium with that of a centralized optimal solution.

1 Introduction

In the last years optics has emerged as a key technology in communication networks due to the promise of data transmission rates several orders of magnitudes higher than conventional technologies [3, 11, 12, 19, 20].

All-optical communications networks exploit photonic technologies for the implementation of both switching and transmission functions [10]. These systems provide all source-destination pairs with end-to-end transparent channels that are identified through a wavelength and a physical path. Maintaining the signal in optical form allows for high data transmission rates in these networks since there is no conversion to and from the electronic form.

The high bandwidth of the optical fiber is utilized through the *wavelength-division multiplexing* (WDM) [4] approach in which two signals connecting source-destination pairs may share a link, provided they are transmitted on carriers having different wavelengths (or colors) of light, see [1, 9] for a survey of the main related results.

We consider all-optical networks in which a service provider has to satisfy a given set of communication patterns, where a pattern can be seen as a request for a connection made by a source-destination pair. Each pair is assigned a path on the network and is charged a cost depending on its wavelength and on the wavelengths of the other requests met along its path in the network. We assume that each request is issued by a selfish agent who is interested only in the minimization of his own cost. Under this assumption any request is willing

R. Královič and O. Sýkora (Eds.): SIROCCO 2004, LNCS 3104, pp. 13–22, 2004.

to be rerouted each time it may be served by a cheaper path in the network and the evolution of the network can be modelled as a multi-player game. A routing strategy, that is an assignment of paths to the requests, in which no request can lower its cost by choosing a different path is said to be a pure Nash equilibrium. Several games [5, 8, 15, 21, 23] have been shown to possess pure Nash equilibria either by using an inductive argument or by defining a suitable potential function decreasing each time a player performs an improving defection. The latter approach, in particular, is useful to prove a stronger result, that is the convergence of the game, independently from its starting state, to a pure Nash equilibrium, see [6] for an interesting discussion.

Clearly, because of the lack of cooperation among the players, Nash equilibria are known not to always optimize the overall performance, with the Prisoner's Dilemma [17, 18] being the best-known example. To this aim, in [13] such a loss of performance is defined as the ratio between the cost of the worst Nash equilibrium and that of an optimal centralized solution and is called *price of anarchy* or *coordination ratio*. Bounding the price of anarchy of selfish routing in different models is now arising as one of the most interesting research areas lying on the boundary between Computer Science and Game Theory, see for example [14, 22].

In this paper we present four reasonable paying strategies that a service provider can adopt in all-optical networks in order to keep the overall performances of the network (modelled as the number of used wavelengths) acceptable, assuming that users are selfish agents who may not cooperate. We determine which payment strategy induces a game converging to a pure Nash equilibrium. For these such strategies, however, we show instances for which the price of anarchy is arbitrarily high.

The paper is organized as follows. In the next section we give the basic definitions and notation used throughout the paper, in section 3 we present some results for the proposed cost measures and finally, in the last section, we address some interesting open problems.

2   Definitions and Notation

We model a network topology by an undirected graph $G = (V, E)$ where V represents the set of sites and each undirected edge $\{x, y\} \in E$ between two nodes $x, y \in V$ represents a link between the sites.

In all-optical communication networks, sites (nodes) communicate via sending signals of different wavelengths (colors) along optical fiber lines (undirected edges). A point-to-point communication requires to establish a uniquely colored path between the two nodes whose color is different from the colors of all the other paths sharing one of its edges.

In the sequel we will use the following notation:

- $P(x, y)$ denotes a path in G from the node x to y, that is a set of consecutive edges beginning in x and ending in y.

- A request is a source-destination pair of nodes (x, y) in G (corresponding to a communication between x and y).
- An instance I is a collection of requests. Note that a given request (x, y) can appear more than once in an instance.
- A routing R for an instance I in G is a set of paths $R = \{P(x, y) | (x, y) \in I\}$. Let c_R be the color function associated to the routing R, i.e. given $(x, y) \in I$, $c_R((x, y))$ is the color of the path $P(x, y) \in R$ and $c(R)$ is the number of colors used by R.
- The conflict graph associated to a routing R is the undirected graph (R, E) with vertex set R and such that two paths of R are adjacent if and only if they share an edge of G.
- The set of available wavelengths (or colors) is $W = \{1, 2, 3, \ldots\}$. $f : W \to \mathbb{R}$ is a pricing function for the colors. Without loss of generality we assume f non decreasing and positive.
- We denote as p a payment function that, given a pricing function on the colors f and a routing R, associates a cost $p((x, y))$ to any request $(x, y) \in I$.

Let G be a graph and I an instance. The problem (G, I) asks the determination of a routing R for the instance I and assigning each request $(x, y) \in I$ a wavelength, so that no two paths of R sharing an edge have the same wavelength. If we think of wavelength as colors, the problem (G, I) seeks a routing R and a vertex coloring of the conflict graph (R, E), such that two adjacent vertices are colored differently.

A game with $n \geq 2$ players is a finite set of actions S_i for each player, and a payoff function u_i for each player mapping $S_1 \times \ldots \times S_n$ to the integers. The elements of $S_1 \times \ldots \times S_n$ will be called action combinations or states. A pure Nash equilibrium is a state $s = (s_1, \ldots, s_n)$ such that for each i, $u_i(s_1, \ldots, s_i, \ldots, s_n) \geq u_i(s_1, \ldots, s_i', \ldots, s_n)$ for any $s_i' \in S_i$. In general a game may not have pure Nash equilibria, however, Nash [16] proved that if we extend the game to include as strategies for i all possible distributions on S_i, with the obvious extension of the u_i's to capture expectation, then an equilibrium is guaranteed to exist.

Consider a graph with node set $S_1 \times \ldots \times S_n$ and an edge (s, s') whenever s and s' differ only in one component, say the i-th, and $u_i(s') > u_i(s)$. We call such a graph the Nash dynamics and the edge (e, e') a move of player i. The Nash dynamics models the evolution of the game as a sequence of moves. If it admits a sink then the game has a pure Nash equilibrium, while if it is acyclic then the game is always guaranteed to converge to a pure Nash equilibrium after a finite number of moves.

Given a graph $G = (V, E)$, an instance I, a pricing function f and a payment function p we define as (G, I, f, p) the following game. We have $r = |I|$ players, where each player i is a request $(x, y) \in I$, the set of actions for player i, $\forall i = 1, \ldots, r$, is $S_i = P(x, y)$ and the utility function is $u_i = -p((x, y))$. Let R^* denote a routing minimizing the number of colors for the problem (G, I) and let $\mathcal{R}$ denote the set of pure Nash equilibria for the game (G, I, f, p), if $\mathcal{R} \neq \emptyset$ the price of anarchy of the game is defined as $\rho(G, I, f, p) = \sup_{R \in \mathcal{R}} \frac{c(R)}{c(R^*)}$.

3   Results for Different Payment Functions

In this section we present different payment functions p and establish which of them give raise to a game possessing a pure Nash equilibrium. For such functions we also give a lower bound on the corresponding price of anarchy.

The payment functions we will consider in this section are the following:

- $p_1((x,y)) = f(c_R((x,y)))$, that is, a request (x,y) pays for the price of the color used to communicate.
- $p_2((x,y)) = \max_{e \in R((x,y))} \max_{(a,b) \in I | e \in R((a,b))} f(c_R((a,b)))$, that is, a request (x,y) pays for the maximum price of all the colors used by any other request sharing an edge with the path used by (x,y).
- $p_3((x,y)) = \sum_{e \in R((x,y))} \max_{(a,b) \in I | e \in R((a,b))} f(c_R((a,b)))$, that is, a request (x,y) pays for the sum over all the edges belonging to the used path of the maximum price of the colors passing through the edge.
- $p_4((x,y)) = \sum_{e \in R((x,y))} \max_{(a,b) \in I | e \in R((a,b))} \frac{f(c_R((a,b)))}{|\{(a,b) \in I | e \in R((a,b))\}|}$, that is, a request (x,y) pays for the sum over all the edges belonging to the used path of the maximum price of the colors passing through the edge, divided by the number of request sharing the edge.

3.1   The Payment Function p_1

The first payment function we consider is, given $(x,y) \in I$ and a routing R, $p_1((x,y)) = f(c_R((x,y)))$, that is, a request (x,y) pays for the price of the color used to communicate.

The following theorem shows that for this payment function a Nash equilibrium is always achieved in a polynomial number of moves.

Theorem 1. *The game* (G, I, f, p_1) *converges to a Nash equilibrium in at most* r^2 *moves.*

Proof. Since any move of a request (x,y) does not affect the payment of any other request, at most r^2 moves are possible until a Nash equilibrium is achieved. In fact, being the number of needed colors at most equal to r and assuming that no request make stupid moves using a color $w > r$, each request can move at most r times.   □

This payment function, however, in unable to guarantee, in general, good performances to the network. Indeed, there exist graphs G and instances I for which the price of anarchy can be very high, as shown in the following theorem.

Theorem 2. *For any pricing function* f, *there exist* G *and* I *such that* $\rho(G, I, f, p_1) = r$.

Proof. Consider a ring having r nodes and the instance $I = \{(v_i, v_{(i+1)mod\ r}) | i \in \{1, \ldots, r\}\}$. In the routing R^* using the same edge $\{v_i, v_{(i+1)mod\ r}\}$ for each request $(v_i, v_{(i+1)mod\ r})$, we have that $c_{R^*} = 1$, that is only one color is used. On the other hand, the routing R that uses, for each request $(v_i, v_{(i+1)mod\ r})$, the path $\{v_i, v_{(i-1)mod\ r}\}, \ldots, \{v_{(i+2)mod\ r}, v_{(i+1)mod\ r}\}$, we have that $c_R = r$. Since the routing R is a Nash equilibrium, the claim follows.   □

Such a bad result is justified by the following consideration. Given a routing R, finding a minimum assignment of wavelengths to the paths in R is equivalent to finding a minimum coloring of the conflict graph induced by R. Since the problem of coloring a graph $G = (V, E)$ cannot be approximated within $|V|^{\frac{1}{7}}$ [2] and a Nash equilibrium achieved in polynomial time is an approximated solution to graph coloring, being $|R| = |I| = r$, the price of anarchy cannot be better than $r^{\frac{1}{7}}$.

One of the drawbacks of the payment functions p_1 is that each request (x, y) is charged a cost depending only on the color it uses, without considering the fact that all the other requests sharing an edge with the path used by (x, y) are not allowed to use the same color. For such a reason one can expect that designing payment functions taking into account all the colors passing through a particular path may yield games with a better price of anarchy.

3.2   The Payment Function p_2

The second payment function we consider is, given $(x, y) \in I$ and a routing R, $p_2((x, y)) = \max_{e \in R((x,y))} \max_{(a,b) \in I | e \in R((a,b))} f(c_R((a, b)))$, that is, a request (x, y) pays for the maximum price of all the colors used by any other request sharing an edge with the path used by (x, y).

For the payment function p_2 is also possible to prove the convergence to a Nash equilibrium even though the number of moves may be exponential.

Theorem 3. *The game (G, I, f, p_2) converges to Nash equilibrium in a finite number of moves.*

Proof. Let β be the sequence of colors $c_R((x, y))$, $\forall (x, y) \in I$ listed in non increasing order. Any move of a player i, from a routing R to a routing R' lowering his payment from p to p', can make the payments of the other players risen to reach at most the value p'. Thus, letting β' be the sequence of colors $c_{R'}((x, y))$, $\forall (x, y) \in I$ listed in non increasing order, we have that β' is lexicographically smaller than β. Thus, being the set of the possible sequences finite, an equilibrium must be achieved.   □

It would be interesting determining a graph G and an instance I creating the situation described in the above theorem, thus effectively implying the non polynomial time convergence in general.

However, also for this payment function, we cannot expect in general a low price of anarchy. In fact, using the same problem (G, I) defined in the proof of Theorem 2, it is possible to prove the following result.

Theorem 4. *For any pricing function f, there exist G and I such that $\rho(G, I, f, p_2) = r$.*

3.3 The Payment Function p_3

The third payment function we consider is, given $(x, y) \in I$ and a routing R, $p_3((x, y)) = \sum_{e \in R((x,y))} \max_{(a,b) \in I | e \in R((a,b))} f(c_R((a, b)))$, that is, a request (x, y) pays for the sum over all the edges belonging to the used path of the maximum price of the colors passing through the edge.

Adding a summation in the payment function clearly tightens the interaction between the requests. If from one side this is expected to improve on the price of anarchy, from the other one the game could become more complex. The next theorem captures this last situation.

Theorem 5. *For any unbounded pricing function of the colors f the game (G, I, f, p_3) in general does not converge.*

Proof. Consider the graph $G = (V, E)$ in which $V = \{a, b, c, v_1, v_2, \ldots, v_{h-1}, w_1, w_2, \ldots, w_{k-1}\}$ and $E = \{\{a, b\}, \{b, c\}, \{a, v_1\}, \{v_1, v_2\}, \ldots, \{v_{h-1}, b\}, \{b, w_1\}, \{w_1, w_2\}, \ldots, \{w_{k-1}, c\}\}$ and the instance $I = \{(a, c), \underbrace{(a, b)}_{n \text{ times}}, \underbrace{(b, c)}_{m \text{ times}}\}$ as shown in Figure1. The initial routing R assigns the colors $\{1, 2, \ldots, n\}$ to the edge $\{a, b\}$ for the n requests (a, b), the colors $\{1, 2, \ldots, m\}$ to the edge $\{b, c\}$ for the m requests (b, c) and the color $n + 1$ to the path $\{a, b\}, \{b, c\}$ for the request (a, c). If the conditions outlined below hold, the following cycle will occur in the Nash dynamics graph: all the m requests (b, c) move on the path containing the "w" nodes (i.e. the path $\{b, w_1\}, \{w_1, w_2\}, \ldots, \{w_{k-1}, c\}$) using the colors $\{1, 2, \ldots, m\}$; the requests (a, c) moves on the path containing the "v" nodes (i.e. the path $\{a, v_1\}, \{v_1, v_2\}, \ldots, \{v_{h-1}, b\}, \{b, c\}$) using the color 1; the m requests (b, c) move on the path containing the only edge $\{b, c\}$ using the colors $\{2, 3, \ldots, m, m + 1\}$; the request (a, c) moves on the path $\{a, b\}, \{b, c\}$ using the color $n + 1$. The first technical condition is $n \geq m \geq 1$. The condition 2 implies that a request (a, b) never moves on the path containing the "v" nodes: $f(n+1) < h \cdot f(1)$. The condition 3 implies that a request (b, c) moves on the path $\{b, c\}$ if it pays at most for the color $m + 1$: $f(m + 1) < k \cdot f(1)$. The condition 4 ensures that the request (a, c) never moves on the path containing the "w" nodes: $2f(n + 1) < (h + k)f(1)$. The condition 5 implies that the request (a, c) can move on the path containing the "v" nodes only if it pays for the color 1: $f(n+1) < (h+1)f(m+1)$. The condition 6 implies that a request (b, c) moves on the path containing the "w" nodes if it pays for the color $n+1$ on the path $\{b, c\}$: $k \cdot f(m) < f(n+1)$. The condition 7 implies that the request (a, c) moves on the path containing the "v" nodes when it can pay for the color 1 and the edge $\{b, c\}$ is free: $(h + 1)f(1) < 2f(n + 1)$. The last condition 8 implies that the request (a, c) moves on the path $\{a, b\}, \{b, c\}$ when the requests (b, c) are routed on the path $\{b, c\}$: $2f(n+1) < h \cdot f(1) + f(m+1)$. All the conditions hold if m is chosen such that $f(m+1) > 2f(1)$, n such that $f(n+1) > \frac{f(m)[f(m+1)+f(1)]}{f(1)}$, h such that $\frac{2f(n+1)-f(m+1)}{f(1)} < h < \frac{2f(n+1)-f(1)}{f(1)}$ and k such that $\frac{f(m+1)}{f(1)} < k < \frac{f(n+1)}{f(m)}$. $\qquad\square$

Even though it may seem that the above theorem proves the non-existence of a Nash equilibrium in general, it must be noted that this is not the case.

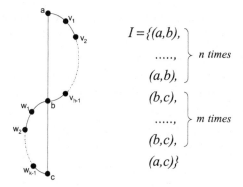

$$I = \{(a,b),$$
$$....., \quad \} \; n \; times$$
$$(a,b),$$
$$(b,c),$$
$$....., \quad \} \; m \; times$$
$$(b,c),$$
$$(a,c)\}$$

Fig. 1.

The following solution, in fact, is a Nash equilibrium: the request (a, c) is on the path $\{a, b\}, \{b, c\}$ with color 1, the n requests (a, b) are on the edge $\{a, b\}$ with colors $\{2, \ldots, n + 1\}$ and the m requests (b, c) are on the edge $\{b, c\}$ with colors $\{2, \ldots, m + 1\}$. However, we conjecture the non-existence of a Nash equilibrium in the general case. Our opinion is based on the fact that the majority of the existence proofs known in the literature exploits the acyclicity of the Nash dynamics graph. Moreover, very often the presence of a cycle in such a graph has been used to suitably redefine the game in order to show the non-existence of the Nash equilibrium.

3.4   The Payment Function p_4

The last payment function we consider is, given $(x, y) \in I$ and a routing R,
$$p_4((x,y)) \;=\; \sum_{e \in R((x,y))} \max_{(a,b) \in I \,|\, e \in R((a,b))} \frac{f(c_R((a,b)))}{|\{(a,b) \in I \,|\, e \in R((a,b))\}|}, \text{ that is, a re-}$$
quest (x, y) pays for the sum over all the edges belonging to the used path of the maximum price of the colors passing through the edge, divided by the number of request sharing the edge.

Theorem 6. *For the pricing function of colors $f(w) = w^2$, the game (G, I, f, p_4) in general does not converge.*

Proof. Consider the graph $G = (V, E)$ in which $V = \{a, b, c, v_1, v_2, \ldots, v_{59}, w_1, w_2\}$ and $E = \{\{a, b\}, \{b, c\}, \{a, v_1\}, \{v_1, v_2\}, \ldots, \{v_{59}, b\}, \{b, w_1\}, \{w_1, w_2\}, \{w_2, c\}\}$ and the instance $I = \{ \underbrace{(a, b)}_{9 \; times}, (b, c), (a, c)\}$ as shown in Figure2. The ini-
tial routing R assigns the colors $\{1, 2, \ldots, 9\}$ to the edge $\{a, b\}$ for the 9 requests (a, b), the color 1 to the edge $\{b, c\}$ for the request (b, c) and the color 10 to the path $\{a, b\}, \{b, c\}$ for the request (a, c). The game will be stuck in the following cycle: the request (b, c) moves on the path containing the "w" nodes (i.e. the path $\{b, w_1\}, \{w_1, w_2\}, \{w_2, c\}$) using the color 1; the request (a, c) moves on the path containing the "v" nodes (i.e. the path $\{a, v_1\}, \{v_1, v_2\}, \ldots, \{v_9, b\}, \{b, c\}$)

using the color 1; the request (b, c) moves on the path containing the only edge $\{b, c\}$ using the color 2; the request (a, c) moves on the path $\{a, b\}, \{b, c\}$ using the color 10. □

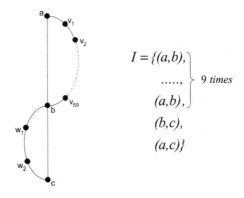

$$I = \{(a,b),$$
$$....., \quad \left.\right\} \; 9 \; times$$
$$(a,b),$$
$$(b,c),$$
$$(a,c)\}$$

Fig. 2.

Clearly, the same observations proposed with respect to Theorem 5, hold also in this case, even though it would be nice to extend the result to any pricing function f.

4   Conclusions

In this paper we have presented four reasonable paying strategies that a service provider can adopt in all-optical networks in order to keep the overall performances of the network (modelled as the number of used wavelengths) acceptable, assuming that users are selfish agents who may not cooperate. Our results are essentially negative, since we prove that only two functions guarantee the convergence of the evolutionary game to a pure Nash equilibrium. Moreover, in both cases the price of anarchy can be made arbitrarily high. However, it must be noted that a good coordination ratio cannot be achieved if the game converges in polynomial time since otherwise this would contradict the inapproximability results of the relative graph coloring problem. We leave open the problem of finding a pricing function yielding a game having a low price of anarchy, in the case that the convergence to a Nash equilibrium cannot be achieved in polynomial time. Another interesting issue if that of proving the non-existence of a Nash equilibrium for the cases in which we prove the non-convergence of the game.

Considering the fact that all of our results are negative, a reasonable step could be that of restricting the graph topology (considering for example only rings, trees, tori and meshes) or the communication patterns (broadcasting, gossiping). Finally, our model could be extended to cover scenarios in which the network performances are measured considering also different metrics as, for example, the hardware costs (see [7]).

Acknowledgments

We would like to thank Michele Flammini for introducing us to the problem and for very useful discussions and comments. A special thank also to the anonymous referees for useful suggestions and improvements.

References

1. B. Beauquier, J. C. Bermond, L. Gargano, P. Hell, S. Perennes, and U. Vaccaro. Graph problems arising from wavelength-routing in all- optical networks. In Proceedings of the 2nd Workshop on Optics and Computer Science, part of IPPS'97, 1997.
2. M. Bellare, O. Goldreich, and M. Sudan. Free bits, PCPs and non-approximability - towards tight results. SIAM Journal of Computing, 27:804–915, 1998.
3. C. A. Brackett. Dense wavelength division multiplexing networks: principles and applications. IEEE Journal on Selected Areas in Communications, 8:948–964, 1990.
4. N. K. Chung, K. Nosu, and G. Winzer. Special issue on dense WDM networks. IEEE Journal on Selected Areas in Communications, 8, 1990.
5. A. Fabrikant, A. Luthra, E. Maneva, C. H. Papadimitriou, and S. Shenker. On a network creation game. In Proceedings of the 22nd ACM Symposium on Principles of Distributed Computing (PODC), pages 347–351, 2003.
6. A. Fabrikant, C. H. Papadimitriou, and K. Talwar. The complexity of pure nash equilibria. Manuscript.
7. M. Flammini, A. Navarra, and A. Proskurowski. On routing of wavebands for gossiping in all-optical paths and cycles. In Proceedings of the 10th Colloquium on Structural Information and Communication Complexity (SIROCCO), pages 133–146, 2003.
8. D. Fotakis, S. Kontogiannis, E. Koutsoupias, M. Mavronicolas, and P. Spirakis. The structure and complexity of nash equilibria for a selfish routing game. In Proceedings of the 29th International Colloquium on Automata, Languages and Programming (ICALP), pages 123–134, 2002.
9. L. Gargano and U. Vaccaro. "Routing in All–Optical Networks: Algorithmic and Graph–Theoretic Problems"in: Numbers, Information and Complexity. Kluwer Academic, 2000.
10. P. E. Green. Fiber-Optic Communication Networks. Prentice Hall, 1992.
11. H. S. Hinton. Architectural considerations for photonic switching networks. IEEE Journal on Selected Areas in Communications, 6:1209–1226, 1988.
12. R. Klasing. Methods and problems of wavelength-routing in all-optical networks. In Proceedings of the MFCS'98 Workshop on Communication, August 24-25, Brno, Czech Republic, pages 1–9, 1998.
13. E. Koutsoupias and C.H. Papadimitriou. Worst-case equilibria. In Proceedings of the 16th Annual Symposium on Theoretical Aspects of Computer Science (STACS), volume 1563 of LNCS, pages 387–396, 1999.
14. M. Mavronicolas and P. Spirakis. The price of selfish routing. In Proceedings of the 33rd Annual ACM Symposium on the Theory of Computing (STOC), pages 510–519, 2001.
15. I. Milchtaich. Congestion games with player-specific payoff functions. Games and Economic Behavior, 13:111–124, 1996.

16. J. F. Nash. Equilibrium points in n-person games. In Proceedings of the National Academy of Sciences, volume 36, pages 48–49, 1950.
17. G. Owen. Game theory. Academic Press, 3rd edition, 1995.
18. C. H. Papadimitriou and M. Yannakakis. On complexity as bounded rationality. In Proceedings of the 26th Annual ACM Symposium on the Theory of Computing (STOC), pages 726–733, 1994.
19. S. Personik. Review of fundamentals of optical fiber systems. IEEE Journal on Selected Areas in Communications, 3:373–380, 1983.
20. R. Ramaswami. Multi-wavelength lightwave networks for computer communication. IEEE Communications Magazine, 31:78–88, 1993.
21. R. W. Rosenthal. A class of games possessing pure-strategy nash equilibria. International Journal of Game Theory, 2:65–67, 1973.
22. T. Roughgarden and E. Tardos. How bad is selfish routing? Journal of ACM, 49(2):236–259, 2002.
23. A. Vetta. Nash equilibria in competitive societies, with applications to facility location, traffic routing and auctions. In Proceedings of the 43rd Annual IEEE Symposium on Foundations of Computer Science (FOCS), pages 416–425, 2002.

Morelia Test:
Improving the Efficiency of the Gabriel Test and Face Routing in Ad-Hoc Networks

Paul Boone[1], Edgar Chavez[2], Lev Gleitzky[3], Evangelos Kranakis[1], Jaroslav Opatrny[4], Gelasio Salazar[3], and Jorge Urrutia[5]

[1] School of Computer Science, Carleton University, Ottawa*
[2] Escuela de Ciencias Físico-Matemáticas
de la Universidad Michoacana de San Nicolás de Hidalgo, México
[3] IICO-UASLP Av. Karakorum 1470. Lomas 4ta Seccion San Luis Potosi, SLP
México 78210
[4] Department of Computer Science, Concordia University, Montréal**
[5] Instituto de Matemáticas, Universidad Nacional Autónoma de México, D.F.,
México City***

Abstract. An important technique for discovering routes between two nodes in an ad-hoc network involves applying the face routing algorithm on a planar spanner of the network. Face routing guarantees message delivery in networks that contains large holes, where greedy algorithms fail. Existing techniques for constructing a suitable planar subgraph involve local tests that eliminate crossings between existing links by deleting some links. They do not test whether the deleted links actually create some crossings and some of the links are deleted needlessly. As a result, some of the routes found in face routing will have an unnecessarily large number of hops from source to destination. We consider a new local test for preprocessing a wireless network that produces a planar subgraph. The test is relatively simple, requires low overhead and does not eliminate existing links unless it is needed to eliminate a crossing, thus reducing overhead associated with multiple hops.

1   Introduction

An *ad-hoc network* is a network consisting of transmitters, often called *hosts*, that is established as needed, typically without any assistance from a fixed infrastructure. Using wireless broadcasts, each host can communicate with other hosts within its transmission range. Typically, not all hosts are within the transmission range of each other. Thus, communication between two hosts in the network is in general achieved by multi-hop routing along a route where intermediate nodes cooperate by forwarding packets. Examples of such networks

---

* Research supported in part by NSERC and MITACS.
** Research supported in part by NSERC.
*** Research partially supported by CONACYT grant no. 37540-A, and PAPIIT Unam.

include sensor, piconet, bluetooth, and home/office networks, and routes in these networks have to be constructed on the fly.

In this paper we consider networks that have the following properties: 1) All hosts know the geometric coordinates (x, y) of their location; 2) All hosts have the same transmission range R, *i.e.* any two hosts at distance $d \leq R$ are able to communicate directly; and 3) Communication links are *bidirectional*. Ad-hoc networks satisfying the conditions above are the most common type of ad-hoc networks considered in literature.

A network can be represented by a geometric undirected graph, $G = (V, E)$, with nodes representing the hosts of the network, and an edge (link) connecting any pair of nodes that can communicate directly.

Discovering a route between two nodes in an ad-hoc network is an important component of current research. In such systems it is vital that route discovery uses only local information and is adaptable to the network connectivity An important technique for discovering routes between any two nodes in a ad-hoc network without the use of flooding is the face routing [2, 7] algorithm The face routing algorithm succeeds in discovering a route in a networr providing that the underlying graph is planar. Since in practice, th original ad-hoc wireless network is never planar with many links crossin each other, before using the face routing we need to extract from th original network, using a local method, a planar connected network spanning the entire underlying network.

There are two important goals we should be concerned with in our reduction from the original wireless network to the geometric planar graph. The first is to **keep long links** and is required so that we can prevent an unnecessary large number of hops from source to destination that require extra processing at the nodes and may cause failures. Secondly, we must **eliminate crossings** of links. This is required in order to create a planar underlying graph. From a practical point of view, the method employed must be efficient and be based on local tests and the resulting graph must be a connected planar spanning subgraph of the original network.

1.1   Results of the Paper

In this paper we consider a method for reducing the overhead of an unnecessarily large number of hops from source to destination.

The *Morelia test* is a new local test for preprocessing a wireless network that produces a planar spanning subgraph of the original wireless network on which we can apply face routing. The Morelia test is a generalization of the Gabriel test and is relatively simple, requires low overhead and does not unnecessarily eliminate existing links if they do not create any crossing, thus reducing the overhead associated with multiple hops. The resulting graph is planar and is a supergraph of the Gabriel graph [5].

In addition to the theoretical justifications of the Morelia test presented in this paper, we conducted simulations of our algorithms on randomly generated networks so that we can quantify the improvements given by the new method.

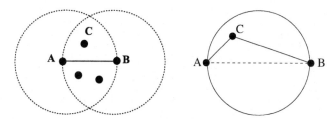

Fig. 1. Left: The RNG test for producing a planar spanner. Right: Eliminating an unnecessary link (dashed line AB) with Gabriel Test

2 Tests for Reduction to Planarity

Here we discuss existing tests for planar reduction in wireless networks. In particular we consider the Relative Neighbourhood Graph and the Gabriel Graph.

A planar spanner of a network can be obtained by applying the *Relative Neighbourhood Graph* (RNG) [10] test to every pair of nodes which are within the transmission range of each other. Let A, B be two nodes whose distance is at most the transmission range R of the network. Consider the region delimited by the intersection of the circles centered at A and B, respectively, where the radius of the circles is R, the power of the stations at A and B, see Figure 1 (left). If there is no node in the region then the link between A and B is kept. If however there is a node C in the region depicted in Figure 1 (left), then nodes A and B remove their direct link. In particular, when A (respectively, B) is queried on routing data to B (respectively, A) the routing table at A (respectively, B) forwards the data through C (or some other similar node if more than one node is in the specified region.) The RNG test suffers from the multiple hop effect because the elimination of crossings is done by elimination of longer links.

One of the most important tests for eliminating crossings in a wireless network is called the *Gabriel test* [5] which, similarly to the RNG test is applied to every link of the network. The main difference between the Gabriel test and the RNG test is the smaller size of the region considered for a link elimination.

Let A, B be two nodes whose distance is less than the transmission range R of the network. In the Gabriel test, if there is no node in the circle with diameter AB then the link between A and B is kept. If however there is a node C in the circle with diameter AB, as depicted in Figure 1 (right), then nodes A and B remove their direct link. In particular, when A (respectively, B) is queried on routing data to B (respectively, A) A (respectively, B) forwards the data through C (or some other similar node if more than one node is in the circle with diameter AB). The advantage of doing this rerouting of data is that the resulting graph is a planar spanner on which we can apply the face routing algorithm for discovering a route from source to destination.

However, the test merely shrinks the "test region" and creates a planar spanner that keeps some of the links that would be eliminated by the RNG test. However, like the RNG test, it does not in any way prevent the multiple hop effect.

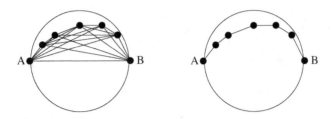

Fig. 2. Multiple hop effect when eliminating a link (line segment AB) via the Gabriel test

For example, consider a set of nodes as depicted in the left-hand side of Figure 2. All nodes are mutually reachable. However, when we apply the Gabriel test the configuration in the right-hand side of Figure 2 results. We can see that although nodes A and B could have reached each other directly in a single hop instead they must direct their data through a sequence of many hops.

We note that the multiple hop effect arises when many nodes are clustering in regions. We should note that while the multi-hop effect may result in slower message delivery, it may have also a positive effect, since it can decrease the power consumption. For example, if instead of using a direct link between A and B an intermediate node, say C is used (see Figure 1 (right)) which lies in the circle with diameter AB then the power consumption decreases from $d(A,B)^{-p}$ to $d(A,C)^{-p} + d(C,B)^{-p}$, for some constant $2 \leq p$. However, this is at the cost of additional overhead implied by intermediate hopping.

3   Planarity and the Multiple Hop Effect

The purpose of this section is to give an algorithm that eliminates crossings but at the same time maintains some "long" links between stations whenever possible, thus reducing the number of hops in face-routing. We introduce the *Morelia test*. The Morelia test is an extension of the Gabriel test where the algorithm checks for the presence of a crossing before eliminating a link.

3.1   Morelia Test

As mentioned in the introduction, we are concerned with the problem of routing in networks with complex topology and containing many holes. Both the RNG and Gabriel tests eliminate some links not because they create crossing of links, but merely for the potential of being involved in crossing. As indicated in the example above, if the Gabriel or RNG test is applied to a complex network, the spanning planar subgraph that is obtained will contain holes of even larger size.

The Morelia test attempts to preserve links whenever possible and as a consequence the resulting planar graph (on which face routing is to be applied) will most likely keep the contour very similar to the original contour and the holes in the network would not grow much. Thus the resulting planar network will have smaller diameter and routes from source to destination will require fewer hops.

The Morelia test is similar to the Gabriel test in that given two nodes A and B it eliminates links based on the inspection of the circle with diameter AB. Unlike the Gabriel test it does not necessarily eliminate the direct link AB when it finds another node inside the circle with diameter AB. Instead, it verifies whether the nodes inside the circle create any crossing of the line AB. If no crossing is created the line AB is kept, otherwise it is removed. The verification of the existence of crossing is done in most cases by inspecting only the neighborhood of nodes A and B at the transmission distance R. In a few cases, the neighborhood of some of the nodes in the circle around AB is inspected. In all cases it is a local test that computes the neighborhood of nodes at distance at most two hops of each node A and B.

In the Morelia test of a link AB we subdivide the area of the circle with diameter AB into four regions, X_1, X_2, Y_1 and Y_2 as in Figure 3. The areas X_2 and Y_2 are determined by an arc of a circle of radius R through A and B. Furthermore, in the testing we use areas X_3 and Y_3 as indicated in Figure 3 that are outside the transmission radius R of the nodes A and B and within distance R from the link AB. For each node A let $N(A)$ be the set of nodes Z such that $d(A, Z) \leq R$.

The precise specification of the Morelia test applied to a link AB is given below (refer to Figure 3).

Morelia Test Rules

1. If there is at least one node in $X_1 \cup X_2$ and at least one node in $Y_1 \cup Y_2$ then the link AB is removed.
2. If there is at least one node in X_1 and no node in $X_2 \cup Y_1 \cup Y_2$ then the node A checks whether any node in $N(A)$ creates a link with nodes in X_1 that crosses AB. If such a crossing occurs, link AB is removed and A sends a message to B to remove the link as well. Similarly node B performs a check of nodes in $N(B)$ for a crossing of the link AB and informs node A if a crossing is found and AB is to be removed.

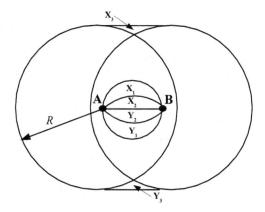

Fig. 3. Morelia Test

3. If there is at least one node in Y_1 and no node in $Y_2 \cup X_1 \cup X_2$ (symmetric to Rule 2) then the node A checks whether any node in $N(A)$ creates a link with nodes in Y_1 that crosses AB. If such a crossing occurs, link AB is removed and A sends a message to B to remove the link as well. Similarly node B performs a check of nodes in $N(B)$ for a crossing of the link AB and informs node A if a crossing is found and AB is to be removed.

4. If there is at least one node in X_2 and no node in $Y_1 \cup Y_2$ then the node A inspects the nodes in $N(A)$ to check whether any node there creates a link with nodes in $X_1 \cup X_2$ that crosses AB. If such a crossing occurs, link AB is removed and A sends a message to B to remove the link as well. Furthermore A sends a message to nodes in X_2 with a request to send back information whether there is a node in the region Y_3. If A receives a message that a node exits in Y_3 then AB is removed and node B is informed to remove the link as well.

5. If there is at least one node in Y_2 and no node in $X_1 \cup X_2$ (symmetric to Rule 4), the node A inspects the nodes in $N(A)$ for a possible crossing of AB. If such a crossing occurs, link AB is removed and A sends a message to nodes in Y_2 with a request to send back information whether there is a node in the region X_3. If A receives a message that a node exits in X_3 then AB is removed and node B is informed to remove the link as well.

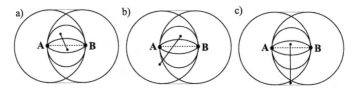

Fig. 4. Examples of the Morelia test situations: a) AB deleted in Rule 1, b) AB deleted in Rule 2, c) AB deleted in Rule 4

Figure 4 illustrates how, unlike the Gabriel test, the Morelia test will check for crossings prior to eliminating a link. It will eliminate link AB because it detects crossings, but it will not eliminate it when no crossing exists.

Notice that in the Rule 1 of the Morelia test, nodes A and B look only at nodes that are in $N(A)$ and $N(B)$ respectively, i.e., nodes that are one hop away. In Rule 2, 3, and 4 node A checks nodes in $N(A)$, node B checks nodes in $N(B)$ and both A and B possibly check $N(x)$ for nodes x in $\in X_2 \cup Y_2$. Thus A or B are checking nodes that are at most two hops away. Therefore, the Morelia test is a local test.

Theorem 1. *If network N is connected then the application of the Morelia test to all links of N produces network N' which is a connected planar spanner of N. Furthermore, N' contains the Gabriel graph of N as its subgraph.*

Proof. Every edge in N that is kept by the Gabriel test is also kept by the Morelia test. Thus, N' contains the Gabriel graph of N as its subgraph. Since the Gabriel test produces a connected spanning subgraph of N, the subnetwork N' is connected.

Assume that there is a link e in N' that crosses a link AB of N'. Since the length of any link in N is at most R, both ends of e are in $N(A) \cup N(B) \cup X_3 \cup Y_3$. If one of the ends of e is in the circle with diameter AB then AB would be deleted by the Morelia test. If both ends of e are outside the circle with diameter AB then one of the ends of the edge AB must be inside the circle with diameter e, since the edges cross each other. However, the Morelia test applied to e would eliminate e because of it being crossed by AB. Thus there can be no crossing in N'.

In Figure 5 we show how the Morelia test keeps some of the long links. The left-hand side shows a link AB and the nodes inside the circle with the diameter AB and the right-hand side shows what edges are kept by the Morelia test.

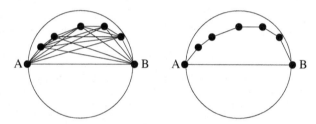

Fig. 5. Multiple hop elimination (line segment AB) via the Morelia test

3.2   The Overhead of the Morelia Test

In Rule 1 of the Morelia test node A or B needs to determine what nodes are in the circle with AB as its diameter and the complexity of it is the same of that of the Gabriel test.

Rule 2 for A involves nodes that are in $N(A)$ and each node of the network needs to know its neighbors anyway. To check for crossings of two line segments involves simple geometrical computation of complexity $O(1)$, and the same applies to node B. The exchange of messages between A and B confirming deletion or retention of the edge is also simple.

Rule 3 and 4 of the test involves the existence of nodes in X_3 and Y_3 that are outside of $N(A) \cup N(B)$. However, all that is needed for A or B is to send a message to the nodes in region X_2 or Y_2 asking the question "is there any node in Y_3 or X_3?" respectively. The region X_3 or Y_3 can be specified by the three corners of the region. Thus, although these rules involve nodes that are two hops away from A and B, it does not create a significant overhead or delay. We show now that the size of the region X_2 or Y_2 is a smaller part of the circle with diameter AB. Since Rule 3 and 4 are used only when X_2 or Y_2 is the only

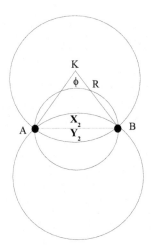

Fig. 6. The arc segment Y_2 (and its symmetric one X_2) used in Rule 5

region of the circle containing a node of the network, the probability of using these rules is also smaller.

We can easily calculate an upper bound on the ratio of $|X_2|$, the area of X_2 to $|X_1|$, the area of X_1 to get an indication on how often Rule 4 is used in comparison to the other rules.

It is easy to see that the ratio $|X_2|/|X_1|$ is getting smaller when the angle ϕ is getting smaller. Thus we get an upper bound on $|X_2|/|X_1|$ by setting $\phi = \pi/3$. In this case the length of AB equals R and the area of the circle determined by the diameter AB is exactly $\pi R^2/4$, and the area of the circular sector X_1 is equal to $\pi R^2 - \sqrt{(3)}R^2/4 = R^2(\frac{2\pi-3\sqrt{(3)}}{12}$. Thus $\frac{|X_2|}{|X_1|} = \frac{2\pi-3\sqrt{(3)}/12}{\pi R^2/8-2\pi-3\sqrt{(3)}/12} = \frac{4\pi-6\sqrt{(3)}}{3\pi-4\pi-6\sqrt{(3)}} < 0.3$.

Thus if a node is located in the circle with diameter AB and its placement is random, it will be in the area $X_2 \cup Y_2$ with probability less than 0.3.

4   Experimental Results

In this section we present our simulations designed to test the Morelia Test in comparison with the Gabriel Test. The purpose of the simulations was to give an indication on how our algorithm actually performs in the real world. We study the operation and performance of our algorithm on randomly generated graphs.

4.1   Goals

There were two main goals to achieve with our simulations. The first was to compare the average link length in the planar spanning subgraph created using the Morelia and Gabriel tests. The second, to compare the number of hops required to deliver messages with face routing on the planar spanning subgraph created by the Morelia and Gabriel tests.

4.2   Network Model

A simulation was designed that allowed the creation of a special type of random graph. Since an ad-hoc network does not assume any specific structure, we considered network models obtained as follows. A square area (called a *grid*) could be defined, nodes could be created with a specific transmission range and placed at random coordinates on the grid. We use the unit disk graph (UDG) model, where all nodes have the same transmission range and the nodes were uniformly distributed.

Each network simulation consisted of 50 nodes each with a transmission range of 250 units. The test area was a square grid on which the 50 nodes were randomly placed. This network setup allowed us to test the performance with different link densities by simply varying the grid size (300 units square to 1300 units square).

4.3   Analysis of Morelia Graph vs. Gabriel Graph

The Gabriel Test and the Morelia Test were run on the *same* series of random graphs. The following are the main metrics that were traced in the simulation: 1) Average length of link on each planar spanning subgraph; and 2) Number of links in each planar spanning subgraph.

Scenario 1 – Uniform Random Graph. The first test, was to generate a series of 30 random networks. The grid size was the only parameter that was altered for a series of trials. This had the effect of increasing or decreasing the network density (the smaller the grid size, the closer together the nodes, the greater number of links, the higher degree vertices).

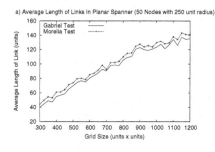
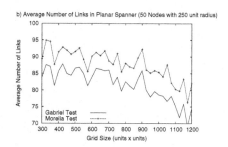

Fig. 7. Average Number and Length of Links in Morelia and Gabriel Graphs

Figure 7 a), shows the average link length of the Gabriel and Morelia graphs. The Morelia graph indeed does keep slightly longer links on average. The average increase in length was 4.86 units or 5.76%.

Figure 7 b), shows the average number of links kept in the planar spanning subgraph for both tests. The Morelia test keeps more links on average in the spanner than the Gabriel test. The average increase in links kept in the Morelia graph was 6.07 or 7.49% over the Gabriel graph.

Scenario 2 – Random Graph with a Sparse Region. This simulation was designed to test the planar spanning subgraph creation when there exists an area of the graph which has a less dense topology. A region of 500 units within the test grid was defined to have a node density of one-half that of the rest of the network.

Here we report, as in the first test, the resulting average link length and number of nodes in the planar spanning subgraph. In Figure 8 a), we see the average length of links (edges) kept in both the Gabriel and Morelia graph. The average Morelia link length is now 5.57 units or 4.5% longer than the average Gabriel link.

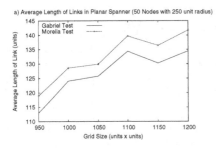 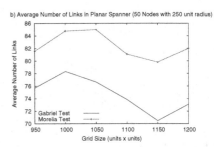

Fig. 8. Average Number and Length of Links in Morelia and Gabriel Graphs

Finally, we show in Figure 8 b), the average number of extra links kept in the two graphs. The average number of additional links kept increased by 7.67, which i an increase of 10.33%

4.4   Face Routing Performance Study

In this section we present the simulation results of the face routing algorithm *Face-2* [2] on the same series of random graphs on which the Morelia and Gabriel tests have produced a planar spanning subgraph. The *Face-2* algorithm is a modified face routing algorithm that at each iteration of a face traversal makes the decision to move to the next face when it determines a link is about to cross the line from source to target, instead of traversing the whole face and keeping track of all crossings that occur [2].

We want to show that, in a network with some sparse areas, the face routing algorithm will perform routing with fewer number of hops on average. Two scenarios were simulated. The first was on a uniform random graph, the latter

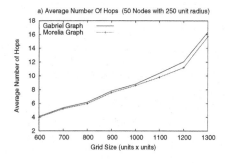 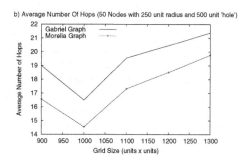

Fig. 9. Face routing applied to Morelia and Gabriel Graphs

scenario was on graphs which had a predetermined sparse region. In these simulations, the only metric studied was the *average number of hops* required for the face routing algorithm.

In the first scenario, a uniform random network of 50 nodes was built. Each node had a transmission range of 250 units. As before, the grid size was the only parameter changed to vary the network density.

We see in Figure 9 a), that in a uniformly distributed network, only minor gains are made in terms of decreasing the number of hops. The average number of hops saved was 0.28 per route.

In the last scenario, we define a region 500 units square and set the number of nodes with the region to one-half that of the first test. Routes were chosen that would traverse the region. All other parameters remained as in the previous scenario.

In Figure 9 b) we can see that there is further improvement of the face routing algorithm. The average number of hops saved with face routing on the Morelia graph has increased to 2.03 hops per route or 10.6%.

5   Conclusion

We have shown that with the application of the Morelia test to the underlying structure of a wireless ad-hoc network we can achieve face routing with a fewer number of hops on average over the Gabriel test when the network has a sparse region Thus the use of the Morelia test would be beneficial in any wireles ad-hoc network having a complex topology.

There are several open problems for future work in this area. One such problem is the extension of the algorithm to a network which is changing dynamically. Another area to investigate would be the improvement of the application of the Morelia test in networks which are very dense. If a network is very dense, i.e. the number of nodes inside a circle of transmission radius R is very high, the Morelia test introduces more overhead than the Gabriel test, the increase in th number of long links kept is not significant, an and thus the number of hops from source to destination remains high. Finally, a study of the Morelia test in combination with other tests that reduce the initial density of links in a network.

References

1. L. Barriere, P. Fraignaud, L. Narayanan, and J. Opatrny, Robust position-based routing in wireless ad hoc networks with unstable transmission ranges, In the Proceedings of the 5th International Workshop on Discrete Algorithms and Methods for Mobile Computing and Communications (DIALM), pp. 19–27, 2001.
2. P. Bose, P. Morin, I. Stojmenovic, and J. Urrutia. Routing with guaranteed delivery in ad hoc wireless networks. In the Proceedings of the 3rd International Workshop on Discrete Algorithms and Methods for Mobile Computing and Communications (DIALM), pp. 48–55, 1999.
3. J. Broch, D. Maltz, D. Johnson, Y.-C. Hu, and J. Jetcheva. A performance comparison of multi-hop wireless ad hoc network routing protocols. In Mobile Computing and Networking, pp. 85–97, 1998.
4. S. Datta, I. Stojmenovic, and J. Wu. Internal node and shortcut based routing with guaranteed delivery in wireless networks. In Proc. IEEE Int. Conf. on Distributed Computing and Systems Workshops; Cluster Computing, pp. 461–466, 2001.
5. K. Gabriel and R. Sokal. A new statistical approach to geographic variation analysis. Systematic Zoology, 18:259–278, 1969.
6. J. Gao, L. Guibas, J. Hershberger, L. Zhang, A. Zhu, Geometric Spanner for Routing in Mobile Networks, Proceedings of the 2nd ACM International Symposium on Mobile Ad hoc Networking and Computing, pp. 45–55, 2001.
7. E. Kranakis, H. Singh, and J. Urrutia, Compass Routing on Geometric Networks, In Proceedings of 11th Canadian Conference on Computational Geometry, pp. 51–54, Vancouver, August, 1999.
8. F. Kuhn, R. Wattenhofer, and A. Zollinger, Asymptotically Optimal Geometric Mobile Ad-Hoc Routing. In Proc. of the 6th International Workshop on Discrete Algorithms and Methods for Mobile Computing and Communications (DIALM), pp. 24–33, 2002.
9. C. E. Perkins, editor, Ad Hoc Networking, Addison Wesley, 2001.
10. G. Toussaint, The relative neighbourhood graph of a finite planar set, Pattern Recognition, 12(4):261–268, 1980.

Path Layout on Tree Networks:
Bounds in Different Label Switching Models

Anat Bremler-Barr and Leah Epstein*

School of Computer Science, The Interdisciplinary Center, Herzliya, Israel
{bremler,lea}@idc.ac.il

Abstract. Path Layout is a fundamental graph problem in label switching protocols. This problem is raised in various protocols such as the traditional ATM protocol and MPLS which is a new label switching protocol standardized recently by the IETF. Path layout is essentially the problem of reducing the size of the label-table in a router. The size is equivalent to the number of different paths that pass through the router, or start from it. A reduction in the size can be achieved by choosing a relatively small number of paths, from which a larger set is composed using concatenation.

In this paper we deal with three variations of the Path Layout Problem according to the special characteristics of paths in three label switching protocols, MPLS, ATM and TRAINET. We focus on tree networks and show an algorithm which finds label-tables of small size while permitting concatenation of at most k paths. We prove that this algorithm gives worst case tight bounds (up to constant factor) for all three models. The bounds are given as a function of the size of the tree, and the maximum degree.

1 Introduction

Improving label switching technologies is a step towards better QoS (Quality of Service) in high speed and IP networks. Label switching is done by using a quite simple idea of label based switching decisions, a packet is switched in the network according to a label it carries. The concept of Label Switching is not new. Its origins are in virtual circuit networks such as ATM networks. However, only recently the combination of switching technology and IP routing became a popular avenue of research and development. The large number of different label switching protocols [7] (tag-switching [18], CSR [15], IP-Switching [17], and threaded indices [6]) motivated the IETF to develop MPLS (Multi Protocol Label Switching), a standard approach to label switching [5]. MPLS was designed to combine the best of several approaches.

MPLS was developed to overcome some drawbacks of the standard IP routing. IP routing is basically a simple method of hop by hop routing. When a packet reaches a router, the routing decision (on the next router in the path) is based only on the destination address of the packet. At any given point in time all packets with the same destination address would be routed in exactly the same way, regardless of any other parameter of the packet. At each router along the packet path, the destination address

---

* Research supported in part by Israel Science Foundation (grant no. 250/01).

R. Královič and O. Sýkora (Eds.): SIROCCO 2004, LNCS 3104, pp. 35–46, 2004.

of the packet is examined, and the longest prefix that matches the address, out of all the prefixes (usually several tens of thousands) in a forwarding table is found. The packet is then routed according to the information associated with the Best Matching Prefix (BMP) in the forwarding table.

The high cost of computing the BMP at each router combined with the inability to distinguish between different flows with the same BMP motivated the development of MPLS. All the packets of the same flow are tagged with a unique label upon entering the network layer. At each switch (router) on the path the label is used to switch (rather than route) the packet to its next hop. The label is used as an index in the switching table. That is, label switching supports the concept of "route once and switch many". In other words, for all the packets in a flow, routing is performed once when setting up the label path, and the label is then used for all the packets at all the intermediate switches. Essentially MPLS establishes a circuit between the source and destination which in the terminology of MPLS is called Label Switch Path (LSP).

The basic advantages of MPLS are as follows. Elimination of the time consuming BMP enables higher rates of packet processing. The second advantage is support of QoS forwarding by assigning different labels to flows that require different services even if destined to the same destination (we do not use such options in this paper). Finally, MPLS enables explicit routing. The source router in the path establishes the path by some setup mechanism (such as RSVP). Henceforth the forwarding is performed by switching on the label which was associated with the explicit route. The explicit routing feature of MPLS enables the support of good traffic engineering (design of paths in the network), load-balancing, and support for resource reservations.

In order to support hierarchical routing (between networks and inside them), MPLS supports multi label paths by placing a stack of labels on each packet of a corresponding flow. In addition, in MPLS each router may use operations on the stack of labels, it can be instructed to push or pop one or more labels. This mechanism is also useful for path concatenation [1, 3]. A path is described by concatenation of LSPs and each LSP described by a label in the stack label. One of the applications of path concatenation is using a small set of LSPs to compose a larger set. The main advantages of this application is reducing the number of labels necessary in the network thus enabling more label switched routes with the same number of labels. Labels in MPLS are a scarce resource. They are the key elements in the scalability of MPLS, and the more labels we need the larger the switching tables are. Since the switching has to operate at high speeds, it is built out of expensive memories and its size is critical.

This paper deals with the following fundamental graph problem. How to choose a small set of routes in order to compose a larger set of routes by using concatenation. We mainly focus on tree networks, which are relevant to many server client applications, such as multicast, and can also be used as a building block for solutions in general graphs [9].

This question is not new. This is the basically **Virtual Path Layout** in ATM, that was considered in the literature [21, 19, 13, 2, 20, 8, 4, 12, 9, 10] in the blossoming year's of ATM protocol. ATM [14] is also protocol based on circuits, where a circuit is a path defined between a source and a destination. In ATM, a VC or a Virtual Channel connects between users and may concatenate many VPs (virtual paths). The main rule of VP is an internal network usage, of reducing the cost of management the establishment of VCs. The concatenation of paths in ATM is done by using the VCI/VPI labels, where VCI is the index of a VC and VPI is the index of a VP. Basically the VCI is the glue that concatenates several VPIs.

Our first contribution in this paper, is giving tight (up to a constant multiplicative factor) upper and lower bounds for the worst case maximum load on any router in a tree network. This result also shades some light on the corresponding open issue in ATM networks.

The second contribution in this paper, is to model the problem of path layout in three different label switching models, ATM, MPLS and a new suggested extended version of MPLS, Trainet. We examine the different power of the models by demonstrating them on the problem of path layout on tree networks.

One of the key differences between MPLS and ATM is caused by the path merging capability in MPLS. MPLS permits tagging all the packets with the same remaining path with the same label, even if there initial path was different [5]. Hence when a label switch path is built in the network, one may use all its, with no extra charge of building and tagging these paths with different labels.

The third model, the Trainet Model, is a model suggested by [1]. The main idea, of this extended MPLS model, is to increase the utilization of each label by using it to define each sub-path of a labeled switched path (LSP) as a legal route. It is achieved by replacing the label with a <label,counter> pair. At each switch the counter shows how many more hops the corresponding packet should take on this LSP. This is much like taking a subway, where the packet has to go off the train after a specified number of stations. This model resembles the idea of Manhattan networks [16].

The outline of the rest of the paper is as follows. The different models are formalized in Section 2. We discuss related work in Section 3. The main result of the paper is presented in Section 4. The algorithms for different cases are given in Section 5. The lower bounds for different cases are given in Section 6. Section 7 explores some results for the different models, which are derived from the main result. We finish the paper with some conclusions.

2   Model

All label switching models establish switch-paths. A packet moving in a network, may move only along a switch-path. Each packet is assigned a label, corresponding to the switch path it is supposed to use . A packet traverses its path using a switching technique. We begin with defining switch-paths. Virtual Paths (in ATM), Label Switch Paths (in MPLS) and Trainet Switch Paths (in Trainet) are all instances of a switch-paths.

We model the network, as a bidirectional graph $G = (V, E)$, where the nodes are routers and the edges are physical connections between the routers. In some cases (where we are interested in paths from a single source) we consider directed graphs.

Definition 2.1. *A switch-path SP in graph G is a simple directed path.*

The label switching models differ in the number and the type of different routes (sub-paths of switch-paths) that a packet may traverse using a specific switch-path of the specific model. Metaphorically, the switch-path resembles a directed express train-line, which the packets can use for fast switching. The routes that packets may use are all sub-paths of the train line. The ATM switch-paths, called also Virtual Paths, define only one route each. In order to use the path, a packet must begin its trip in the first station, the source of the switch-path, and may leave only in the last station, the destination. In MPLS, the packet may enter the train-line at any station, but must stay till the end of the switch-path. A label switch path in MPLS defines a subset of routes, each of which corresponds to a suffix of the label switch-path. In the Trainet model, a packet may enter the train-line at every station, and get off at any later station. A Trainet switch-path defines a set of routes, each of them is some sub-path of the switch-path i.e., each path in the Trainet model introduces all its sub-paths as routes. Next, we give definitions, which formally define the routes that are induced by each type of switch-path.

Definition 2.2. *Let V be a Virtual Switch Path. We define the set of routes that are induced by VSP in the ATM model, simply as $\{V\}$.*

Definition 2.3. *Let L be a Label Switch Path. We define the set of routes that are induced by L in the MPLS model, as the set of routes $\{r|r$ is suffix of $L\}$.*

Definition 2.4. *Let T be a Trainet Switch Path. We define the set of routes that are induced by T in the Trainet model, as the set of routes $\{r|r$ is (a non-empty) sub-path of $T\}$.*

In the next definition, we define the **path layout** in a Model M (where M can be ATM, MPLS or Trainet).

Definition 2.5. *Let $G = (V, E)$ be a directed graph. Let P be a collection of routes in G. A collection of switch-paths P' in G is **a k-hop path layout** of P in the model M, iff for every $r \in P$, it can be created by concatenating of at most k valid routes induced by P' in the model M.*

Our paper focuses on finding a path layout that minimizes the size of the largest label-table (or the switching table) of any router. A label-table of a router is the table that contains the database of the labels. The size of this table for a given router vertex is also called **the vertex load**. Next we give a formal definition of vertex load in the different Models. Similarly to the load of a vertex, one can define the load on an edge. We prefer to discuss the vertex loads due to their direct connection to the sizes of the label-tables. The vertex load of node v is denoted $L(v)$.

Definition 2.6. *Let VSP be the set of all the virtual paths in an ATM networks. Let $VSP(v)$ be the subset of paths $p \in VP$ such that $v \in p$. $L(v) = |VSP(v)|$.*

Definition 2.7. *Let LSP be the set of all label switch paths p in a MPLS network. Let LSP' be the set of all routes induced by LSP. Let $LSP'(v)$ be the set of all routes p in LSP', such that v is the first vertex in p. $L(v) = |LSP'(v)|$. Note that this definition captures the merging of labels capabilities of the MPLS model.*

Definition 2.8. *Let* TSP *be the set of Trainet switch paths in a Trainet Network. Let* TSP' *be the set of all routes induced by* TSP. *Let* $TSP'(v)$ *be the subset of all routes in* TSP' *such that* v *is the first vertex in* p. *Let* $TSP''(v)$ *be a minimal subset of* $TSP'(v)$ *such that for each* $p' \in TSP'(v)$ *there exists a route* $p'' \in TSP''(v)$ *such that* p' *is a prefix of* p''. $L(v) = |TSP''(v)|$. *Note again, that this captures the merging of labels capabilities of the Trainet model.*

We say that a path layout has vertex load m if the maximum load on any vertex in the graph is m. Our paper deals with finding a k-hop path layout, that requires minimum vertex load, and hence minimizes the memory used for the incoming label-table at the router. The paper focuses on the case of a tree network where the basic set of paths P is naturally a set of shortest paths. In the following definitions, we define three problems of path layout in a tree, one-to-all path layout, all-to-one path layout, and all-to-all path layout.

Definition 2.9. *Let* r *be a designated vertex in graph* G. *A* **one-to-all path layout** *is a path layout where* r *has shortest path routes to all other nodes in* G.

Definition 2.10. *Let* r *be a designated vertex in graph* G. *A* **all-to-one path layout** *is a path layout where all the vertices in* G *have shortest path routes to* r.

Definition 2.11. *A* **all-to-all path layout** *is a path layout in* G, *where every node has a shortest path route to every other node in the graph.*

Throughout the paper, the symbol n denotes the number of nodes in the network. The **size** of a tree is defined to be the number of nodes in it.

3   Related Work

While there is extended literature about the virtual path layout in general ATM networks [2, 20, 8, 4, 12, 9–11], there is no work, to the best of our knowledge, that deals with the new model of label switch path layout.

The paper of Gerstel, Cidon and Zaks deals with the problem of virtual path layout in a tree network [9]. In that paper, an optimal polynomial algorithm for finding a virtual path layout (in ATM model) with minimum edge load was introduced. Gerstel and Zaks [13] provided a greedy algorithm that constructs a layout with smallest possible vertex load. However, to the best of our knowledge, there is no tight bound known for vertex loads in tree networks. The paper [13] does not analyze the resulting optimal vertex load found by the algorithm. The results of [9, 13] are important and raise many further questions. There is practical need in knowing the maximum resulting load, the reason to study the vertex load is that it is equivalent to the maximum size of the label-table of any vertex. Our paper, fills this gap, and gives a lower bound for the minimum vertex load that any algorithm can find. We feel that our lower bound, is an important step in reaching a real understanding and estimating the gain by using path concatenation.

Our paper also introduces polynomial algorithms, which give a tight upper bound. The paper [10] gives an upper bound of $O(\sqrt{n})$ of edge load for the 2-hop path layout, which leads to vertex load of $O(d\sqrt{n})$. In this paper we give algorithms for the general

k-hop path layout problem. Specifically, for the special case of the 2-hop path layout, our algorithm gives a vertex load of only $O(\sqrt{dn})$.

To the best of our knowledge, our work is also the first paper to model the virtual path problem in the MPLS and Trainet models.

4   Main Results

In the rest of the paper, we prove the following theorem, and derive some results on the different models from it. We consider a network which is a rooted directed tree. We define d to be the maximum out-degree of any node in the directed tree. It is also possible to see the tree as undirected. This does not change the result since the maximum degree in that case is at least d and at most $d + 1$ and a change in d by a multiplicative factor of at most two does not affect the results. I.e., the following theorem gives tight bounds on the worst case vertex load for the one-to-all k-hop path layout in tree networks. The theorem holds in all three models.

Theorem 4.1. *Given a rooted directed tree with n nodes and maximum degree d. For an integer $k > 1$, the following function f_k describes the maximum size of the switching table of any node (i.e, the vertex load), for a fixed value of k (maximum number of concatenated paths).*
If k is even then

$$f_k(n,d) = \begin{cases} \Theta(n^{\frac{1}{k}} d^{\frac{k-1}{k}}), & \text{for } d \in [1, n^{\frac{1}{k+1}}]; \\ \Theta((nd)^{\frac{2}{k+2}}), & \text{for } d \in [n^{\frac{1}{k+1}}, n^{\frac{2}{k}}]; \\ \Theta(d), & \text{for } d \in [n^{\frac{2}{k}}, n]. \end{cases}$$

If k is odd then

$$f_k(n,d) = \begin{cases} \Theta(n^{\frac{1}{k}} d^{\frac{k-1}{k}}), & \text{for } d \in [1, n^{\frac{1}{k+1}}]; \\ \Theta(n^{\frac{2}{k+1}}), & \text{for } d \in [n^{\frac{1}{k+1}}, n^{\frac{2}{k+1}}]; \\ \Theta(d), & \text{for } d \in [n^{\frac{2}{k+1}}, n]. \end{cases}$$

There is one exception which is the case $d = 1$ in the Trainet model, for which $f_k(n, d) = 1$.

To prove the theorem, we need to show algorithms for all cases, and to give negative examples to all cases as well. We do that in the next sections.

5   Algorithms

We start with designing the algorithms, and construct the lower bounds in the next section. We define a procedure DECOMPOSE. This procedure is used by the algorithms for all the different cases. The procedure receives a rooted directed tree T' of size m, and an integer $\ell \le m$ and returns a node v. v is the root of a minimal sub-tree of size at least ℓ (we require the sub-tree to contain all the descendants of v in T'). Note that since the sub-tree is minimal, the size of the sub-trees of the children of v are smaller than ℓ. The operation of DECOMPOSE is iterative and simple; It moves a pointer along a descending path in the tree. The pointer starts at the root of T, and goes along a directed edge to a child whose sub-tree is of size at least ℓ, as long as such a child (whose sub-tree is large enough) still exists. We later show how to combine several iterations of decompose into a single run of DFS.

All algorithms have a similar general structure. The original tree T is decomposed recursively into smaller and smaller trees, until the trees are small enough. In each decomposition step, some switch-paths are defined.

Case 1: An algorithm for even or odd k and $d \in [1, n^{\frac{1}{k+1}}]$

Let $\ell_0, \ldots, \ell_{k-1}$ be a monotonically decreasing sequence such that $\ell_0 = n$. The values ℓ_i serve as thresholds for sizes of trees (numbers of nodes). We define sets of roots $R_0, \ldots R_{k-1}$ in the following way. $R_0 = \{r\}$, where r is the root of T. R_i contains nodes, which are defined to be roots of sub-trees (those are sub-trees which contain a subset of their descendants). such sub-trees have sizes of at most ℓ_i. Note that the sets are not necessarily disjoint. Moreover, each vertex $r_i \in R_i$, has an associated vertex in R_{i-1}, which is expressed by a function $f_i : R_i \rightarrow R_{i-1}$. We explain below how to create the sets and the functions.

The algorithm uses the term "trees of type i". Those are trees whose root is in R_i. Clearly the size of a tree of type i ($0 \le i \le k - 1$) is of size at most ℓ_i. The original tree T is a type 0 tree. A root of a tree of type i, i.e. a node of R_i, is called a root of type i. The root of T is also called *the main root*.

Decomposition into smaller trees: The algorithm uses the set of thresholds to recursively define sets of trees of type i for each i. Starting with T (which is the only tree of type 0), trees of type $i + 1$ are simply defined by decomposing trees of type i.

The structure of switch-paths: Switch-paths are defined in a way that for each root r of type $i + 1$, either there exists a root of type i ($f_{i+1}(r)$) with a switch-path going to r, or r itself is a type i root. The exact way to do this is defined together with the defining the decomposition algorithm.

Finally each root of type $k - 1$ is defined to have switch paths to all the nodes in its sub-tree (of type $k - 1$). The value of ℓ_{k-1} has to be relatively small so that roots of type $k - 1$ do not have a large vertex load.

Using the above set of paths we get that the set of all concatenations of at most $k-1$ switch-paths from the main root, allow us to get to every root of type $k - 1$. Using the switch paths from roots of type $k - 1$ we get that it is possible to use a concatenation of at most k switch-paths to get from the main root to any node.

An algorithm for decomposing a tree of type i into trees of type $i + 1$: Given a tree T_i of type i, fix $\ell_{i+1} = n^{\frac{k-i-1}{k}} d^{\frac{i+1}{k}}$ and do the following until the tree T_i is left with at most ℓ_{i+1} nodes: (If already $|T_i| \le \ell_{i+1}$, then the tree is not decomposed further). Run DECOMPOSE with the parameter ℓ_{i+1} on the tree, and get a sub-tree rooted at x. Add all paths from the root of T_i to x and its direct children to the set of switch-paths, and remove this sub-tree from T_i. Let x be the root of the removed sub-tree. Each direct child y of x is defined to be a root of type $i + 1$ (the tree of type $i + 1$ contains all the descendants of y that were not removed previously from the tree). Since the sub-tree of x is minimal, the size of the sub-tree of each child of x is of size at most ℓ_{i+1}. The remainder of T_i (after all removals) is also defined to be a tree of type $i + 1$. Recall that by definition, the remainder is of size at most ℓ_{i+1}.

Analysis of the vertex load: The number of times that DECOMPOSE was run (for a given i) is at most $\frac{\ell_i}{\ell_{i+1}}$. Consider the labels given only while decomposing trees of type i into trees of type $i + 1$. For each sub-tree of type $i + 1$, at most $d + 1$ new switch-paths are defined, hence the root of T_i is assigned at most $(d + 1)\frac{\ell_i}{\ell_{i+1}}$ new labels. Using $\ell_i = n^{\frac{k-i}{k}} d^{\frac{i}{k}}$, we get $\frac{\ell_i}{\ell_{i+1}} = n^{\frac{1}{k}} d^{-\frac{1}{k}}$ and the number of new labels assigned to each root of type $i < k - 1$ is at most $(d + 1)n^{\frac{1}{k}} d^{-\frac{1}{k}} = \Theta(n^{\frac{1}{k}} d^{\frac{k-1}{k}})$ labels. For $i = k - 1$, the size of each tree of type $k - 1$ is at most $\ell_{k-1} = n^{\frac{1}{k}} d^{\frac{k-1}{k}}$, hence each root of type $k - 1$ also has $O(n^{\frac{1}{k}} d^{\frac{k-1}{k}})$ new labels defined for switch-paths to its descendants. To calculate the total number of labels that any node has received, note that each node in the tree T belongs to at most one tree of each type. While considering trees of type i, all switch-paths were defined inside the trees. In the worst case, a node in a tree of type i belongs to all switch-paths starting at its root. Since there are k types of trees, and k is constant, each node has at most $\Theta(n^{\frac{1}{k}} d^{\frac{k-1}{k}})$ labels of different paths.

Case 2: Algorithm for odd k and $d \geq n^{\frac{1}{k+1}}$

The algorithm is similar to the previous algorithm. We specify the differences between this case and case 1. Instead of k types of trees, there are only $\frac{k+1}{2}$ types. The number of thresholds is naturally also $\frac{k+1}{2}$, and again $\ell_0 = n$. Clearly, the number of recursive decomposition phases is also smaller.

The structure of switch-paths: For each root of type $i + 1$ ($i > 0$) there exists a concatenation of at most two switch-paths to it from some root of type i (the root of type $i + 1$ may be already a root of type i, this can be seen as a special case). In the trees of type $\frac{k-1}{2}$, there is a switch-path from each root to all the nodes in the sub-tree.

The maximum number of switch-paths that need to be concatenated to be able to get from the root to some node is two times the number of phases of decomposition, plus 1 (for the switch-paths from every root of type $\frac{k+1}{2}$ to its descendants). This number is exactly k.

An algorithm for decomposing a tree of type i into trees of type $i + 1$: We set $\ell_{i+1} = n^{\frac{k-2i-1}{k+1}}$. To decompose a tree of type i, T_i, into trees of type $i + 1$, we again run DECOMPOSE with the parameter ℓ_{i+1}, removing each tree until $|T_i| \leq \ell_{i+1}$. Let x be a root of a removed sub-tree. A switch-path is fixed from the root of T_i to x, and switch-paths are also fixed from x to all its direct children. The sub-tree of each direct child is defined to be a tree of type $i + 1$. The remainder of T_i after all removals is also defined to be a tree of type T_{i+1}.

Analysis of the vertex load: To calculate the number of labels given to a node while considering trees of type i, note that there are two cases. Nodes got labels either on a switch-path from a root of type i, or on a switch-path of length 1 (to roots of type $i + 1$ from their direct parents). In the first case, a root of type i is assigned at most $\frac{\ell_i}{\ell_{i+1}}$ labels for $i < \frac{k-1}{2}$ and at most ℓ_i for $i = \frac{k-1}{2}$. This equals to $n^{\frac{2}{k+1}}$. In the second case, a node is defined to have switch-paths going to its children, which gives at most d labels for a node. Hence the maximum number of labels given to one node is $O(\max(d, n^{\frac{2}{k+1}}))$.

Using similar reasoning as in the first case, the total number of labels given to one node in total is also $O(\max(d, n^{\frac{2}{k+1}}))$.

Case 3: Algorithm for even k, $d \in [n^{\frac{1}{k+1}}, n^{\frac{2}{k}}]$

We use a combination of the first and the second algorithms. We omit the details

Case 4: Algorithm for even k, $d \geq n^{\frac{2}{k}}$

Let $a = \lceil \frac{\log n}{\log d} \rceil - 1$. We run the second algorithm with $a + 1$ types of trees $\ell_i = \frac{n}{d^i}$, for $0 \leq i \leq a$. We omit the details

Running time: We assume that the algorithm needs to gather information, but does not need to construct the paths. The decomposition of a tree of type i into trees of type $i+1$ can be done e.g. while running a Depth First Search on the tree. Each node may have a counter of the size of its sub-tree. As soon as the search returns to a node with enough descendants, switch-paths from the root are defined, and the counter is set to zero. We run this once for each tree (may be done recursively, running a new DFS when a node is declared to be a root of some type), The running time is $O(n)$ for each type of trees (each node or edge belongs to at most one tree of each type). The total running time is $O(kn)$. We can conclude that the running time is linear in n.

The upper bounds above assume that k is a constant, which is the case in reality. However, even if k is non-constant, it is possible to get the same upper bound in the Trainet model. The details are omitted.

6 Lower Bounds

The lower bounds presented in this section are valid for all three models. Whenever, we use routes, in the proof, the routes are induced by the switching path according to any of the models. To prove *lower bounds* we introduce several trees in which at least one node has a large switching table.

If $d = 1$, then the tree is a chain. We need to show that there exists a node with $\Omega(n^{\frac{1}{k}})$ labels in the ATM and in the MPLS models. Assume to the contrary that all nodes have a table of size at most $\frac{1}{3}n^{\frac{1}{k}}$. Starting at the root of T, using a concatenation of exactly i routes, the root has access to at most $(\frac{1}{3}n^{\frac{1}{k}})^i$ different nodes. Using a concatenation of at most k routes we get access to at most $1 + (\frac{1}{3}n^{\frac{1}{k}}) + (\frac{1}{3}n^{\frac{1}{k}})^2 + \ldots (\frac{1}{3}n^{\frac{1}{k}})^k \leq 1 + \frac{1}{3}n + (\frac{1}{3})^2n + \ldots \leq 1 + \frac{n}{2} < n$ (for $n \geq 3$). Contradiction.

For other cases, we build trees of sizes at most n that give negative examples.

Case a: Lower bound for $d \in [2, n^{\frac{1}{k+1}}]$

We define the following tree types. Let $n' = n/2$. Let T_1 be a maximum complete binary tree with at most $(\frac{n'}{d})^{\frac{1}{k}}$ leaves and T_2 a maximum complete binary tree with at most $(\frac{n'}{d^{k+1}})^{\frac{1}{k}}$ leaves (if $\frac{n'}{d^{k+1}} < 1$, or $(\frac{n'}{d})^{\frac{1}{k}} < 1$ use a tree which consists of a single node instead of a binary tree. Note that $d^k < n$).

Let T'_1 (resp. T'_2) be a T_1 (resp. T_2) tree such that each leaf has d children. We define a concatenation of two trees S_1 and S_2 as follows : $S_1 S_2$ is rooted at the root of S_1 and each leaf of S_1 has a S_2 as a sub-tree (the leaf of the S_1 is a root of an S_2). Let $T = T'_1(T'_2)^{k-1}$ (See Figure 1). Note that T has at most n' leaves, and at most $2n' = n$

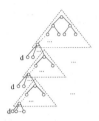

Fig. 1. Illustration of the lower bound tree of Case a

nodes. We need to show that some node has $\Omega(n^{\frac{1}{k}}d^{\frac{k-1}{k}})$ labels. For a set of vertices Q in a tree T, let $D(Q)$ denote the set of all descendants of nodes in Q (including Q i.e. $Q \subseteq D(Q)$). For a tree T with a given switching table, where the root can reach any node concatenating k routes or less, let $T(i)$ ($0 \le i \le k$) denote the set of nodes that are reachable from the root by at concatenating at most i routes, or some descendant of the node is reachable. It is easy to see that $T(0)$ is the root and $T(k)$ is the set of all nodes.

Assume that at most $\frac{1}{4}(n'^{\frac{1}{k}}d^{\frac{k-1}{k}})$ labels for the root and at most $\frac{1}{8}(n'^{\frac{1}{k}}d^{\frac{k-1}{k}})$ labels for any other node are enough. We will reach a contradiction which would mean that some node needs $\Omega(n^{\frac{1}{k}}d^{\frac{k-1}{k}})$ labels. Let I_0 be the set of leaves of the tree T_1', and let I_i be the set of leaves of $T_1'(T_2')^i$ for $i = 1, .., k - 1$.

We can prove the following claim using induction if we assume that the root has at most $\frac{1}{4}(n'^{\frac{1}{k}}d^{\frac{k-1}{k}})$ labels, and each other node has at most $\frac{1}{8}(n'^{\frac{1}{k}}d^{\frac{k-1}{k}})$ labels.

Claim. For every $0 \le i \le k - 1$, the set I_i contains a subset I_i' which satisfies the following properties: 1. $|I_i'| \ge \frac{d}{2}$. 2. All vertices in I_i' have the same parent node. 3. $T(i+1) \cap D(I_i') = \emptyset$ (i.e. it is impossible to get from the root of T to any of the nodes in the subset or any nodes in their sub-trees by concatenating at most $i + 1$ routes).

The claim shows that there exists a leaf node of T which does not have a concatenation of k routes from the root which leads to this leaf. This is a contradiction. (Actually we showed that there are at least $\frac{d}{2}$ such leaves).

Case b: Lower bound for even k and $d \in [n^{\frac{1}{k+1}}, n^{\frac{2}{k}}]$

We build the following tree: Let $n' = n/2$. Let T_3 be a maximum complete binary tree with at most $(\frac{n}{d^{\frac{k}{2}}})^{\frac{2}{k+2}}$ leaves (this is well defined since $d \le n^{\frac{2}{k}}$). Let T_3' be a defined analogously to T_i' in previous proof. Let $T = (T_3')^{\frac{k}{2}}T_3$. Note that T has at most n' leaves, and at most $2n' = n$ nodes. Define J_{2i-1} to be the leaves of $(T_3')^i$ ($1 \le i \le \frac{k}{2}$), J_k be the leaves of T and let J_{2i} be the parents of J_{2i+1} ($0 \le i \le \frac{k}{2} - 1$).

Assume to the contrary that at most $\frac{1}{4}(n'd)^{\frac{2}{k+2}}$ labels for the root and at most $\frac{1}{8}(n'd)^{\frac{2}{k+2}}$ labels for any other node are enough.

If the root has at most $\frac{1}{4}(n'd)^{\frac{2}{k+2}}$ labels and each other node has at most $\frac{1}{8}(n'd)^{\frac{2}{k+2}}$ labels, we can prove the following claim, which is similar to the claim for case a. The proof is omitted.

Claim. For all odd i, there exists a set $J_i' \subset J_i$ which satisfies the following properties:
1. $|J_i'| \geq \frac{d}{2}$. 2. All vertices in J_i' have the same parent node. 3. $T(i) \cap D(J_i') = \emptyset$ (i.e. it is impossible to get from the root of T to any of the nodes in the subset J_i' or any nodes in their sub-trees by concatenating at most i routes).

For even $i > 0$, there exists a set $J_i' \subset J_i$ which satisfies the following properties 1. All vertices in J_i' have the same ancestor node z in J_{i-1}, i.e. they belong to a single T_3 tree. 2. $|J_i'| \geq \frac{1}{2}|D(\{z\}) \cap J_i|$ (i.e. J_i' contains at least half of the leaves of some T_3 tree). 3. $T(i) \cap D(J_i') = \emptyset$ (i.e. it is impossible to get from the root of T to any of the nodes in the subset J_i' or any nodes in their sub-trees by concatenating at most i routes).

Using the claim for $i = k$ we get that there exists a leaf of T which is not reachable from the root by concatenating k routes. Contradiction.

Case c: Lower bound for odd k, $d \in [n^{\frac{1}{k+1}}, n^{\frac{2}{k+1}}]$

Let T_4 be a maximum complete binary tree with at most $\frac{n^{\frac{2}{k+1}}}{d}$ leaves (note that this value is larger or equal 1). Let T_4' be defined analogously to T_1'. Let $T = (T_4')^{\frac{k+1}{2}}$. The proof for this tree is very similar to the previous proof.

We omit the lower bound proof of the other cases.

7   Results in the Different Models

In this section we show two corollaries which adapt the basic algorithm to give an all-to-all path layout in general tree networks for the Trainet Model, and for directed tree networks in the MPLS model. The proofs of the claims in this section are omitted.

The special properties of the Trainet model allow us construct an all-to-all path layout with no extra cost.

Claim. Given a bidirectional tree. In the Trainet model, it is possible to build a k-hop all-to-all path layout with the same vertex load as a k-hop one-to-all path layout in the same model.

In the MPLS model, it is also possible to extend the reachability with no extra cost. This can be done in the directed rooted tree.

Claim. Given a directed tree, in the MPLS model it is possible to build a k-hop all-to-all path layout using label-tables of the same size as are used for a k-hop one-to-all path layout in the same model.

8   Conclusions

We have given tight bounds on the one-to-all path layout in trees. This leaves several open questions. Is it possible to give tight bounds for general graphs? Is it possible to give tight bounds for the all-to-all path layout? Note that for the Trainet model this paper gives a tight bound for the all-to-all path layout, but the same question for the MPLS and ATM models is still open.

Acknowledgments: We gratefully acknowledge the fruitful discussions with Haim Kaplan and Yehuda Afek from Tel-Aviv University. We also thank Rob van Stee from CWI for reading the paper and commenting on it.

References

1. Y. Afek and A. Bremler-Barr. Trainet: A new label switching scheme. In Proc. of The Conference on Computer Communications (INFOCOM 2000), pages 874–883, 2000.
2. S. Ahn, R. P. Tsang, S.-R. Tong, and David H.-C. Du. Virtual path layout design on ATM networks. In Proc. of The Conference on Computer Communications (INFOCOM 1994), pages 192–200, 1994.
3. D. O. Awduche and Y. Rekhter. Multi-protocol lambda switching: Combining MPLS traffic engineering control with optical crossconnects. IEEE Comm. Mag., 39(3):111–116, 2001.
4. J.-C. Bermond, N. Marlin, D. Peleg, and S. Perennes. Directed virtual path layouts in atm networks. Theor. Comput. Sci., 291(1):3–28, 2003.
5. R. Callon, P. Doolan, N.Feldman, A. Fredette, and G. Swallow. A framework for multiprotocol label switching, 1997. manuscript.
6. G. Chandranmenon and G. Varghese. Trading packet headers for packet processing. IEEE Transactions on Networking, 4(2):141–152, 1996.
7. B. Davie, P. Doolan, and Y. Rekhter. Switching in IP Networks. Morgan Kaufmann Publishers Inc., 1998.
8. T. Eilam, M. Flammini, and S. Zaks. A complete characterization of the path layout construction problem for ATM networks with given hop count and load. Parallel Processing Letters, 8(2):207–220, 1998.
9. O. Gerstel, I. Cidon, and S. Zaks. The layout of virtual paths in ATM networks. Transactions on Networking, 4(6):873–884, 1996.
10. O. Gerstel, I. Cidon, and S. Zaks. Optimal virtual path layout in ATM networks with shared routing table switches. The Chicago Journal of Theoretical Computer Science, 3, December 1996.
11. O. Gerstel, A. Wool, and S. Zaks. Optimal layouts on a chain ATM network. DAMATH: Discrete Applied Mathematics and Combinatorial Operations Research and Computer Science, 83, 1998.
12. O. Gerstel and S. Zaks. The virtual path layout problem in ATM ring and mesh networks. In Proc. of the First Conf. on Structure, Information and Communication Complexity (SIROCCO 1994), 1994.
13. O. Gerstel and S. Zaks. The virtual path layout problem in fast networks. In Symposium on Principles of Distributed Computing (PODC '94), pages 235–243, 1994.
14. R. Handler and M.N. Huber. Integrated Broadband Networks:an introduction to ATM-based networks. Addison-Wesley, 1991.
15. Y. Katsube, K. Nagami, and H. Esaki. Toshiba's router architecture extensions for atm: Overview. rfc 2098. Technical report, 1997.
16. N. F. Maxemchuk. Routing in the manhattan street network. Transactions on Communications, 5:503–512, 1987.
17. P. Newman, G. Minshall, and L. Huston. Ip switching and gigabit routers. IEEE Communications Magazine, 35(1):64–69, 1997.
18. Y. Rekhter, B. Davie, D. Katz, E. Rosen, G. Swallow, and D. Farinacci. Tag switching architecture overview, 1996. manuscript.
19. L. Stacho and I. Vrťo. Virtual path layouts in ATM networks. SIAM Journal on Computing, 29(5):1621–1629, 2000.
20. A. Vakhutinsky and M. Ball. Fault-tolerant virtual path layout in atm networks. In Proceedings of the 15th International Teletraffic Conference (ITC 15), pages 1031–1041, 1997.
21. S. Zaks. Path layout in ATM networks. In Theory and Practice of Informatics, Seminar on Current Trends in Theory and Practice of Informatics (SOFSEM'97), volume 24, pages 144–160. 1997.

On Approximability of the Independent Set Problem for Low Degree Graphs

Miroslav Chlebík[1],[*] and Janka Chlebíková[2],[*],[**]

[1] Max Planck Institute for Mathematics in the Sciences
Inselstraße 22-26, D-04103 Leipzig, Germany
chlebik@mis.mpg.de
[2] Christian-Albrechts-Universität zu Kiel
Institut für Informatik und Praktische Mathematik
Olshausenstraße 40, D-24098 Kiel, Germany
jch@informatik.uni-kiel.de

Abstract. We obtain slightly improved upper bounds on efficient approximability of the MAXIMUM INDEPENDENT SET problem in graphs of maximum degree at most B (shortly, B-MAXIS), for small $B \geq 3$. The degree-three case plays a role of the central problem, as many of the results for the other problems use reductions to it. Our careful analysis of approximation algorithms of Berman and Fujito for 3-MAXIS shows that one can achieve approximation ratio arbitrarily close to $3 - \frac{\sqrt{13}}{2}$. Improvements of an approximation ratio below $\frac{6}{5}$ for this case translate to improvements below $\frac{B+3}{5}$ of approximation factors for B-MAXIS for all odd B. Consequently, for any odd $B \geq 3$, polynomial time algorithms for B-MAXIS exist with approximation ratios arbitrarily close to $\frac{B+3}{5} - \frac{4(5\sqrt{13}-18)}{5} \frac{(B-2)!!}{(B+1)!!}$. This is currently the best upper bound for B-MAXIS for any odd B, $3 \leq B < 613$.

Keywords: maximum independent set, approximation algorithm, bounded degree graphs

1 Introduction

The problem of finding an independent set of maximum cardinality in a graph, the MAXIMUM INDEPENDENT SET problem (MAXIS), is one of fundamental NP-hard combinatorial optimization problems. Recall, that an independent set in a graph is a set of vertices that are mutually non-adjacent. The MAXIS problem is notoriously known for its approximation hardness. In general graphs the best

[*] The permanent address of the authors: Faculty of Mathematics, Physics and Informatics, Comenius University, Mlynská dolina, 842 48 Bratislava, Slovakia, {chlebik, chlebikova}@fmph.uniba.sk

[**] Supported by the EU-Project ARACNE, Approximation and Randomized Algorithms in Communication Networks, HPRN-CT-1999-00112 and EU-Project APPOL II, Approximation and Online Algorithms for Optimization Problems, IST-2001-32007.

R. Královič and O. Sýkora (Eds.): SIROCCO 2004, LNCS 3104, pp. 47–56, 2004.
© Springer-Verlag Berlin Heidelberg 2004

known polynomial time approximation algorithm achieves an approximation ratio $O(\frac{n}{\log^2 n})$ [3]. The result of Håstad [7] explains this phenomenon by providing a strong lower bound of $n^{1-\varepsilon}$ for any $\varepsilon > 0$, under a reasonable complexity theoretic assumption that NP-complete problems do not have randomized polynomial time algorithms.

The restricted versions of B-MaxIS with very small degree bound B (like $B = 3$, 4, 5) have turned up to be important as intermediate steps to many other problems, e.g., the Metric Travelling Salesman problem. Currently the best lower bounds are $\frac{95}{94}$ for 3-MaxIS, $\frac{48}{47}$ for 4-MaxIS, and $\frac{46}{45}$ for 5-MaxIS, obtained in [4]. Namely, it is NP-hard to achieve an approximation within the corresponding multiplicative factor for each of problems. On the other hand, approximation ratios of algorithms have been continuously improved by a number of new techniques and better analysis. The best upper bounds on approximability for B-MaxIS with small degree bound $B \geq 3$ are provided by algorithms of Berman, Fujito, and Fürer ([1], [2]) with approximation ratios arbitrarily close to $\frac{B+3}{5}$. The basic ingredient of their heuristics is a type of a local search algorithm that additionally searches in the complement of a current solution. Such local search algorithm, taking advantage of degree boundedness, searches for an improvement of a current feasible solution in far distance. To accelerate its performance even further, some general reduction methods and tricks are incorporated in the most difficult case, when $B = 3$.

Asymptotically, better lower and upper bounds have been obtained. Trevisan [8] has proved NP-hardness of approximation for B-MaxIS within $\frac{B}{2^{O(\sqrt{\log B})}}$. Halldórsson and Randhakrishnan [6] presented an $O(\frac{B}{\log\log B})$-approximation via subgraph removal technique, and then Vishwanathan (firstly documented in [5]) has observed that an improved approximation factor of $O(\frac{B\log\log B}{\log B})$ can be obtained using semi-definite programming relaxation combining known combinatorial and algorithmic results on Lovász theta function and k-colorable graphs.

Despite this fact, for small values of B (namely, for $3 \leq B < 613$), the upper bound $\frac{B+3}{5} + \varepsilon$ hasn't been improved over the years.

1.1   Our Contribution

In this contribution we improve slightly the upper bounds on efficient approximability for B-MaxIS in case of small degree bound B. We essentially follow the approach of Berman and Fujito and make the careful analysis of their algorithms for 3-MaxIS presented in [1]. This results in approximation ratio arbitrarily close to $3 - \frac{\sqrt{13}}{2}$ for the problem.

As a consequence, the approximation threshold for B-MaxIS is also strictly less than $\frac{B+3}{5}$ for any odd $B \geq 3$. This follows from the recursive method of Berman and Fürer [2] that an efficient r_{B-2}-approximation algorithm for $(B-2)$-MaxIS can be used to design the one for B-MaxIS with an approximation ratio r_B arbitrarily close to $1 + \frac{B-2}{B+1} r_{B-2}$. Consequently, a polynomial time algorithm for B-MaxIS (with odd $B \geq 3$) exists with an approximation ratio arbitrarily close to $\frac{B+3}{5} - \frac{4(5\sqrt{13}-18)}{5}\frac{(B-2)!!}{(B+1)!!}$.

This is now the best upper bound for B-MAXIS for any odd B, $3 \le B < 613$. It is worth pointing out that using Nemhauser-Trotter Theorem this also leads to better than $(2 - \frac{5}{B+3})$-approximation for the MINIMUM VERTEX COVER problem in graphs with maximum degree B.

In [1] Berman and Fujito declare that their algorithms for 3-MAXIS in 3-regular graphs have approximation ratios arbitrarily close to $\frac{7}{6}$. However, we show that the arguments from [1] are flawed and this result can't be taken as guaranteed by their proof.

1.2 Preliminaries and Notation

For a simple graph $G = (V, E)$, let $\alpha(G)$ denote the size of a maximum independent set in G. For a vertex subset $U \subseteq V$, let $G(U)$ denote the subgraph of G induced by U and $N(U)$ the set of neighbors of U, i.e., $N(U) = \{v \in V : \exists u \in U \text{ such that } (u, v) \in E\}$.

The local search method is based on augmentation of an independent set U by a vertex set called an *improvement*. A subset $I \subseteq V$ is an s-*improvement* for U, if $G(I)$ is a connected graph, the symmetric difference $U \oplus I$ is an independent set larger than U, and $|I \setminus U| \le s$. A solution is s-*optimal* if it has no s-improvement.

The algorithm of Berman and Fürer ([2]) for building a large independent set in low degree graphs $G = (V, E)$ is a local search in two directions. For some constant parameter k, a solution A is repeatedly augmented by applying improvements of size $O(k \log |V|)$. When such augmentation is no longer possible and G was of maximum degree 3, one can find efficiently a maximum independent set A' in its complement $G(V \setminus A)$, because it is a graph of maximum degree at most 2 (otherwise A was not 1-optimal). If $|A'| > |A|$, the local search is started again from A'. The resulting solution A is $O(k \log |V|)$-optimal and $\alpha(G(V \setminus A)) \le |A|$, which together assure $(\frac{5}{4} + \frac{1}{k})|A| \ge \alpha(G)$. Moreover, this inequality is tight ([2]). Berman and Fujito in [1] suggested some additional reductions to improve the performance ratio for 3-MAXIS to $\frac{6}{5} + \varepsilon$.

2 Approximation Algorithm for B-MaxIS in Low Degree Graphs

The algorithm $\mathcal{A}_k$ presented in Fig. 1 is Berman and Fujito's algorithm for 3-MAXIS ([1]) with some minor changes.

The algorithm $\mathcal{A}_k$ consists of four phases: (1) preprocessing, (2) a construction of initial solution, (3) a local search, and (4) postprocessing. A set S and a stack are used in preprocessing and postprocessing phase for the construction of a solution for an input graph G.

(1) In the preprocessing phase Branchy, Nemhauser-Trotter, and Small Commitment reductions are applied to a graph as long as it is possible to reduce the graph.

Branchy reduction: (i) While there is a vertex v of degree 0 or 1 in G, add v to the set S, and remove it (along with its neighbor, if any) from G, (ii) while

Input: A graph $G = (V, E)$ of degree 3
{**preprocessing phase**}
repeat
    oldsize := $|V|$;
    do Branchy reduction;
    do Nemhauser-Trotter reduction;
    do Small Commitment reduction;
until $|V| =$ *oldsize*;
{**construction of initial solution (for reduced graph)**}
$A_1 := \{v \in V : d(v) = 2\}$;
if $A_1 = \emptyset$ **then** $A_2 := \{$any 1-optimal set$\}$
        else $A_2 := \{$an independent set recursively computed by $\mathcal{A}_k$
               in $G(V \setminus (A_1 \cup N(A_1)))\}$;
$A := A_1 \cup A_2$;
{**local search for improvement**}
repeat
    oldsize := $|A|$;
    do all possible improvements of size $\max\{3k \log |V|, 4k + 2k \log |V|\}$ to A;
    do Acyclic Complement procedure;
    Find an optimal solution A_3 in $G(V \setminus A)$;
    if $|A_3| > |A|$ **then** $A := A_3$;
until $|A| =$ *oldsize*;
{**postprocessing phase**}
Reconstruction of a solution for an input graph G

Fig. 1. The algorithm $\mathcal{A}_k$

there is a path $P = [v_1, v_2, v_3, v_4]$ with both vertices v_2 and v_3 of degree 2, push P into the stack, remove v_2 and v_3 from G and, if (v_1, v_4) is not an edge in G, add it to G.

    Small Commitment reduction: If for some vertex subset $U \subseteq V$ of size at most k, U is a maximum independent set in $G(U \cup N(U))$, add U to S, and remove $U \cup N(U)$ from G.

    Nemhauser-Trotter reduction: This reduction in polynomial time partitions a set V into subsets V_1, V_2, V_3 such that (i) there is a maximum independent set containing all the vertices of V_1 but none of V_2, (ii) there is no edge between V_1 and V_3, and (iii) $\alpha(G_{V_3}) \leq \frac{1}{2}|V_3|$. If $V_3 \neq V$, then add V_1 to S and remove V_1 and V_2 from G.

Definition 1. *Let $\mathcal{P}(k)$ denote the set of graphs of maximum degree at most 3 that are not reduced by the preprocessing phase of the algorithm $\mathcal{A}_k$.*

    Obviously, each graph G from $\mathcal{P}(k)$ has the following properties:

(P1) each vertex is of degree 2 or 3,
(P2) any two vertices of degree 2 are nonadjacent,
(P3) $\alpha(G) \leq \frac{1}{2}|V|$, and
(P4) every vertex set U of size at most k is not a maximum independent set in
      $G(U \cup N(U))$.

(2) An initial solution for a local search is constructed with preference given to vertices of degree 2. Let A_1 be the set of all degree 2 vertices in the graph obtained by preprocessing. This is an independent set due to (P2). If A_1 is empty, then we take any 1-optimal independent set as an initial solution A. Otherwise, we first obtain an independent set A_2 applying recursively $\mathcal{A}_k$ to $G(V \setminus (A_1 \cup N(A_1)))$, and take $A := A_1 \cup A_2$ as an initial solution we start with.

(3) The algorithm $\mathcal{A}_k$ makes improvements of A in the following steps:

  (i)   it makes A to be $O(k \log |V|)$-optimal;
 (ii)   without decreasing its size, A is improved by Acyclic Complement procedure to be an independent set with acyclic complement;
(iii)   the MAXIS problem is optimally solved for a graph $G(V \setminus A)$ of degree at most 2, and if the solution is larger, A is updated.

This is repeated in a cycle until no more improvements are achieved after steps (i)–(iii). We start with a local improvement (compare with [1]) to ensure that Acyclic Complement procedure always receives a 2-optimal solution as its input.

  Acyclic Complement procedure: If I is a 2-optimal independent set in a connected graph $G = (V, E)$ of maximum degree 3 (and G is not a complete graph on 4 vertices) this procedure finds in polynomial time an independent set I' of size $|I'| \geq |I|$ such that $G(V \setminus I')$ is an acyclic subgraph (see [1] for details).

(4) In the postprocessing phase the solution for an input graph G is constructed from the solution A found in the phase (3) for its reduced graph (obtained from G by preprocessing phase). The whole set S produced by Branchy reduction (i), Nemhauser-Trotter reduction, and Small Commitment reduction can be added to the solution A. For any path $[v_1, v_2, v_3, v_4]$ pushed into the stack by Branchy reduction (ii), we can add either v_2 or v_3 to A and keep it independent.

Remark 1. It can be seen that the effect of all reductions from the preprocessing phase is no loss in approximation quality. Hence, to prove that $\mathcal{A}_k$ approximates 3-MAXIS within $1 + \delta$, it is enough to do so for graphs from $\mathcal{P}(k)$.

  To find the upper bound on the approximation ratio of the algorithm $\mathcal{A}_k$, we consider a graph $G = (V, E)$ from $\mathcal{P}(k)$ as an input and let A' be the approximate solution provided by $\mathcal{A}_k$. We try to find relations between its size and the size of an optimal solution. By B' denote one fixed optimal solution in G such that $|A' \cap B'|$ is of the maximum size, and among all sets with this property B' contains maximum number of degree 2 vertices. Partition V into four sets, A ("approximative"), B ("best"), C ("common"), and D, such that $C := A' \cap B'$, $A := A' \setminus C$, $B := B' \setminus C$, $D := V \setminus (A \cup B \cup C)$. For the reader's convenience, we use the same notations as in [1].

  Based on this partition of V, every vertex in D will be further classified according to its neighboring subsets. A degree 3 vertex is of type $[XYZ]$ if its neighbors belong to vertex sets X, Y, Z, where each set X, Y, and Z, is one of sets A, B, C, D. In the same way we define for a degree 2 vertex its type $[XY]$. Due to our assumptions that B' has maximum possible intersection with A', and additionally it has maximum number of degree 2 vertices (relatively to the first condition), there are only 4 types for degree 2 vertices from D ($[AC], [BC], [CC], [CD]$) and 13 types for degree 3 vertices from D ($[AAB]$, $[AAC], [ABB], [ABC], [ABD], [ACC], [ACD], [BBC], [BCC], [BCD], [CCC]$, $[CCD], [CDD]$). Denote the cardinalities of sets A, B, C, and D, by a, b, c, and d, respectively, and the number of degree 2 vertices in sets A, B, and C, by a_2, b_2, and c_2, respectively. The number of D-vertices of type $[XYZ]$ (resp., $[XY]$) is denoted by its lower case counterpart, $[xyz]$ (resp., $[xy]$).

The purpose of the preprocessing and the acyclic complement procedures is to normalize a given graph and temporary solutions so that the final approximate solution has properties which allow us to compare its size with the size of an optimal solution. The proof of the following relations (1)–(7) among sizes of vertex subsets of the partition defined above can be found in [1].

$$\frac{k+1}{k}c \geq 3[bc] + 2[bbc] + [bcd] + [cdd] + [cd] + [bcc] + [cc]$$
$$- (2[acc] + [aac] + [abc] + [acd] + [ac]) \tag{1}$$

$$\frac{k+1}{k}c \geq [bbc] + [bc] + [bcc] + [cc] + \frac{1}{2}([bcd] + [cdd] + [cd] + [ccd]) \tag{2}$$

$$c \geq (b - a) + [ccc] + [acc] + [aac] + [ac] + [cc]$$
$$+ \frac{1}{2}([cdd] + [ccd] + [acd] + [cd]) \tag{3}$$

$$\frac{k+1}{k}a \geq 3(b - a) + a_2 + 2[aab] + 2[aac] + [abb] + [abd]$$
$$+ [acc] + [acd] + [abc] + [ac] \tag{4}$$

$$d \geq c + (b - a) \tag{5}$$

$$b_2 = 3(b - a) + a_2 + ([aab] + 2[aac] + [acc] + [acd] + [ac])$$
$$- ([abb] + 2[bbc] + [bcc] + [bcd] + [bc]) \tag{6}$$

$$3c - c_2 = 3[ccc] + 2([acc] + [bcc] + [ccd] + [cc]) + [aac] + [abc] + [bbc]$$
$$+ [acd] + [bcd] + [cdd] + [ac] + [bc] + [cd] \tag{7}$$

In our proof that $\mathcal{A}_k$ has approximation ratio $\leq (1 + \delta)$, for a fixed constant δ, we want argue by contradiction. Assume that for some graphs $\mathcal{A}_k$ does not perform so well, then for minimal counterexamples we can prove some additional relation. This is given in the following Lemma which corresponds to the inequality (8) from [1] (choosing $\delta = \frac{1}{5}$).

Lemma 1. *Let k be a fixed positive integer and $\delta \in (0, 1)$ be such that the approximation ratio of $\mathcal{A}_k$ is $> 1 + \delta$. If G is a graph with minimum number of vertices for which $\mathcal{A}_k$ returns an independent set of size $< \frac{1}{1+\delta}\alpha(G)$, then G belongs to $\mathcal{P}(k)$ and*

$$a_2 + [bc] + [cc] \geq \frac{\delta}{1 - \delta}(b_2 + c_2 + [ac] + [cd]). \tag{$8_{1+\delta}$}$$

Proof. As follows from Remark 1, G belongs to $\mathcal{P}(k)$. Hence G is not changed by preprocessing steps of $\mathcal{A}_k$ and the algorithm effectively starts with the construction of an initial solution. If G is 3-regular, then the inequality trivially holds with both sides equal to 0. Hence, we can assume in what follows that the set A_1 of degree 2 vertices of G is nonempty. $\mathcal{A}_k$ takes as initial solution the set $A := A_1 \cup A_2$, where A_2 is the solution that $\mathcal{A}_k$ finds recursively for the graph $G(V \setminus (A_1 \cup N(A_1)))$ as an input.

From the set A_1, opt_0 vertices belong to $B \cup C$ and opt_i vertices have exactly i neighbors in $B \cup C$, $i = 1, 2$. Removing the set $A_1 \cup N(A_1)$ from G, we can lose at most $opt_0 + opt_1 + 2opt_2$ vertices from $B \cup C$ and hence $G(V \setminus (A_1 \cup N(A_1)))$ contains an independent set of size at least $(b + c) - (opt_0 + opt_1 + 2opt_2)$. The assumption of minimality of G implies that $\mathcal{A}_k$ finds an independent set A_2 of size at least $\frac{1}{1+\delta}(b + c - opt_0 - opt_1 - 2opt_2)$ in this smaller graph, and hence

$$|A| \geq (opt_0 + opt_1 + opt_2) + \frac{1}{1+\delta}(b + c - opt_0 - opt_1 - 2opt_2)$$
$$= \frac{1}{1+\delta}(b + c) + \frac{1-\delta}{1+\delta}\left(\frac{\delta}{1-\delta}(opt_0 + opt_1) - opt_2\right).$$

On the other hand, $|A| < \frac{1}{1+\delta}\alpha(G)$ by our assumptions. Combining the inequalities we get $opt_2 > \frac{\delta}{1-\delta}(opt_0 + opt_1)$, that can be easily translated as $(8_{1+\delta})$, even with the strict inequality in this case.$\square$

Theorem 1. *The algorithm $\mathcal{A}_k$ approximates the* 3-MaxIS *problem within* $3 - \frac{\sqrt{13}}{2} + \frac{13-\sqrt{13}}{52k}$ *in time* $O(n^{O(k)})$, *where n is the number of vertices of an input graph.*

Proof. Denote $\delta_* := 2 - \frac{\sqrt{13}}{2}$, $\varepsilon := \frac{13-\sqrt{13}}{52k}$, and $\delta := \delta_* + \varepsilon$. We will prove that an approximation ratio of $\mathcal{A}_k$ is at most $1 + \delta$ via contradiction. Let G be a graph with minimum number of vertices for which $\mathcal{A}_k$ returns an independent set of size less than $\frac{1}{1+\delta}\alpha(G)$.

By Lemma 1, G belongs to $\mathcal{P}(k)$ and $(8_{1+\delta})$ holds together with (1)–(7). Summing up inequalities and equations (1)–(7), $(8_{1+\delta})$ with respective multiplicative factors of $1+2\delta$, $2(1-2\delta)$, $4(1-2\delta)$, $3-2\delta$, $1-2\delta$, 4δ, $-1+2\delta$, $4(1-\delta)$ yields the inequality

$$\frac{k+1}{k}(a + c)(3 - 2\delta) \geq (14 - 4\delta)(b - a) + (6\delta - 1)(a_2 + c_2 + [bc]) + (2 + 4\delta)[cd]$$
$$+ (5 + 2\delta)[aab] + (7 - 2\delta)[aac] + (2 - 4\delta)([abb] + [acc] + [bbc])$$
$$+ 2([abd] + [acd] + [cdd]).$$

As clearly $\delta \in [\frac{1}{6}, \frac{1}{2}]$, all the terms in brackets are of nonnegative value. Thus the inequality holds with the right hand side simplified to $(14 - 4\delta)(b - a)$, and we rewrite it as $\frac{b-a}{a+c} \leq \frac{k+1}{k}\frac{3-2\delta}{14-4\delta}$. On the other hand, our assumption $a + c < \frac{b+c}{1+\delta}$ implies $\delta < \frac{b-a}{a+c}$. Combining these inequalities we get $\delta < \frac{k+1}{k}\frac{3-2\delta}{14-4\delta}$, which we rewrite as $k < \frac{3-2\delta}{16\delta-4\delta^2-3}$. Substituting $\delta := \delta_* + \varepsilon$, we get equivalently $k <$

$\frac{1}{4\varepsilon}\frac{3-2\varepsilon-2\delta_*}{4-\varepsilon-2\delta_*}$, where the right hand side is clearly less than $\frac{1}{4\varepsilon}\frac{3-2\delta^*}{4-2\delta^*} = \frac{13-\sqrt{13}}{52\varepsilon} = k$, a contradiction. This contradiction shows that the approximation factor of the algorithm $\mathcal{A}_k$ is at most $1 + \delta$, as required.

The time complexity of $\mathcal{A}_k$ is dominated by that of its local neighborhood search. A search space consists of all connected subgraphs of vertex set size at most $\max\{3k\log n, 4k + 2k\log n\}$. One can show that the number of such subgraphs of size at most s in a degree 3 graph with n vertices is smaller that $n3^s$. Hence, it takes $O(n^{1+3k\log 3})$ time to find a small improvement. As a solution is augmented at most n times, $\mathcal{A}_k$ runs in time $O(n^{O(k)})$. $\square$

In [2] Berman and Fürer have shown how one can use an algorithm with the approximation ratio r_{B-2} for $(B-2)$-MaxIS to obtain an algorithm for B-MaxIS with the approximation ratio r_B (for a fixed constant k) such that

$$r_B \le 1 + \frac{1}{k(B+1)} + \frac{B-2}{B+1}r_{B-2}.$$

For the approximation threshold t_B on polynomial time approximability of B-MaxIS (defined as $t_B = \inf\{r > 1 : \text{there is a polynomial time algorithm that approximates } B\text{-MaxIS within } r\}$) this implies the recurrent relation $t_B \le 1 + \frac{B-2}{B+1}t_{B-2}$.

As follows from the results of [2] and [1], $t_B \le \frac{B+3}{5}$ for any $B \ge 2$. Our Theorem 1 shows that $t_3 \le 3 - \frac{\sqrt{13}}{2}$, which in turn implies better upper bound on t_B than $\frac{B+3}{5}$ for any odd $B \ge 3$. If we rewrite t_B in the form $t_B = \frac{B+3}{5} - \varepsilon_B$, then we have that $\varepsilon_3 \ge \frac{5\sqrt{13}-18}{10}$, and the above recurrent relation reads as $\varepsilon_B \ge \frac{B-2}{B+1}\varepsilon_{B-2}$. Therefore, for any odd $B \ge 3$,

$$\varepsilon_B \ge \frac{B-2}{B+1}\cdot\frac{B-4}{B-1}\cdots\frac{5}{8}\cdot\frac{3}{6}\cdot\varepsilon_3 = 8\frac{(B-2)!!}{(B+1)!!}\varepsilon_3 \ge \frac{4(5\sqrt{13}-18)}{5}\frac{(B-2)!!}{(B+1)!!}.$$

Finally, $t_B \le \frac{B+3}{5} - \frac{4(5\sqrt{13}-18)}{5}\frac{(B-2)!!}{(B+1)!!}$ for any odd $B \ge 3$. To our knowledge, this is currently the best upper bound on polynomial time approximability of B-MaxIS for any odd B, $3 \le B < 613$. See the paper by Halldórsson and Randhakrishnan [6] for comparison of results and for the range of B, when methods used to prove asymptotic result $t_B = o(B)$ achieve upper bounds below $\frac{B+3}{5}$.

Hence, we can summarize the results as follows

Corollary 1. *The B-MaxIS problem can be approximated in polynomial time with an approximation ratio arbitrarily close to $\frac{B+3}{5} - \frac{4(5\sqrt{13}-18)}{5}\frac{(B-2)!!}{(B+1)!!}$, for every odd $B \ge 3$.*

Remark 2. Berman and Fujito in [1] claimed that their algorithm $\mathcal{A}_k$ achieve for 3-regular graphs approximation factors arbitrarily close to $\frac{7}{6}$ (depending on k). However, they proved such approximation ratio of $\mathcal{A}_k$ only for 3-regular graphs belonging to $\mathcal{P}(k)$. In the following we observe that an input 3-regular graph can be reduced by the preprocessing phase of $\mathcal{A}_k$ to the one with vertices of degree 2 as well. Hence, the fact that 3-regular graphs from $\mathcal{P}(k)$ have better approximation ratio cannot be simply extended to all 3-regular graphs.

Let us consider a gadget H with 6 vertices a, b, c, a', b', and c', and 8 edges defined as follows: the triple a, b, c, and similarly a', b', c', is a triangle in H and the triangles are connected by edges (a, b') and (a', b). If an input 3-regular graph G contains a copy of H as an induced subgraph, then Small Commitment reduction (with $k \geq 2$) can commit itself to $U := \{a, a'\}$ (respectively to $U := \{b, b'\}$) and remove all vertices H of the gadget from G. If there are many copies of H (or similar gadgets) in G, then the graph G' obtained from G by the preprocessing phase can have many vertices of degree 2.

However, we can obtain slightly better estimate $r \in (\frac{7}{6}, 3 - \frac{\sqrt{13}}{2})$ on the approximation ratio of $\mathcal{A}_k$ for 3-regular graphs, than for general 3-MAXIS. We do not focus on estimating the precise value of r in this paper and provide just a sketch how such result can be obtained.

Let G be a 3-regular graph and $G' \in \mathcal{P}(k)$ be the graph obtained from G after preprocessing phase of $\mathcal{A}_k$.

(1) If a relatively small fraction of vertices of G' is of degree 2, we can apply arguments similar to those of Berman and Fujito [1] in the proof of roughly $\frac{7}{6}$ approximation factor for 3-regular graphs in $\mathcal{P}(k)$. The method is robust enough to show that in case of "few" degree 2 vertices in G', $\mathcal{A}_k$ performs within r on G', and hence on G as well.

(2) On the other hand, if the fraction of degree 2 vertices of G' is above some fixed level, then one can prove that the number of stored vertices in S is significant, and $\alpha(G) - \alpha(G')$ is at least a fixed fraction of $\alpha(G')$. It implies, that even if $\mathcal{A}_k(G')$ is guaranteed to be only roughly within $3 - \frac{\sqrt{13}}{2}$ of $\alpha(G')$, the postprocessing phase of $\mathcal{A}_k$ will improve this approximation ratio for G to r.

References

1. P. Berman and T. Fujito. On approximation properties of the independent set problem for low degree graphs. Theory of Computing Systems, 32:115–132, 1999. (also in Proccedings of 4th Workshop on Algorithms and Data Structure, WADS 1995, LNCS 955, 449–460).
2. P. Berman and M. Fürer. Approximating maximum independent set in bounded degree graphs. In Proceedings of the 5th Annual ACM-SIAM Symposium on Discrete Algorithms, SODA 1994, pages 365–371, 1994.
3. R. B. Boppana and M. M. Halldórsson. Approximating maximum independent sets by excluding subgraphs. BIT, 32(2):180–196, 1992.
4. M. Chlebík and J. Chlebíková. Inapproximability results for bounded variants of optimization problems. In Proceedings of the 14th International Symposium on Fundamentals of Computation Theory, FCT 2003, LNCS 2751, pages 27–38, 2003.
5. M. M. Halldórson. A survey on independent set approximations. In Proceedings of the 1st International Workshop on Approximation Algorithms for Combinatorial Optimization, APPROX 1998, LNCS 1444, pages 1–14.

6. M. M. Halldórsson and J. Radhakrishnan. Improved approximations of independent sets in bounded-degree graphs. Nordic Journal of Computing, 1(4):475–492, 1994.
7. J. Håstad. Clique is hard to approximate within $n^{1-\varepsilon}$. Acta Mathematica, 182(1):627–636, 1999.
8. L. Trevisan. Non-approximability results for optimization problems on bounded degree instances. In Proceedings of the 33rd Symposium on Theory of Computing, STOC 2001, pages 453–461.

Asynchronous Broadcast in Radio Networks[*]

Bogdan S. Chlebus and Mariusz A. Rokicki

Department of Computer Science and Engineering
University of Colorado at Denver
Denver, CO 80217, USA

Abstract. We study asynchronous packet radio networks in which transmissions among nodes may be delayed. We consider the task of broadcasting a message generated by the source node. The timing of arrivals of messages is controlled by adversaries. We consider three different adversaries. The edge adversary can have a transmitted message delivered at different times to different recipients. The crash adversary is the edge one augmented by the ability to crash nodes. The node adversary can have a message received at arbitrary times, but simultaneously by all the recipients. A protocol specifies for each node how many times the message is retransmitted by this node, after it has been received. The total number of transmissions of nodes is defined to be a measure of performance of a broadcast protocol, and is called its work. The radio network is modeled as a graph and is given as input to a centralized algorithm. An aim of the algorithm could be either to find a broadcast protocol, possibly with additional properties, or to verify correctness of a given protocol. We give an algorithm to find a protocol correct against the edge adversary. The obtained protocol is work-exponential in general. This is an inherent property of the problem, as is justified by a lower bound. We develop a polynomial algorithm to verify correctness of a given protocol for a given network against the edge adversary. We show that a problem to decide if there exists a protocol, for a given network and of a specified work performance, is NP-hard. We extend some of these results to the remaining two adversaries.

1   Introduction

In this paper we consider broadcast protocols in packet radio networks. One node is designated as a source and all the other nodes are reachable from the source along directed paths. The source is an origin of a message that needs to be disseminated among all the other nodes in the network. This is achieved by relaying the message from node to node.

Radio networks studied in the literature are typically synchronous, with local clocks ticking at the same rate and producing the same round numbers. It is also usually assumed that a transmission by a node occurs within one round and occupies the whole round.

The standard mode of radio communication has the following properties:

---

[*] This work is supported by the NSF Grant 0310503.

R. Královič and O. Sýkora (Eds.): SIROCCO 2004, LNCS 3104, pp. 57–68, 2004.

(i) if a message is sent by a node, then it is simultaneously sent to *all* the out-neighbors;

(ii) a message sent at a round is delivered to the recipients in the same round, but is not necessarily *heard*, that is received successfully, by all of them;

(iii) a message is heard by a node when it is the *only* message arriving at the node in a round.

There are also the following two tacitly made assumptions. First is that the nodes given in a network specification are the only ones participating in communication protocols. Second is that communication protocols, when actually run, can control the behavior of the nodes in the sense that a protocol is specified as a schedule that determines for each step whether a transmission occurs or not in the step, the nodes follow the protocol precisely as stipulated. In this paper we introduce models of asynchronous radio communication. Our approach is by way of relaxing these two tacit assumptions, and possibly some of the ones listed as (i-iii) above.

Previous Work. Broadcasting problems have been considered in various computational settings, including centralized [1, 5–7, 15], distributed deterministic [4, 8, 9, 11, 13, 14, 21–23], and distributed randomized [2, 14, 23, 25]. A lower bound $\Omega(\log^2 n)$ for broadcasting was given by Alon, Bar-Noy, Linial and Peleg [1]; it holds even for networks with a constant-bounded diameter and centralized protocols with complete knowledge. A lower bound $\Omega(D \log n)$ was shown by Bruschi and Del Pinto [4]. It was later improved to $\Omega(n \log D)$ by Clementi, Monti and Silvestri [13]. Fault-tolerant broadcasting in radio networks was studied by Kushilevitz and Mansour [26] and by Kranakis, Krizanc and Pelc [24].

Gossiping is a natural extension of broadcasting, in which each node is a source of an individual message, and a goal is to have all the nodes learn all the messages, see [11, 27, 28]. Other problems of multiple communication were studied by Clementi, Monti and Silvestri [12] and by Bar-Yehuda, Israeli, and Itai [3]. Radio protocols that are efficient in terms of energy use were given for the wakeup-like problem by Zheng, Hou and Sha [30], and for the leader-election problem by Jurdziński, Kutyłowski and Zatopiański [19].

Broadcasting in radio networks is a special case of the wakeup problem. In that problem, initially only a source node is awoken, while the remaining nodes join in disseminating the source message after having received a wakeup message from the source. The wakeup problem for radio networks was first considered by Gąsieniec, Pelc and Peleg [17], in the case of single-hop networks, who developed deterministic and randomized protocols. A general version was considered by Chrobak, Gąsieniec and Kowalski [10]. The best known deterministic wakeup algorithm for general ad-hoc networks was given in [10] and is of time $\mathcal{O}(n^{5/3} \log n)$ performance. The paper [10] also showed how to synchronize local clocks and how to solve leader election in radio networks, relying on a solution to the wakeup problem, with only a logarithmic-time overhead. Jurdziński and Stachowiak [20] studied randomized wakeup protocols for single-hop radio networks. Indyk [18] considered constructive wakeup protocols.

2   Technical Preliminaries

A radio network is modeled as a directed graph $G = (V, A)$, where V is a set of nodes, and A is a set of directed edges. An edge $\langle x, y \rangle \in A$ is also denoted as $x \to y$. When $x \to y$ is an edge, then node x is its *start point,* and node y is its *end point,* while x is an *in-neighbor* of y and y an *out-neighbor* of x. The set of all the in-neighbors of y is denoted as $\text{IN}(y)$, and the set of all the out-neighbors of x is denoted as $\text{OUT}(x)$. We assume that a network is represented by each vertex having a list of pointers to its in- and out-neighbors.

For a node x and each outgoing edge $x \to y$, there is a *outgoing buffer* $\text{out}_x(y)$ at x associated with the edge. When the node x *sends* a message, then the message is put into *each* such a buffer $\text{out}_x(z)$ at x, for any outgoing edge $x \to z$. For a node y and an incoming edge $x \to y$, there is an *incoming buffer* $\text{in}_y(x)$ at y associated with the edge. The message in an outgoing buffer $\text{out}_x(y)$ is eventually *delivered* from the sender x to the recipient y. This is achieved by removing the message from the outgoing buffer $\text{out}_x(y)$ and putting it into the incoming buffer $\text{in}_y(x)$. At most one message can traverse an edge in a step from an out-going to the in-coming buffer. We distinguish a message being delivered from it being heard, that is successfully received. A message delivered at node x at a round is *heard* at x if it is the only message delivered to x at this round. If more than one messages arrive at a node at a round, then none is received successfully. An interpretation is that all the messages are transmitted on radio waves of the same frequency and so many arriving simultaneously interfere with each other. A message delivered to a node y is removed from the incoming buffer in the same round.

A broadcast protocol $\mathcal{B}$ is determined by a function $\mathcal{B} : V \to \mathbb{N}$ that assigns a non-negative integer $\mathcal{B}(v)$ to each node $v \in V$. Such a function is called a *repeat-broadcast function* and $\mathcal{B}(v)$ is a *repeat-broadcast value of v*. The protocol $\mathcal{B}$ determined by such a function operates as follows: once a node v receives a message, then it sends the message $\mathcal{B}(v)$ times.

Timings of deliveries of messages are controlled by an adversary, once they have been put into the outgoing buffer. The adversary aims to disrupt the flow of copies of the broadcast message through the network. The only *general constraint* imposed on adversaries is that, during each step, if some buffer is nonempty, then at least one copy of the message is transmitted in this step. Transmitting a message is done by moving it from the outgoing buffer to the in-coming one. We consider three types of adversaries, each has to act subject to the general constraint.

Edge adversary: It acts subject only to the general constraint.
Crash adversary: It is the edge adversary augmented with the power to fail nodes by crashing. The adversary chooses the vertices to crash prior to the start of an execution. The faulty nodes do not participate in executions, and may be considered as removed from the graph. This may make some nodes unreachable; such nodes are also considered to be crashed.

Node adversary: It acts subject to an additional constraint: if a message is received at node y from some other node x, then at the same step the message is received at node z for any edge $x \rightarrow z$.

The *work* $\mathcal{W}(\mathcal{B})$ of protocol $\mathcal{B}$ is defined to be equal to the sum of the numbers of all the broadcasts at the nodes of G, that is $\mathcal{W}(\mathcal{B}) = \sum_{v \in V} \mathcal{B}(v)$. We can observe that the time after which the broadcast is completed is at most $\mathcal{W}(\mathcal{B}) = \sum_{v \in V} \mathcal{B}(v)$, since the adversary has to transmit at least one message in each step. Protocol $\mathcal{B}$ for a given graph G is *correct with respect to adversary* $\mathcal{A}$ if the message is heard eventually by each node of G, when the repeat-broadcast function of $\mathcal{B}$ is applied, and while it is $\mathcal{A}$ who decides on the timing of arrivals of each message broadcast.

3 Protocol by Enforcing Dominance

A node x is called a *relay for node* y if $x \rightarrow y$ is an edge and the distance from the source to x is smaller than to y. Let $S \subseteq V$ be a set of nodes. A node $x \in S$ is said to *dominate* S if the inequality

$$\mathcal{B}(x) > \sum_{y \in S \setminus \{x\}} \mathcal{B}(y)$$

holds, while the empty sum is assumed to be equal to zero. Observe that each node different from the source has at least one relay node, because all the nodes are reachable. If there is an edge $x \rightarrow y$ and node x dominates all the in-neighbors of y, then x is said to *guard* node y, which is also denoted $x = L(y)$. A node x is *guarded* if one of its neighbors guards it.

Lemma 1. *If, for each node y different from the source, there is a relay node that guards y, then the protocol is correct with respect to the edge adversary.*

Next we describe an algorithm DISTANCE-DOMINATION to find a protocol correct against the edge adversary. The algorithm assigns values of repeat-broadcast function to the nodes. In general, the work of the obtained protocol is exponential in the size of the network.

Let us fix an enumeration $\langle v_i \rangle_{1 \leq i \leq n}$ of all the nodes so that if the distance from the source s to v_k is smaller than the distance from the source to v_ℓ, then $\ell < k$. In particular we have that $s = v_n$. Initialize the broadcast function $\mathcal{B}(v) = 0$, for each node $v \in V$. It is next modified for each node, in the increasing order of indices. For a current setting of the repeat-broadcast function, define $C_j = 1 + \sum_{x \in \text{IN}(v_j)} \mathcal{B}(x)$. The repeat-broadcast function is set to its final value by considering all the nodes in increasing order of their indices. Take an index i, such that $1 \leq i \leq n$ and all $1 \leq k < i$ have been processed. Set $\mathcal{B}(v_i)$ equal to the maximum value among C_j over all edges $v_i \rightarrow v_j$, if at least one such an edge exists, and $\mathcal{B}(v_i) = 0$ otherwise.

Theorem 1. *Algorithm DISTANCE-DOMINATION produces a protocol that is correct against the edge adversary. Its work is always less than 2^{n-1}, where $|V| = n$.*

The work-performance of a protocol produced by the algorithm DISTANCE-DOMINATION is exponential, which cannot be improved in general, as we show in Section 4. One could also observe that ordering the nodes on the distance to them from the source in increasing order of indices, and then assigning 2^i to a node v_i as a value of a repeat-broadcast function, results in a correct protocol. This is because each set of nodes has a dominating element, hence Lemma 1 applies. If this is so simple, why to develop the algorithm DISTANCE-DOMINATION? An answer is that for some classes of graphs the work of a protocol obtained by this algorithm could be $o(2^n)$ for networks of n nodes.

Theorem 2. *The work of a protocol obtained by the algorithm* DISTANCE-DO-MINATION, *on networks with in-degrees at most* d, *is* $\mathcal{O}(C^n)$, *where* $1 < C < 2(1 - 2^{-d})$.

Proof. The numbers $\mathcal{B}(v_m)$ satisfy an inequality $\mathcal{B}(v_{m+1}) \leq \sum_{k=m-d}^{m} \mathcal{B}(v_k)$, for $m \geq d$. Hence an inequality $\mathcal{B}(v_k) \leq A_k$ holds, where the numbers A_k are defined by the recurrence relation $A_{k+d+1} = A_{k+1} + A_{k+2} + \ldots + A_{k+d}$ and suitable initial values. This is a definition of d-step Fibonacci numbers. The characteristic equation of this recurrence is $x^d - x^{d-1} - x^{d-2} - \ldots - x - 1$. It has one positive simple root C in the interval $[1, 2]$, which satisfies the inequality $1 < C < 2(1 - 2^{-d})$, see [29]. □

4   Lower Bound

In this section we show that there are networks requiring exponential work from broadcast protocols. More precisely, we construct a family of networks G_n, each with n nodes, such that for any broadcast protocol on G_n correct against the edge adversary, its work has to be at least $c^{n^{1/3}}$, for some constant $c > 1$.

We define a family of networks F_k, for each integer $k > 0$. Network F_k has $\binom{k}{3} + k + 1$ nodes, partitioned into three layers. The *top layer* L_0 consists of only the source node. The *middle layer* L_1 consists of some k nodes, and the *bottom layer* L_2 that includes the remaining $\binom{k}{3}$ nodes. Directed edges connect the source with each element of layer L_1. There is a one-to-one correspondence between the three-element subsets of L_1 and nodes of L_2: for each such a subset X there is exactly one node v in the layer L_2 that X consists of the in-neighbors of v.

Lemma 2. *Let* $\mathcal{B}$ *be a broadcast protocol for* F_k *correct against the edge adversary. Then the following inequalities hold:* $\mathcal{B}(v_1) \geq 0$ *and* $\mathcal{B}(v_i) \geq f_{i-2}$, *for* $k \geq i \geq 2$, *where* f_i *is the ith Fibonacci number.*

Theorem 3. *There is a sequence* $\langle G_n \rangle_{n \geq 1}$ *of networks, such that* G_n *has* n *nodes and any broadcast protocol for this network, that is correct against the edge adversary, requires work* $\Omega(\phi^{\sqrt[3]{n}})$, *where* $\phi = (1 + \sqrt{5})/2$ *is the golden ratio.*

Proof. Take $k = \lfloor n^{1/3} \rfloor$ and consider F_k. The network F_k is of a size $\Theta(n)$. By Lemma 2, and properties of Fibonacci numbers, the work of the algorithm is at least $\sum_{i=1}^{k} \mathcal{B}(v_i) \geq \sum_{i=2}^{k} f_{i-2} = f_k - 1$. Since f_k is equal to $\phi^k / \sqrt{5}$ rounded to a closest integer, the work of the algorithm is $\Omega(\phi^{\sqrt[3]{n}})$. □

5   Verifying Correctness

Given a network G with a source s, and a repeat-broadcast function $\mathcal{B}$, we consider a problem to verify if the broadcast protocol determined by $\mathcal{B}$ is correct against the edge adversary.

For two nodes u and v, we say that *node u depends on v* if each path from the source s to u traverses either v or a node w with $\mathcal{B}(w) = 0$. If an edge $u \to v$ has a property that u depends on v, then such an edge is said to be *redundant*. An edge that is not redundant is called *significant*. If edge $u \to v$ is redundant, then transmissions by u cannot block v from receiving a message, since node v receives the message before u. Redundant edges can be removed without affecting the set of nodes that receive the source message, for any behavior of the adversary.

We may start with removing all the redundant edges. To identify redundant edges, do the following. For each vertex v, first remove v and all the nodes w with $\mathcal{B}(w) = 0$, and then identify all the nodes than cannot be reached from the source by a directed path. If node u is among them and $u \to v$ is an edge of G then this edge is redundant. In the remaining part of this section we assume that any edge is significant.

Lemma 3. *If node v does not have a guard, then the adversary can block v from receiving the message.*

Lemma 3 implies that every vertex has to have a guard for the protocol to be correct. This is a necessary but not a sufficient condition however. At least one transmission by $L(v)$ cannot be blocked by one of a different neighbor of v, because $L(v)$ dominates them all. This shows:

Lemma 4. *If there is a guard of vertex v, and the guard receives the message, then v also will receive the message eventually.*

Now we develop a routine to verify if a vertex v can be blocked from receiving a message. A pseudocode for this routine is in Figure 1, where $L(v)$ denotes a function that returns a guard of vertex v. Notice that guards may change as the repeat-broadcast function evolves.

The routine maintains a set SBV, which stands for set-of-blocked-vertices, of vertices that cannot receive the message to block vertex v. Initialize SBV $= \{v\}$ and modify $\mathcal{B}$ so that $\mathcal{B}(v) = 0$ since v will never broadcast as a blocked vertex. Remove all redundant edges for such a modified $\mathcal{B}$. Lemma 4 implies that $L(v)$ cannot receive a message to block v. Add $L(v)$ to SBV and modify $\mathcal{B}$ so that $\mathcal{B}(L(v)) = 0$. Now remove all redundant edges for such a modified $\mathcal{B}$. Next, for each vertex x in SBV, find its guard $L(x)$ and add the guard to the SBV while

```
procedure CHECK-VERTEX-BLOCKED ( v )
     if  v = s  then return false
     SBV := {v}  ;  B(v) := 0
     remove all the redundant edges
     repeat
            NewVerticesToBlock := ∅
            for each vertex u ∈ SBV
                 if L(u) exists then
                       NewVerticesToBlock := NewVerticesToBlock ∪ {L(u)}
                       B(L(u)) := 0
                       remove all redundant edges
            SBV := SBV ∪ NewVerticesToBlock
     until   NewVerticesToBlock = ∅
     if s ∈ SBV    then return true    else return false
```

Fig. 1. A procedure to verify if node v can be blocked by the edge adversary.

modifying B to be equal to zero on $L(x)$. Keep iterating this, until at some point the set SBV stops increasing, which occurs when there are no more guards to be blocked. If the source s is in SBV, then v cannot be blocked, otherwise it can.

Lemma 5. *The procedure* CHECK-VERTEX-BLOCKED *is correct.*

An algorithm VERIFY-CORRECT(G) operates as follows: for each node $v \in G$ verify if v is blockable by calling CHECK-VERTEX-BLOCKED(v).

Theorem 4. *Given a network G of n nodes, and a repeat-broadcast function B, the algorithm* VERIFY-CORRECT *checks if the protocol B is correct against the edge adversary in time $\mathcal{O}(n^6)$.*

Proof. Each node v is added to the set SBV at most once, since after v has been put in SBV, its repeat-broadcast value is set to zero and v is never a guard. The until loop can be executed up to n times. The inner for loop takes $|SBV|$ iterations. Each iteration consist of finding $L(u)$, which takes time $\mathcal{O}(|\text{IN}(u)|)$ per call, followed by removing the redundant edges in time $\mathcal{O}(n^3)$. The runtime of CHECK-VERTEX-BLOCKED is $\mathcal{O}((1 + 2 + \cdots + n)n^3) = \mathcal{O}(n^5)$, which yields a total runtime $\mathcal{O}(n^6)$.   □

6   Hard Problems for Edge Adversary

We present two decision problems regarding broadcasting in asynchronous networks, with timing of arrivals of messages controlled by the edge adversary, that are NP-complete. The first problem is about existence of a broadcast protocol with work bounded locally.

Problem: **0-1 Asynchronous Radio Broadcast**
Instance: A network G with a distinguished source.
Question: Is there a broadcast protocol for G correct against the edge
adversary and such that each vertex broadcasts at most once?

Theorem 5. *The problem* 0-1 ASYNCHRONOUS RADIO BROADCAST *is NP-complete.*

The next problem is about existence of a broadcast protocol with work
bounded globally.

Problem: **Work-Bounded Asynchronous Radio Broadcast**
Instance: A network G with a distinguished source, and a positive integer W.
Question: Is there a broadcast protocol for G correct against the edge
adversary and such that its work is at most W?

Theorem 6. *The problem* WORK-BOUNDED ASYNCHRONOUS RADIO BROADCAST *is NP-complete.*

7   Crash Adversary

The crash adversary has all the power of the edge adversary, but additionally
can crash nodes. We assume that the source s is never failed. For a broadcast
protocol $\mathcal{B}$, the adversary chooses nodes to be crashed prior to the start of an
execution. The protocol $\mathcal{B}$ is correct if the message reaches all non-faulty nodes
that are reachable from the source by a directed path of non-faulty nodes.

Verification of Correctness of a Broadcast Protocol. We would like to
verify if a protocol $\mathcal{B}$ is correct for the given input network G. Node v is said to
be *blockable for* $\mathcal{B}$ if the crash adversary can crash some nodes, possibly none,
so that there is path $P = (s, u_1, \ldots, u_k, v)$ from s to v consisting of non-faulty
nodes, and node v never receives the message in some execution of $\mathcal{B}$.

Lemma 6. *If there exists a node v that is blockable by the crash adversary, then
there exist a node u that is blockable by crashing only some neighbors of u.*

Next we consider an algorithm VERIFY-AGAINST-CRASHES to verify correctness of a protocol, by way of checking if there exists a blockable node. For each
node v check if there is a subset $S \subseteq \text{IN}(v)$ such that S does not contain a guard.
If for some node v such a subset S is found, then this v is blockable, and the
protocol is not correct, otherwise $\mathcal{B}$ is correct. The algorithm is in Figure 2.

Theorem 7. *The algorithm* VERIFY-AGAINST-CRASHES*($\mathcal{B}$) checks if $\mathcal{B}$ is correct in time* $\mathcal{O}(n^5)$.

Proof. Correctness follows from Lemma 6. The outer for loop takes n iterations.
The inner while loop takes at most n iterations. Removing redundant edges costs
$\mathcal{O}(n^3)$. The total runtime is thus $\mathcal{O}(n^5)$. $\square$

```
algorithm VERIFY-AGAINST-CRASHES ( B )
    remove all the redundant edges
    for each node v do
        C := B
        while  L(v) exists for C  do
            C(L(v)) := 0
            remove redundant edges for C
            if IN(v) is nonempty then
                return "v is blockable"
    return "the protocol is correct"
```

Fig. 2. Algorithm to verify correctness of B against the crash adversary.

The algorithm VERIFY-AGAINST-CRASHES checks if there *exist* a blockable node, which can be done in polynomial time. Interestingly enough, the problem to check if *a specific node* is blockable is NP-complete.

Optimal Work against the Crash Adversary. We show that work $\Omega(2^n)$ is sometimes necessary to have for a protocol that is correct against the crash adversary. This amount of work is then optimal, since it is always achievable, as observed in Section 3.

Lemma 7. *Let B be a broadcast protocol. If any subset of IN(v) contains a guard, for each node v, then the protocol B is correct against the crash adversary.*

We say that vertex v is *vulnerable* if a removal of an arbitrary subset $S \subset$ IN(v) of nodes of v and also of the node v does not disconnect the nodes in IN$(v) \backslash S$ from the source s. Intuitively, a node v is vulnerable if in a situation when an arbitrary proper subset of neighbors of v have been crashed, the remaining in-neighbors of v can receive the message before v receives it.

Lemma 8. *Let $v \in G$ be a vulnerable node. If protocol B is correct for G, then each subset of IN(v) contains a guard.*

Lemma 9. *Let $V = \{v_0, \ldots, v_{n-1}\}$ and $\mathcal{F} : V \rightarrow \mathbb{N}$. If $\mathcal{F}$ has the property that for each subset $S \subseteq V$ there exists an element $v \in S$ such that v dominates S, then the inequality $\mathcal{F}(v_i) \geq 2^i$ holds.*

Next we define a network H. It consists of n nodes partitioned into three layers. The top layer L_0 consists of one source vertex s. The middle layer L_1 consists of $n - 2$ nodes. The bottom layer L_2 consists of one vertex v. The vertex s is the in-neighbor of each vertex from layer L_1 and each vertex from L_1 is the in neighbor of the vertex v. We can see that vertex v is vulnerable.

Theorem 8. *There exists a network G of n nodes for which any protocol B correct against the crash adversary has work $\Omega(2^n)$.*

Proof. Consider the network H. Vertex v is vulnerable, so by Lemma 8 there exists a guard in each subset of L_1. By Lemma 9, the smallest possible amount of work is achieved when vertices in L_1 transmit $2^0, 2^1, \ldots, 2^{n-3}$ times respectively. Hence the work is $\Omega(2^n)$. $\square$

NP Complete Problem for Crash Adversary. Next we describe a decision problem regarding broadcast protocols correct against the crash adversary that is NP-complete.

> Problem: **Asynchronous Radio 3-Broadcast with Crashes**
> Instance: A network G.
> Question: Is there a repeat-broadcast function for G that assigns at most three transmissions to each node so that the resulting broadcast protocol is correct against the crash adversary?

Theorem 9.   *The problem* ASYNCHRONOUS RADIO 3-BROADCAST WITH CRASHES *is NP-complete.*

Broadcast Protocols and Colorings. We show that a vertex coloring yields a broadcast protocol. Let F be the graph and $\chi(F)$ its chromatic number. We assume that colors used to color F are the numbers $\{0, \ldots, \chi(F)-1\}$. The graph is colored so that the color 0 represents to most frequently used color, color 1 represents the second most frequently used color, and so on, with color $\chi(F)-1$ representing a least frequently used color.

Given network G, build graph $F(G)$ as follows. The nodes are those of G. For each pair of edges $u \to w$ and $v \to w$ of G, create edge (u, v) in $F(G)$.

Theorem 10. *If $\chi(F(G))$ is a chromatic number of the graph $F(G)$, then there is a broadcast protocol $\mathcal{B}$ for network G with work $\mathcal{O}(n\, 2^{\chi(F(G))}/\chi(F(G)))$.*

Proof. The coloring guarantees that, for each $u \in G$, all the in-neighbors of u have different colors. For each node v, set $\mathcal{B}(v)$ to be equal to $2^{\mathcal{C}(v)}$. Such an assignment guarantees that each node will have a guard node in each subset of its in-neighbors, so it is correct by Lemma 7. The total work of the protocol is at most $\mathcal{O}(n\, 2^{\chi(F(G))}/\chi(F(G)))$, since the maximal size of a clique in $F(G)$ is $\chi(F(G))$. $\square$

8   Node Adversary

The node adversary is weaker than the edge one, in the sense that each algorithm correct against the edge adversary is also correct against the node one.

Theorem 11. *There is a sequence $\langle G_n \rangle_{n \geq 1}$ of networks, each of n nodes and any broadcast protocol for this network, that is correct against the node adversary, requires work $\Omega(c^{\sqrt[3]{n}})$, for a constant $c > 1$.*

Consider a counterpart of the problem WORK-BOUNDED ASYNCHRONOUS RADIO BROADCAST, which was shown to be NP-hard in Section 6. The problem is as follows: Given a network and an integer $w > 0$: is there a protocol for this network that is correct against the node adversary and of work at most w? We conjecture that this problem does not even belong to NP. If true, this would imply that there is no deterministic polynomial algorithm to verify if a given broadcast protocol is correct against the node adversary.

9   Conclusion and Open Problems

We showed that sometimes the work of broadcast protocols for asynchronous networks has to be exponential in the size of the underlying networks. A reason of this phenomenon is that the adversaries considered are very strong. It would be interesting to identify weaker but natural adversaries, for which polynomial-work protocols always exist.

The problems considered in this paper concern centralized algorithms. Introducing a viable model of asynchronous radio networks and developing distributed communication protocols for them is an interesting topic to pursue.

References

1. N. Alon, A. Bar-Noy, N. Linial and D. Peleg, A lower bound for radio broadcast, Journal of Computer and System Sciences, 43 (1991) 290 - 298.
2. R. Bar-Yehuda, O. Goldreich, and A. Itai, On the time complexity of broadcast in radio networks: an exponential gap between determinism and randomization, Journal of Computer and System Sciences, 45 (1992) 104 - 126.
3. R. Bar-Yehuda, A. Israeli, and A. Itai, Multiple communication in multihop radio networks, SIAM Journal on Computing, 22 (1993) 875 - 887.
4. D. Bruschi, and M. Del Pinto, Lower bounds for the broadcast problem in mobile radio networks, Distributed Computing, 10 (1997) 129 - 135.
5. I. Chlamtac, and S. Kutten, On broadcasting in radio networks - problem analysis and protocol design, IEEE Transactions on Communications, 33 (1985) 1240 - 1246.
6. I. Chlamtac, and S. Kutten, Tree based broadcasting in multihop radio networks, IEEE Transactions on Computers, 36 (1987) 1209 - 1223.
7. I. Chlamtac, and O. Weinstein, The wave expansion approach to broadcasting in multihop radio networks, IEEE Transactions on Communications, 39 (1991) 426 - 433.
8. B.S. Chlebus, L. Gąsieniec, A. Gibbons, A. Pelc, and W. Rytter, Deterministic broadcasting in unknown radio networks, Distributed Computing, 15 (2002) 27 - 38.
9. B.S. Chlebus, L. Gąsieniec, A. Östlin, and J.M. Robson, Deterministic radio broadcasting, in Proc., 27th Colloquium on Automata, Languages and Programming (ICALP), 2000, LNCS 1853, pp. 717-728.
10. M. Chrobak, L. Gąsieniec, and D.R. Kowalski, The wake-up problem in multi-hop radio networks, in Proc., 15th ACM-SIAM Symposium on Discrete Algorithms (SODA), 2004, pp. 985 - 993.

11. M. Chrobak, L. Gąsieniec, and W. Rytter, Fast broadcasting and gossiping in radio networks, Journal of Algorithms, 43 (2002) 177 - 189.

12. A. Clementi, A. Monti, and R. Silvestri, Distributed multi-broadcast in unknown radio networks, in Proc., 20th ACM Symposium on Principles of Distributed Computing (PODC), 2001, pp. 255 - 264.

13. A. Clementi, A. Monti, and R. Silvestri, Selective families, superimposed codes, and broadcasting on unknown radio networks, in Proc., 12th ACM-SIAM Symposium on Discrete Algorithms (SODA), 2001, pp. 709 - 718.

14. A. Czumaj, and W. Rytter, Broadcasting algorithms in radio networks with unknown topology, in Proc., 44th IEEE Symposium on Foundations of Computer Science (FOCS), 2003, pp. 492 - 501.

15. I. Gaber, and Y. Mansour, Broadcast in radio networks, J. Algorithms, 46 (2003) 1 - 20.

16. M.R. Garey, and D.S. Johnson, "Computers and Intractability: A Guide to the Theory of NP-Completeness," 1979, W.H. Freeman, New York.

17. L. Gąsieniec, A. Pelc, and D. Peleg, The wakeup problem in synchronous broadcast systems, SIAM Journal on Discrete Mathematics, 14 (2001) 207 - 222.

18. P. Indyk, Explicit constructions of selectors and related combinatorial structures, with applications, in Proc., 13th ACM-SIAM Symposium on Discrete Algorithms (SODA), 2002, pp. 697 - 704.

19. T. Jurdziński, M. Kutyłowski, and J. Zatopiański, Efficient algorithms for leader election in radio networks, in Proc., 21st ACM Symposium on Principles of Distributed Computing (PODC), 2002, pp. 51 - 57.

20. T. Jurdziński, and G. Stachowiak, Probabilistic algorithms for the wakeup problem in single-hop radio networks, in Proc., 13th International Symposium on Algorithms and Computation (ISAAC), 2002, LNCS 2518, pp. 535 - 549.

21. D.R. Kowalski, and A. Pelc, Deterministic broadcasting time in radio networks of unknown topology, in Proc., 43rd IEEE Symposium of Foundations of Computer Science (FOCS), 2002, pp. 63 - 72.

22. D.R. Kowalski, and A. Pelc, Faster deterministic broadcasting in ad-hoc radio networks, in Proc., 20th Symposium on Theoretical Aspects of Computer Science (STACS), 2003, LNCS 2607, pp. 109 - 120.

23. D.R. Kowalski, and A. Pelc, Deterministic broadcasting time in radio networks of unknown topology, in Proc., 22nd ACM Symposium on Principles of Distributed Computing (PODC), 2003, pp. 73 - 82.

24. E. Kranakis, D. Krizanc, and A. Pelc, Fault-tolerant broadcasting in radio networks, Journal of Algorithms, 39 (2001), 47-67.

25. E. Kushilevitz, and Y. Mansour, An $\Omega(D \log(N/D))$ lower bound for broadcast in radio networks, SIAM Journal on Computing, 27 (1998) 702 - 712.

26. E. Kushilevitz, and Y. Mansour, Computation in noisy radio networks, in Proc., 9th ACM-SIAM Symposium on Discrete Algorithms (SODA), 1998, pp. 236 - 243.

27. D. Liu, and M. Prabhakaran, On randomized broadcasting and gossiping in radio networks, in Proc., 8th Computing Combinatorics Conference (COCOON), 2002, LNCS 2387, pp. 340 - 349.

28. Y. Xu, An $\mathcal{O}(n^{1.5})$ deterministic gossiping algorithm for radio networks, Algorithmica, 36 (2003) 93 - 96.

29. D.A. Wolfram, Solving generalized Fibonacci recurrences, The Fibonacci Quarterly, 36 (1998) 129 - 145.

30. R. Zheng, J.C. Hou, and L. Sha, Asynchronous wakeup for power management in ad hoc networks, in Proc., 4th ACM Symposium on Mobile Ad Hoc Networking and Computing (MOBIHOC), 2003, pp. 35 - 45.

Two-Hop Virtual Path Layout in Tori[*]

Sébastien Choplin[1], Lata Narayanan[2], and Jaroslav Opatrny[2]

[1] Laboratoire de Recherche en Informatique d'Amiens,
Université de Picardie J. Verne, France
sebastien.choplin@u-picardie.fr
[2] Department of Computer Science,
Concordia University, Montréal, Canada
{lata,opatrny}@cs.concordia.ca

Abstract. We consider the problem of D-hop virtual path layout in ATM (Asynchronous Transfer Mode) networks. Given a physical network and an all-to-all traffic pattern, the problem consists of designing a virtual network with a given diameter D, which can be embedded in the physical one with a minimum congestion (the congestion is the maximum load of a physical link). Here we propose a method to solve this problem when the diameter is 2. We use this method to give an asymptotically optimal solution for the 2-hop virtual path layout problem for all-to-all traffic when the physical network is a mesh, a torus or a chordal ring.

1 Introduction

Asynchronous Transfer Mode (ATM) is the world's most widely deployed backbone technology. This standards-based transport medium is widely used in all kinds of telecommunications networks to send data, video and voice at ultra high speeds. An ATM network consists of nodes and physical links connecting the nodes of the network. There are two kinds of routes in ATM network, called *virtual paths* and *virtual channels*. A virtual channel is used to create a connection between two nodes that wish to communicate, while a virtual path serves an internal network role and is used to build several virtual channels that share part of their route. Each virtual path corresponds to a simple path in the network while a virtual channel is a concatenation of several virtual paths. A *D-hop* virtual channel is the concatenation of D virtual paths. Given a set R of pairs of nodes in a network, the *D-hop virtual path layout (VPL)* problem is the problem of determining a set of virtual paths and virtual channels so that any pair of nodes in R is assigned an ℓ-hop virtual channel with $\ell \leq D$. The cost of implementing a D-hop virtual path layout depends on the *maximum load* of any physical link, *i.e.*, the number of paths that share any physical link. A layout that minimizes this maximum load will have the smallest cost. At the same time, the transmission cost on a virtual channel is proportional to the hop count D, and thus it is useful to consider virtual path layouts for small values of D.

---

[*] The work was supported partially by NSERC, Canada and was done while the first author was visiting Concordia.

The VPL problem for a set of communicating nodes R in a given graph G and fixed hop count D can be seen as finding a *virtual graph* with the vertex set R and with diameter D, and then finding an embedding of the virtual graph in G that minimizes the load. For instance, if we consider the all-to-all traffic pattern in a graph G and are interested in only one-hop paths, the only possible virtual graph is the complete graph. Thus the problem reduces to finding an embedding of the complete graph in G with minimum load. When the number of hops is more than one however, one has to choose both the virtual graph and the embedding. We formalize this model below.

Let $G = (V, E)$ and $H = (V, E')$ be two undirected graphs with the same set of vertices V. An embedding P of H in G consists in associating with each edge e of H an elementary path $P(e)$ in G. So P is a mapping of E' into the set of paths in G. The pair (H, P) is called a **virtual path layout** (VPL for short) on G. G is called the **physical graph** and H is called the **virtual graph**. The edges of G are called **physical edges** and the edges of H are called **virtual edges**. The **load of a physical edge** e of the graph $G = (V, E)$ for the VPL (H, P), denoted by $\pi(G, H, P, e)$ is the number of paths of $P(E')$ which use the physical edge $e : \pi(G, H, P, e) = |\{ e' \in E' \text{ s.t. } e \in P(e') \}|$. The **maximum load** of the embedding is denoted by $\pi(G, H, P) = \max_{e \in E} \{ \pi(G, H, P, e) \}$. A virtual graph H on G can be embedded in several ways; we denote by $\pi(G, H)$ the minimum of $\pi(G, H, P)$ over all embeddings P. As stated in the introduction, we are interested in virtual graphs H of diameter at most D. Our aim is to find a virtual graph H which has the **minimum value** $\pi(G, H)$ **among all the graphs with diameter at most** D. We will denote this **minimum value** $\pi(G, D)$.

In [2], authors give a general lower bound for this problem:

Theorem 1 ([2]). *For any instance (G, D), where G is an n-vertex graph of bounded genus and bounded degree, $\pi(G, D) = \Omega\left(\sqrt[2D]{n^3} \right)$.*

As mentioned earlier, if $D = 1$ there is only one possible virtual graph, namely the complete graph and $\pi(G, 1)$ is related to the edge-forwarding index [7]. For $D \geq 1$, the problem has been solved exactly when the physical graph G is a chain [3] and for $D = 2$ when G is a cycle [4]. For $D = 3i$, where i is an integer, the authors of [2] give constructions for meshes which are asymptotically optimal. Results concerning other physical graphs and set of requests can be found in [5, 6]. As far as we know, there are no previous asymptotically optimal results for tori and meshes for $D = 2$.

In this paper, we give an asymptotically optimal solution to the problem of 2-hop virtual path layout for all-to-all traffic in meshes and tori. In order to achieve a 2-hop layout, our method is to designate some nodes in the graph to be switches, which are used to interconnect the remaining nodes. A similar approach was used in [4]. The novel idea we use here is that the connections between nodes are specified according to a suitable projective plane. This enables us to achieve a diameter of two, as well as to minimize the load on the links adjacent to switches. By a careful placement of the switches and assignment of paths to switches, we are then able to achieve an asymptotically optimal load for meshes, tori, and chordal rings. The constant factor in the leading term for the load in our construction is much smaller than that in the result of [2].

The rest of the paper is organized as follows. The next section describes the construction of our family of virtual graphs and its properties. Section 3 gives our results on tori, meshes and chordal rings. We conclude with a discussion of open problems in Section 4.

2   The Virtual Graph

One way of constructing a virtual graph of diameter 2 is to designate a set of nodes to be so-called *switches* such that every node is connected to some subset of switches in such a way that for every pair of nodes, there is always one switch that they are both connected to. This can be achieved by a star topology, where all nodes are connected to the same central switch. However, the congestion at the central switch is obviously very high if its degree is bounded. To distribute the task of this central switch, the nodes are divided into subsets, which are interconnected by means of several switches. To specify which subsets of nodes are connected to which switches, we use the notion of projective planes.

2.1   Projective Planes and Difference Sets

Definition 1 (Projective Plane [8]). *For q a power of a prime, a finite projective plane of order q denoted $PP(q)$, consists of a set X of q^2+q+1 elements called **points**, and a set $\mathcal{L}$ of $(q+1)-$element subsets of X called **lines** having the property that any two points lie on a unique line.*

The existence of a projective plane $PP(q)$ is related to the existence of a field of order q, and we have:

Lemma 1. *[9] There is a finite projective plane $PP(q)$ whenever q is a power of a prime number.*

We use the properties of projective planes given by the following lemma:

Proposition 1. *In a projective plane of order q, the following holds:*

- *Every point lies on $q+1$ lines;*
- *Two lines meet in a unique point;*
- *There are q^2+q+1 lines.*

One way of constructing projective planes is using the notion of difference sets. Since this construction is helpful to us in calculating the load of the embedding of the virtual graph created by using the projective plane, we describe it below:

Definition 2 (Difference set [1]). *A difference set* modulo n *is a set of* $q + 1$ *positive integers*
$\{d_0, d_1, d_2, d_3, ..., d_q\}$, *where* $n > q + 1$, *with the property that all differences* $d_i - d_j$ *modulo* n *for* $i \neq j$ *are different. The numbers* n *and* $q + 1$ *are called the modulus and size of the difference set respectively.*

The operation of adding 1 (modulo n) to each integer in a difference set is called a *shift*, and by performing a shift, one may obtain a new difference set. Thus, one can perform $q^2 + q$ successive shifts to obtain $q^2 + q + 1$ distinct difference sets. In other words, if D is the difference set $\{d_0, \ldots, d_q\}$, the difference set $D_i = \{d_0 + i(\bmod\ n), \ldots, d_q + i(\bmod\ n)\}$ for $i \in \{0, \ldots, n - 1\}$. For example, for $q = 2, n = 7$, the set $\{1, 2, 4\}$ can be verified to be a difference set. By performing shifts modulo 7, we can obtain the difference sets $\{2, 3, 5\}, \{3, 4, 6\}, \{4, 5, 7\}$, $\{1, 5, 6\}, \{2, 6, 7\}$, and $\{1, 3, 7\}$.

Difference sets are closely linked to projective planes as shown by the following theorem:

Theorem 2. *Let* D *be a difference set of size* $q + 1$ *where* $q \geq 2$ *is a power of a prime and modulus* $n = q^2 + q + 1$. *Then* D *induces a projective plane* $PP(q)$ *of order* q *which has the following properties:*

- *the points of* $PP(q)$ *are the integers* $\{0, 1, 2, 3, ..., q^2 + q\}$.
- *the lines of* $PP(q)$ *are the difference sets obtained by performing* n *successive shifts on* D.

The projective plane given by the difference set $\{1, 2, 4\}$ modulo 7 and its shifts is the so-called Fano plane and is shown in Figure 1.

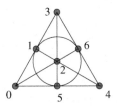

Fig. 1. The Fano plane

2.2   Virtual Graph Construction

We now describe the key idea of our construction of a family of virtual graphs.

Definition 3 (Projective Graph). *Let* q *be a power of a prime, and let the number of vertices* n *be fixed such that* $n \geq q^2 + q + 1$. *Then a projective graph* $H(n, q)$ *is constructed as follows: divide the vertex set* V *into* $q^2 + q + 1$ *sets,* $V_0, V_1, \ldots, V_{q^2+q}$, *corresponding to the points of* $PP(q)$. *Each set* V_j *has a special designated vertex called the switch, denoted* s_i. *If the line* i *of* $PP(q)$ *consists of the points* $i_1, i_2, \ldots, i_{q+1}$, *then each vertex in the set* V_j *is connected to the switches* $s_{i_1}, s_{i_2}, \ldots, s_{i_{q+1}}$, *barring any self-loops.*

Lemma 2. *For q a power of a prime and $n \geq q^2 + q + 1$, a projective graph $H(n,q)$ has diameter 2.*

Proof. Let x and y be two vertices in the graph $H(n,q)$. We show that in each of the cases below that $dist(x,y) \leq 2$.

- $x, y \in V_i$ for some i: From the definition of a projective graph, it follows that x and y are connected to the same set of switches, which is a non-empty set, and so $dist(x,y) \leq 2$.
- $x \in V_i$, $y \in V_j$ where $i \neq j$: Since any two points of $PP(q)$ lie on a line, there is a switch k such that either $x = s_k$ or $y = s_k$ and they are directly connected, or x and y are both connected to s_k, in which case $dist(x,y) = 2$.

Lemma 3. *Let G be an n-vertex graph of maximum degree Δ and $H(n,q)$ a projective graph,*

$$\pi(G,H) \geq \frac{n(q+1)}{\Delta(q^2 + q + 1)} - \frac{q+1}{\Delta}$$

Proof. Let $\{V_i\}$ be the partition of the vertices such that each V_i corresponds to a line of $PP(q)$ of a projective graph $H(n,q)$. As each vertex of V_i is connected to $q+1$ switches, the sum of the degree in H of the switches is at least $(q+1) \times \sum_{0 \leq i \leq q^2 + q} |V_i|$. So one of the switches has a degree in H of at least $\frac{q+1}{q^2+q+1}(n - (q^2 + q + 1))$ and the virtual edges connected to this switch can be balanced on at most Δ physical edges.

Corollary 1. *Let G be an n-vertex planar graph of bounded genus and bounded degree. If $H(n,q)$ is a projective graph and $\pi(G,H) = \Theta\left(\sqrt[4]{n^3}\right)$ then $q = \Theta\left(\sqrt[4]{n}\right)$.*

Proof. Theorem 1 shows that, if G is of bounded genus and bounded degree, $\pi(G,2) = \Omega\left(\sqrt[4]{n^3}\right)$. Lemma 3 indicates that, to achieve this bound, q must be $\Theta\left(\sqrt[4]{n}\right)$.

3   Embedding the Virtual Graph

In this section, we describe some ways in which a virtual projective graph as described above can be embedded in a torus, mesh, and some chordal rings. We give bounds on the load of these embeddings.

3.1   Torus and Mesh

We denote by $T_{L \times \ell}$ a torus with L rows and ℓ columns, vertex $x_{i,j}$ denotes the vertex of the torus in row i and column j, where $0 \leq i < L, 0 \leq j < \ell$. The following lemma shows that one can define a projective graph as described in the last section that can be embedded in some square tori such that the load is asymptotically optimal.

Lemma 4. *Let q be a power of a prime and $L = \ell = q^2 + q + 1$. Then there exists an embedding of a projective graph in $T_{L \times \ell}$ such that the load on any horizontal edge of the torus is at most $\frac{\sqrt[4]{n^3}}{2} + o\left(\sqrt[4]{n^3}\right)$ and the load on any vertical edge of the torus is at most $\frac{5\sqrt[4]{n^3}}{12} + o\left(\sqrt[4]{n^3}\right)$ where n is the number of vertices in the torus.*

Proof. Consider the torus $T_{\sqrt{n} \times \sqrt{n}}$ such that $q^2 + q + 1 = \sqrt{n}$ with q a power of a prime. We specify below a projective graph and its embedding (or routing) in $T_{\sqrt{n} \times \sqrt{n}}$ with an asymptotically optimal load. Using the difference set $\{d_0, d_1, \ldots, d_q\}$ (with $d_0 < d_1 < \ldots < d_q$), construct a virtual projective graph $H(n, q)$ as follows:

- Partition the set of vertices V into the subsets V_i ($0 \le i \le q^2 + q$) such that V_i is the set of vertices in the i-th row, that is, $V_i = \{x_{i,j} : 0 \le j \le q^2 + q\}$.
- Designate the middle vertex of row i, that is, $x_{i,(q^2+q)/2}$ to be the switch s_i of the vertex set V_i.
- The edges of H are, for $v \in V_i$, $[v, s_{i+d \mod q^2+q+1}]$ for $0 \le i \le q^2 + q$ and $d \in \{d_0, \ldots, d_q\}$ (barring self-loops).

We will associate a path in $T_{\sqrt{n} \times \sqrt{n}}$ with each edge of H such that the maximum load is $\Theta\left(\sqrt[4]{n^3}\right)$. Let i, j and k be integers such that s_k is connected to $x_{i,j}$ in H. The associated path in $T_{\sqrt{n} \times \sqrt{n}}$ will follow the shortest path along the horizontal edges from s_k to $x_{k,j}$ followed by the shortest path along the vertical edges from $x_{k,j}$ to $x_{i,j}$.

Figure 2 shows the groups, switches and paths connecting group V_0 to switches s_0, s_1 and s_3 with the construction using a projective plane of order 2 with the difference set $\{0, 1, 3\}$.

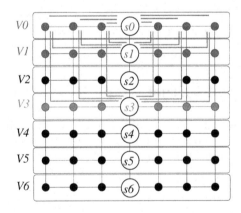

Fig. 2. Virtual edges between V_0 and switches s_0, s_1 and s_3 and their routing in $T_{7 \times 7}$

Using this routing, the load on the horizontal edges of $T_{\sqrt{n} \times \sqrt{n}}$ of type $(x_{i,j}, x_{i,j\pm1})$ is only induced by edges connecting s_i to non-switch vertices and its maximum is reached on the edges $(x_{i,((q^2+q)/2)-1}, s_i)$ and $(s_i, x_{i,((q^2+q)/2)+1})$. Since s_i is connected to $(q+1) \times (q^2+q)$ non-switch vertices and the load is balanced on these two edges, the maximum load on any horizontal edge (called *maximum horizontal load*) is equal to $\frac{(q+1)(q^2+q)}{2} = \frac{\sqrt[4]{n^3}}{2} + o\left(\sqrt[4]{n^3}\right)$.

The maximum load on a vertical edge of $T_{\sqrt{n} \times \sqrt{n}}$ (called *maximum vertical load*) is determined by the chosen difference set $\{d_0, d_1, \ldots, d_q\}$. Each vertex of the torus is the endpoint of $q+1$ paths between the vertex and switches and these paths start with vertical parts of length d_i if $d_i \leq (q^2+q)/2$ or of length $d_i - q^2 + q$ if $d_i > (q^2+q)/2$, since the vertical part does not need to be more than the half of the number of rows. Since the vertical parts are the same for each vertex, it is easy to observe that the load on any vertical edge of the torus is the same and is the sum of the lengths of the vertical parts which is equal to $\sum_{i=0}^{q} \min\{d_i, |q^2+q-d_i|\}$. To obtain an upper bound on the value of the sum, observe first that any element of the sum is at most equal to $(q^2+q)/2$ and there can be at most two such values. Secondly, since $\{d_0, d_1, \ldots, d_q\}$ is a difference set, the differences between any two consecutive elements in the difference set must be distinct. Using these two facts, we can obtain the required bound. When q is even, the required sum would be at most obtained by taking $d_{q/2} = (q^2+q)/2$, $d_{q/2+1} = (q^2+q)/2+1$ and $d_{q/2-i} = d_{q/2-i+1} - 2i$, $1 \leq i \leq q/2$, and $d_{q/2+1+i} = d_{q/2+1+i-1}+2i+1$, for $1 \leq i \leq q/2-1$. Similarly when q is odd, the upper bound can be obtained by taking $d_{(q-1)/2} = (q^2+q)/2$, $d_{(q-1)/2+1} = (q^2+q)/2+1$ and $d_{(q-1)/2-i} = d_{(q-1)/2-i+1} - 2i$, and $d_{(q-1)/2+1+i} = d_{(q-1)/2+1+i-1} + 2i + 1$, for $1 \leq i \leq (q-1)/2$. Thus $\sum_{i=0}^{q} \min\{d_i, q^2+q-d_i\} \leq \sum_{i=0}^{q/2}(q^2+q)/2 - i(i+1) + \sum_{i=0}^{q/2-1}((q^2+q)/2 - i(i+2)) = 5/12q^3 + 5/8q^2 + 7/12q \leq \frac{5\sqrt[4]{n^3}}{12} + o\left(\sqrt[4]{n^3}\right)$.

In Figure 2, the maximum vertical load given by the embedding is 3 and the maximum horizontal load is 9. We remark here that the multiplicative constant in the above bound can be improved by obtaining a better general bound on the vertical load which is determined by the difference sets. A similar projective graph construction can be embedded into more general cases of tori with optimal load as shown by the following theorem:

Theorem 3. *Let q be a power of a prime and L and ℓ be two integers such that $L = c_1(q^2+q+1)$ and $\ell = c_2(q^2+q+1)$ for some integer constants c_1 and c_2. Then*

$$\pi(T_{L \times \ell}, 2) = \Theta(\sqrt[4]{n^3})$$

where n is the number of vertices in the torus, which is asymptotically optimal.

Proof. Using a construction similar to the one from Lemma 4, we divide the torus into consecutive bands of c_1 rows with one switch in a central position in each band. The vertical load increases by a factor proportional to c_1 and the horizontal load increases by the factor $c_1 c_2$. Thus the maximum load is proportional to q^3. Clearly, $n = \Theta(q^4)$ and the lower bound given by Theorem 1 is $\Omega(q^3)$.

We denote by $M_{L \times \ell}$ the mesh with L rows and ℓ columns. It is easy to prove that $\pi(T_{L \times \ell}, D) \le \pi(M_{L \times \ell}, D) \le 2\pi(T_{L \times \ell}, D)$. It is worth pointing out however, that in the embedding described in Lemma 4, no horizontal wraparound edges are used, and hence the maximum horizontal load for the mesh is identical to that in the torus, while the vertical load in the case of the mesh can be at most double that in the torus.

Theorem 4. *Let q be a power of a prime and and $L = \ell = q^2 + q + 1$. Then there exists an embedding of a projective graph in $T_{L \times \ell}$ such that*

$$\pi(M_{L \times \ell}, 2) \le \frac{5 \sqrt[4]{n^3}}{6} + o\left(\sqrt[4]{n^3}\right)$$

where n is the number of vertices in the mesh, which is asymptotically optimal.

Improving the Result for Torus. As we can see from Figure 2, the horizontal edges are not evenly loaded, and the load on some horizontal edges is greater than the load on any vertical edge. The lower bound of Lemma 3 is $\frac{\sqrt[4]{n^3}}{4}$ and to achieve this bound, the load on all edges incident on each switch must be evenly balanced. This can be done with a better layout of the switches in the torus. In the torus with $q^2 + q + 1$ rows and columns we distribute the switches as follows: Place the first switch in the last column. If the switch in row i is in located in column j, place the switch in row $i+1$ in the column $(j + (q^2+q)/2) \mod (q^2+q+1)$. This layout of switches is shown for a 7×7 torus in Figure 3. With this layout, if we use the same routing as in Lemma 4, the horizontal edges just above and below a switch have very low load. Thus, we can re-route some of the paths coming to the switch from the left and right by using the lightly loaded part of the row above and below the switch and the vertical edges incident on the switch. Using this technique we can improve the bound on the load by a constant factor. For example, for the 7×7 torus, we can achieve a maximum load of 6 which is optimal over all virtual path layouts defined by projective graphs. In general we obtain the following result:

Theorem 5. *Let q be a power of a prime and $L = \ell = q^2 + q + 1$. Then there exists an embedding of a projective graph in $T_{L \times \ell}$ such that*

$$\pi(T_{L \times \ell}, 2) \le \frac{11 \sqrt[4]{n^3}}{24} + o\left(\sqrt[4]{n^3}\right)$$

where n is the number of vertices in the torus, which is asymptotically optimal.

3.2   Chordal Ring

The technique applied to meshes and tori in the previous section is applicable to many cases of the chordal rings. We can divide the chordal ring of length $(q^2 + q + 1)^2$ into segments of length $q^2 + q + 1$. The ring edges can then be treated as horizontal edges in the torus, and the chord edges can play the role of the vertical edges if the chord length is not "too short" or "too long". By using a similar embedding as described in Section 3.1, we obtain the following result:

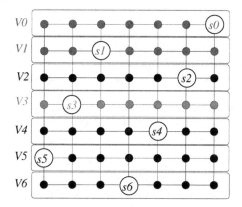

Fig. 3. Layout of switches that balances the load

Theorem 6. *Let q be a power of a prime, $n = (q^2 + q + 1)^2$ and G be a chordal ring on n vertices with chord length k such that $q^2 + q + 1 \leq \alpha k < 2(q^2 + q + 1)$ for some integer α. Then $\pi(G, 2) = \Theta(\sqrt[4]{n^3})$.*

The same bound can be obtained if $n = \Theta(q^4)$.

4 Discussion

We give an asymptotically optimal solution to the virtual path layout problem for all-to-all traffic in meshes, tori and chordal rings with a maximum hop count of 2. Although we stated the results for the case when the number of rows and columns is proportional to $q^2 + q + 1$ for a prime power q, we can easily extend our results to other sizes. For example, a grid having 10 rows can be partitioned into 7 segments such that 6 segments consist of one and half rows, one segment contains one row, and we proceed with $q = 2$.

Our solution uses a construction based on a projective plane structure. It would be interesting to see whether this kind of finite geometry construction gives optimal or near-optimal solutions for other families of graphs. It would be also interesting to see whether some other finite geometries yield an optimal solution when the number of hops is greater than two.

References

1. L.D. Baumert. Cyclic Difference Sets. Lecture Notes in Mathematics 182, Springer-Verlag, 1971.
2. L. Becchetti, P. Bertolazzi, C. Gaibisso, and G. Gambosi. On the design of efficient ATM routing schemes. Theoretical Computer Science, Volume 270, Issues 1-2:341–359, 2002.
3. S. Choplin. Virtual Path Layout in ATM path with given hop count. In International Conference on Networking, ICN01, volume 2094, Part II of LNCS, pages 527–537. Springer, 2001.

4. S. Choplin, A. Jarry, and S. Pérennes. Virtual network embedded in the cycle. Discrete Applied Mathematics, to appear.
5. I. Cidon, O. Gerstel, and S. Zaks. A scalable approach to routing in ATM networks. In Proc. of the 8th International Workshop on Distributed Algorithms, WDAG '94, volume 857 of Lecture Notes in Computer Science, pages 209–222. Springer Verlag, 1994.
6. O. Gerstel and S. Zaks. The Virtual Path Layout problem in fast networks (extended abstract). In Proceedings of the Thirteenth Annual ACM Symposium on Principles of Distributed Computing, pages 235–243, Los Angeles, California, 14–17 August 1994.
7. M-C. Heydemann, J-C. Meyer, and D. Sotteau. On forwarding indices of networks. Discrete Applied Mathematics, 23:103–123, 1989.
8. S. Jukna. Extremal Combinatorics with applications in computer science. Springer-Verlag, 2001.
9. W. D. Wallis. Combinatorial Designs. Marcel Dekker, New York, 1988.

Robot Convergence
via Center-of-Gravity Algorithms

Reuven Cohen and David Peleg

Department of Computer Science and Applied Mathematics,
The Weizmann Institute of Science, Rehovot 76100, Israel
{r.cohen,david.peleg}@weizmann.ac.il

Abstract. Consider a group of N robots aiming to converge towards
a single point. The robots cannot communicate, and their only input is
obtained by visual sensors. A natural algorithm for the problem is based
on requiring each robot to move towards the robots' center of gravity.
The paper proves the correctness of the center-of-gravity algorithm in the
semi-synchronous model for any number of robots, and its correctness in
the fully asynchronous model for two robots.

1 Introduction

1.1 Background and Motivation

In hazardous or hostile environments, it may be desirable to employ large groups
of low cost robots for cooperatively performing various tasks. This approach has
the advantage of being more resilient to malfunction and more configurable than
a single high cost robot. Consequently, autonomous mobile robot systems have
been studied in different contexts, from engineering to artificial intelligence (e.g.,
[1–11]). A survey on the area is presented in [12, 13].

During the last decade, various issues related to the coordination of multiple
robot systems have been studied from the point of view of distributed computing (cf. [14–20]). The focus is on trying to model an environment consisting of
mobile autonomous robots, and studying the capabilities the robots must have
in order to achieve their common goals. A number of computational models were
proposed in the literature for such systems. In this paper we follow the models
of [19–21]. The robots are identical and indistinguishable, cannot communicate
between them, and do not operate continuously. They wake up at unspecified
times, observe their environment using sensors which are capable of identifying
the locations of the other robots, perform a local computation determining their
next move and move accordingly.

To model the behavior of the robots, several activation scheduling models
have been suggested. In the synchronous model, all robots are active at every
cycle. In the semi-synchronous model some robots are active at each cycle. In
the asynchronous model, no cycles exist and no limit is placed on the latency
of each robot movement. The models are presented more formally in the next
section.

R. Královič and O. Sýkora (Eds.): SIROCCO 2004, LNCS 3104, pp. 79–88, 2004.

Two basic coordination tasks in autonomous mobile robot systems that have received considerable attention are *gathering* and *convergence*. The gathering task requires the robots to occupy a single point within a finite number of steps, starting from any initial configuration. The closely related convergence task requires the robots to converge to a single point, rather than reach it (namely, for every $\epsilon > 0$ there must be a time t_ϵ by which all robots are within distance of at most ϵ of each other).

A straightforward approach to these problems is based on requiring the robots to calculate some median position of the group and move towards that position. The current paper focuses on what is arguably the most natural variant of this approach, namely, using the *center of gravity* (also known as the *center of mass*, the *barycenter* or the average position) of the robot group.

The center of gravity approach is easy to analyze in the fully synchronous model. In the semi-synchronous or asynchronous models, however, analyzing the process becomes more involved, since the robots may take their measurements at different times, including while other robots are in movement. This might result in oscillatory effects on the centers of gravity calculated by the various robots, and possibly cause them to pass each other by in certain configurations. Subsequently, the correctness of the center of gravity algorithm has not been proven formally so far (to the best of our knowledge), i.e., it has not been shown that this algorithm guarantees convergence.

Several different and more sophisticated algorithms have been proposed before, some of which also guarantee gathering within finite time. The gathering problem was first discussed in [19, 20] in the semi-synchronous model. It was proven that it is impossible to achieve gathering of *two* oblivious autonomous mobile robots that have no common sense of orientation under the semi-synchronous model. (Convergence is easy to achieve in this setting.) Also, an algorithm was presented in [20] for gathering $N \geq 3$ robots in the semi-synchronous model. In the asynchronous model, an algorithm for gathering $N = 3, 4$ robots is brought in [22, 16], and an algorithm for gathering $N \geq 5$ robots has recently been described in [15]. The gathering problem was also studied in a system where the robots have limited visibility [23, 24].

Nevertheless, the center of gravity approach has several important advantages, making it desirable in many cases.

- It requires a very simple and efficient calculation, which can be carried out on simple hardware, and requires very low computational effort.
- It can be easily applied to 1, 2 or 3 dimensions and to any number of robots.
- It has a bounded and simple to calculate error due to rounding.
- It is oblivious, i.e., it requires no memory of previous actions and positions, rendering it both memory-efficient and self-stabilizing (in the sense that a finite number of transient errors cannot prevent eventual convergence).
- It prevents deadlocks, that is, every robot can move at any given position (unless it is already at the center of gravity).

In the current paper, we address the convergence of the center of gravity algorithm. In Section 2 we prove the correctness of the algorithm in the semi-synchronous model of [19]. The problem appears to be more difficult in the asynchronous model. In Section 3 we make a modest first step by providing a convergence proof in this model for $N = 2$ robots.

Recently, we have been able to extend the result presented in the current paper and show that the center of gravity algorithm converges in the fully asynchronous model for any number of robots [25].

1.2   The Model

The basic model studied in [14–20] can be summarized as follows (with some minor changes). The robots are expected to execute a given algorithm in order to achieve a prespecified mission. Each of the N robots R_i in the system is assumed to operate individually, repeatedly going through simple cycles consisting of four steps:

- **Look:** Identify the locations of all robots in R_i's private coordinate system; the result of this step is a multiset of points $P = \{p_1, \ldots, p_N\}$ defining the current *configuration*. The robots are indistinguishable, so each robot R_i knows its own location p_i, but does not know the identity of the robots at each of the other points.
- **Compute:** Execute the given algorithm, resulting in a goal point p_G.
- **Move:** Move on a straight line towards the point p_G. The robot might stop before reaching its goal point p_G, but is promised to traverse a distance of at least S (unless it has reached the goal).
- **Wait:** The robot sleeps for an indefinite time and the awakens for the next Move.

The "look" and "move" operations are identical in every cycle, and the differences between various algorithms are in the "compute" step. The procedure carried out in the "compute" step is identical for all robots. Notics that for the sake of analysis, the Compute and Wait steps can be absorbed into the Move step since no restriction is placed on the relative rate of movement during this step.

In most papers in this area (cf. [26, 19, 23, 16]), the robots are assumed to be rather limited. To begin with, the robots are assumed to have no means of directly communicating with each other. Moreover, the robots are also assumed to be *oblivious* (or memoryless), namely, they cannot remember their previous states, their previous actions or the previous positions of the other robots. Hence the algorithm employed by the robots for the "compute" step cannot rely on information from previous cycles, and its only input is the current configuration. While this is admittedly an over-restrictive and unrealistic assumption, developing algorithms for the oblivious model still makes sense in various settings, for two reasons. First, solutions that rely on non-obliviousness do not necessarily work in a dynamic environment where the robots are activated in different cycles, or robots might be added to or removed from the system dynamically. Secondly, any algorithm that works correctly for oblivious robots is inherently self-stabilizing, i.e., it withstands transient errors that alter the robots' local states.

We consider two timing models. The first is the well studied *semi-synchronous* model (cf. [19]). This model is partially synchronous, in the sense that all robots operate according to the same fixed length clock cycles. However, not all robots are necessarily active in all cycles (the model is based on the assumption that each cycle is instantaneous). Rather, at each cycle t, any non-predetermined subgroup of the robots may commence the Look–Compute–Move cycle. The activation of the different robots can be thought of as managed by a hypothetical scheduler, whose only fairness obligation is that each robot must be activated and given a chance to operate infinitely often in any infinite execution.

The second model is the *fully asynchronous* model (cf. [15, 16]). In this model, robots operate on their own (time-varying) rates, and no assumptions are made regarding the relative speeds of different robots. In particular, robots may remain inactive for arbitrarily long periods between consecutive operation cycles.

To describe the center of gravity algorithm, hereafter named Alg. Go_to_COG, we use the following notation. Denote by $\bar{r}_i[t]$ the location of robot i at time t. Denote the true center of gravity at time t by $\bar{c}[t] = \frac{1}{N} \sum_{i=1}^{N} \bar{r}_i[t]$. Denote by $\bar{c}_i[t]$ the center of gravity as last calculated by the robot i before time t, i.e., if the last calculation was done at time $t' < t$ then $\bar{c}_i[t] = \bar{c}[t']$. Note that, as mentioned before, robot i calculates this location in its own private coordinate system; however, for the purpose of describing the algorithm and its analysis, it is convenient to represent these locations in a unified global coordinate system (which of course is unknown to the robots themselves). By convention $\bar{c}_i[0] = \bar{r}_i[0]$ for all i.

Algorithm Go_to_COG is very simple. After measuring the current configuration at some time t, the robot i computes the average location of all robot positions (including its own), $\bar{c}_i[t] = \sum_i \bar{r}_i[t]/N$, and then proceeds to move towards the calculated point $\bar{c}_i[t]$. (As mentioned earlier, the move may terminate before the robot actually reaches the point $\bar{c}_i[t]$, but in case the robot has not reached $\bar{c}_i[t]$, it must has traversed a distance of at least S. Also, in the semi-synchronous model, the move operation terminates by the end of the cycle.)

2   Convergence in the Semi-synchronous Model

In this section we prove that in the semi-synchronous model, Algorithm Go_to_COG converges for any number of robots $N \geq 2$.

We start with a technical Lemma. The robots' *moment of inertia* at time t is defined as

$$I[t] = \frac{1}{N} \sum_{i=1}^{N} (\bar{r}_i[t] - \bar{c}[t])^2 .$$

Lemma 1. *Suppose that at some given time, all robots reside in a radius R circle centered at the origin, and let $0 < x_0 \leq R$. If the x coordinate of each robot satisfies $x \geq x_0$, then their moment of inertia satisfies $I \leq 4(R^2 - x_0^2)$.*

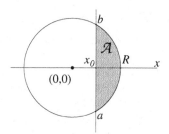

Fig. 1. Illustration of Lemma 1.

Proof. Since the region $\mathcal{A}$ obtained from the intersection of the circle and the half-plane $x \geq x_0$ is convex, the center of gravity $\bar{c}$ must also lie inside $\mathcal{A}$. Since $\mathcal{A}$ contains less than half the circle, the longest chord in $\mathcal{A}$ is the segment between the points $a = (x_0, -\sqrt{R^2 - x_0^2})$ and $b = (x_0, \sqrt{R^2 - x_0^2})$ (see Figure 1). The length of this segment is $L = 2\sqrt{R^2 - x_0^2}$ and therefore the distance between each robot and the center of gravity cannot exceed L. Thus,

$$ I \equiv \frac{1}{N} \sum_{i=1}^{N} (\bar{r}_i - \bar{c})^2 \leq \frac{1}{N} \cdot N \cdot L^2 \leq L^2 . $$

□

Defining the function $I_{\bar{x}}[t] \equiv \frac{1}{N} \sum_j (\bar{r}_j[t] - \bar{x})^2$, we notice the following (cf. [27]).

Lemma 2. $I_{\bar{x}} \geq I_{\bar{c}}$ *for every* $\bar{x}$.

Lemma 3. *For the semi-synchronous model, in any execution of Algorithm* Go_to_COG, $I[t]$ *is a non-increasing function of time.*

Proof. In each cycle t, each robot that is activated, and is not already at the current center of gravity, can get closer to the observed center of gravity $\bar{c}[t]$, hence $I_{\bar{c}[t]}[t] \geq I_{\bar{c}[t]}[t+1]$. By Fact 2, $I_{\bar{c}[t]}[t+1] \geq I_{\bar{c}[t+1]}[t+1]$. Combined, we have

$$ I[t] = I_{\bar{c}[t]}[t] \geq I_{\bar{c}[t+1]}[t+1] = I[t+1] . $$

□

Recall that S is the minimum movement distance of the robots.

Lemma 4. *If at some time* t_0 *the robots' moment of inertia is* $I_0 \equiv I[t_0]$, *then there exists some* $\hat{t} > t_0$, *such that* $I_1 \equiv I[\hat{t}] \leq \max\{I - \frac{S^2}{N}, (1 - \frac{1}{100N^5}) I\}$.

Proof. Assume without loss of generality that $\bar{c}[t_0]$ is at the origin. Take i to be the robot most distant from the origin at time t_0, and assume again that it lies on the positive x axis, at point $(R, 0)$. Since it was the most distant robot, we have $R^2/N \leq I_0 \leq R^2$. Now take t_1 to be the time of this robot's first activation after t_0. By Lemma 3, $I[t_1] \leq I_0$. We separate the analysis into two cases. First, suppose that the distance of robot i from $\bar{c}[t_1]$, the center of gravity at time t_1, is at least $\frac{R}{10N^2}$. Then, in its turn, it can move a distance at least $m = \min\{\frac{R}{10N^2}, S\}$. This decreases I by at least $\frac{m^2}{N}$ as required. Thus taking $\hat{t} = t_1 + 1$, the lemma follows.

If the robot cannot move a distance m in its turn, this implies that the distance of this robot from the center of gravity at time $\hat{t}$ is less than m. Since the circle is convex, no robot can reside at a distance greater than R from the origin. The difference between the x coordinate of robot i and $\bar{c}[t_1]$, the center of gravity at time t_1, cannot exceed $m \leq \frac{R}{10N^2}$. Since the center of gravity is the average of the robots' locations, the x coordinate of each robot j must satisfy $x_j \geq R - Nm$. By Lemma 1, $I[t_1] \leq 4(R^2 - (R - Nm)^2) \leq 8RNm \leq \frac{8R^2}{10N} \leq \left(1 - \frac{1}{100N^5}\right) I[t_0]$. Thus taking $\hat{t} = t_1$, the lemma follows.   □

Theorem 1. *In the semi-synchronous model, for $N \geq 2$ robots in 2-dimensional space, Algorithm Go_to_COG converges.*

3   Asynchronous Convergence of Two Robots

Let us now turn to the fully asynchronous model. We prove that in this model, Algorithm Go_to_COG converges for two robots. Our first observation applies to any $N \geq 2$.

Lemma 5. *If for some time t_0, $\bar{r}_i[t_0]$ and $\bar{c}_i[t_0]$ for all i reside in a closed convex curve, $\mathcal{P}$, then for every time $t > t_0$, $\bar{r}_i[t]$ and $\bar{c}_i[t]$ also reside in $\mathcal{P}$ for every $1 \leq i \leq N$.*

Proof. For the Move operation, it is clear that if for some i, $\bar{r}_i[t_0]$ and $\bar{c}_i[t_0]$ both reside in a convex hull then for the rest of the move operation $\bar{c}_i[t] = \bar{c}_i[t_0]$ does not change and $\bar{r}_i[t]$ is on the segment $[\bar{r}_i[t_0], \bar{c}_i[t_0]]$, which is inside $\mathcal{P}$.

For the Look step, if $N = 2$ then the calculated center of gravity is on the line segment connecting both robots, and therefore respects convexity. For $N > 2$ robots the center of gravity is on the line connecting the center of gravity of $N - 1$ robots and the Nth robot, and the Lemma follows by induction.   □

For the following we assume the robots reside on the x-axis. For every t, let $H[t]$ denote the convex hull of the points $\bar{r}_1[t]$, $\bar{r}_2[t]$, $\bar{c}_1[t]$ and $\bar{c}_2[t]$, namely, the smallest closed interval containing all four points.

Corollary 1. *For $N = 2$ robots, for any times t, t_0, if $t > t_0$ then $H[t] \subseteq H[t_0]$, namely, the points $\bar{r}_1[t]$, $\bar{r}_2[t]$, $\bar{c}_1[t]$ and $\bar{c}_2[t]$ reside in the line segment $H[t_0]$.*

Lemma 6. *If for some time t_0, $\bar{r}_2[t_0] \geq \bar{r}_1[t_0]$, $\bar{c}_2[t_0] \geq \bar{r}_1[t_0]$ and $\bar{r}_1[t]$ is a monotonic non-increasing function for $t \in [t_0, \hat{t}]$, then*

1. *at all times $t \in [t_0, \hat{t}]$, $\bar{r}_2[t] \geq \bar{r}_1[t]$ and $c_2[t] \geq \bar{r}_1[t]$, and*
2. *if at time $t^* \in [t_0, \hat{t}]$ robot 2 performed a Look step, then $\bar{c}_2[t] \in [\bar{r}_1[t], \bar{r}_2[t]]$ at all times $t \in [t^*, \hat{t}]$.*

Proof. Suppose that during the time interval $[t_0, \hat{t}]$, robot 2 performed $k \geq 0$ Look steps. Set $t_{k+1} = \hat{t}$, and if $k \geq 1$ then denote the times of these Look steps by $t_1, t_2, \ldots, t_k \in [t_0, t_{k+1}]$. We now prove by induction that claims 1 and 2 hold also for all times $t \in [t_i, t_{i+1}]$ for $0 \leq i \leq k$.

For $i = 0$, observe that since robot 2 did not perform a Look step throughout the entire time interval $[t_0, t_1]$, its location $\bar{r}_2[t]$ at any time $t \in [t_0, t_1]$ is in the interval $[\bar{r}_2[t_0], \bar{c}_2[t_0]]$, which is entirely to the right of $\bar{r}_1[t_0] \geq \bar{r}_1[t]$, hence claim 1 of the lemma follows. Claim 2 of the lemma holds vacuously, and we are done.

We now assume that claim 1 holds for the interval $[t_{i-1}, t_i]$ and prove that both claims hold at any time in the interval $[t_i, t_{i+1}]$. Indeed, notice that by assumption robot 2 is to the right of robot 1 at time t_i, and since a Look step is performed at that time, $\bar{c}_2[t_i]$ is at the average position of the robots, hence it is also to the right of robot 1. Thus, $\bar{c}_2[t] = \bar{c}_2[t_i] \geq \bar{r}_1[t_i] \geq \bar{r}_1[t]$ for all $t \in [t_i, t_{i+1}]$ and claim 2 follows for that interval. During the time interval $[t_i, t_{i+1}]$, robot 2 traverses the segment $[\bar{c}_2[t_i], \bar{r}_2[t_i]]$, which is entirely to the right of $\bar{r}_1[t_i] \geq \bar{r}_1[t]$, hence claim 1 follows as well. □

Lemma 7. *For $N = 2$ robots, for every time t_0 there exists a time $\hat{t} \geq t_0$ in which $\bar{c}_i[\hat{t}] = \bar{r}_i[\hat{t}]$ and $\bar{c}_j[\hat{t}]$ resides on the line segment $[\bar{r}_1, \bar{r}_2]$ for $i \neq j$ and $i, j \in \{1, 2\}$.*

Proof. Let $[a, b] = H[t_0]$. Take $t_1 > t_0$ to be the first time at which robot 1 completes its next Move operation. Thus, $\bar{r}_1[t_1] = \bar{c}_1[t_1]$. Without loss of generality assume $\bar{r}_2[t_1] > \bar{r}_1[t_1]$, i.e. $\bar{r}_2[t_1]$ falls in the segment $[\bar{r}_1[t_1], b]$. There are now three possible configurations (see Fig. 2).

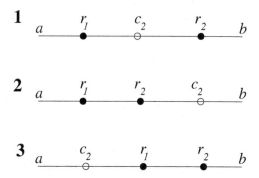

Fig. 2. The three cases in the proof of Lemma 7: ordering of the robots.

1. $\bar{c}_2$ resides on the segment $[\bar{r}_1, \bar{r}_2]$. Then take $\hat{t} = t_1$ and we are done.
2. $\bar{c}_2$ resides on the segment $[\bar{r}_2, b]$. Then let $\hat{t}$ be the time when robot 2 completes its current Move operation. Since robot 2 moves away from robot 1, $\bar{c}_1$ will always reside on the segment $[\bar{r}_1, \bar{r}_2]$ (by Lemma 6). At $\hat{t}$, $\bar{r}_2 = \bar{c}_2$ and c_1 resides on the segment $[\bar{r}_1, \bar{r}_2]$. Thus, we are done.

3. $\bar{c}_2$ resides on the line segment $[a, \bar{r}_1]$. In this case take t_2 to be the time at which robot 2 ends its Move operation. Whence, $\bar{c}_2[t_2] = \bar{r}_2[t_2]$. Since for any $t \in [t_1, t_2]$, $\bar{r}_2[t]$ is on the right of $\bar{c}_2[t]$ and $\bar{c}_1[t] = (\bar{r}_1[t_3] + \bar{r}_2[t_3])/2$ for some $t_3 \in [t_1, t]$, it follows that $\bar{c}_1[t]$ is to the right of $\bar{c}_2[t]$ and therefore also $\bar{r}_1[t]$ stays to the right of $\bar{c}_2[t] = \bar{c}_2[t_1]$. Hence, at time t_2, $\bar{r}_1$ and $\bar{c}_1$ are on the left of $\bar{r}_2 = \bar{c}_2$. Therefore, we reach either case 1 or 2 with the roles of 1 and 2 reversed. This completes the proof.   □

Lemma 8. *For $N = 2$ robots, for every time t_0 there exists a time $\hat{t} > t_0$ such that $|H[\hat{t}]| \leq |H[t_0]|/2$.*

Proof. Let $L = |H[t_0]|$. Take $t_1 > t_0$ to be the time when both robots have completed at least one cycle after t_0. By Corollary 1 they are still in $H[t_0]$. By Lemma 7 there exists $t_2 \geq t_1$ such that at time t_2, $\bar{c}_1[t_2] = \bar{r}_1[t_2]$ and $\bar{c}_2[t_2] \in [\bar{r}_1, \bar{r}_2]$ (where robots 1 and 2 and the left/right directions are chosen to make this true). By Corollary 1, $[\bar{r}_1[t_2], \bar{r}_2[t_2]] \subseteq H[t_0]$. Take $D = d(\bar{r}_1[t_2], \bar{r}_2[t_2]) \leq L$.

Now take t_3 to be the time of the next Look step of robot 1. Clearly $L_1 \equiv d(\bar{c}_1[t_3], \bar{r}_1[t_3]) < L/2$ (since the center of gravity is the average of the robots' locations and $d(\bar{r}_1[t_3], \bar{r}_2[t_3]) \leq |H(t_3)| \leq |H(t_0)| = L$). Now take $\hat{t}$ to be the time of the end of robot 1's Move operation, so $\bar{c}_1[\hat{t}] = \bar{r}_1[\hat{t}]$. Without loss of generality assume at t_3 robot 1 is the rightmost. Either of the following three cases is possible at $\hat{t}$ (see Fig. 3).

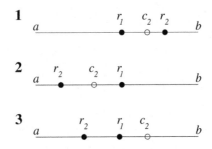

Fig. 3. The three cases in the ordering of the robots in Lemma 8.

1. $\bar{r}_2[\hat{t}]$ is to the right of $\bar{r}_1[\hat{t}]$. This means that for some time $t_4 \in [t_3, \hat{t}]$, $\bar{r}_1[t_4] = \bar{r}_2[t_4]$. At any $t \in [t_2, t_4]$, $\bar{r}_1[t]$ was to the right of $\bar{r}_2[t]$, implying $\bar{c}_2[t]$ was also on the right of $\bar{r}_2[t]$, and specifically $\bar{c}_2[\hat{t}] \geq \bar{r}_2[\hat{t}] \geq \bar{r}_1[\hat{t}] = \bar{c}_1[\hat{t}]$. Therefore they are all on the segment $[\bar{r}_1[\hat{t}], \bar{r}_1[t_3]]$ of length $L_1 \leq L/2$.
2. $\bar{r}_2[\hat{t}]$ and $\bar{c}_2[\hat{t}]$ are to the left of $\bar{r}_1[\hat{t}]$. Since robot 2 must have been moving right at all $t_2 \leq t \leq \hat{t}$ it must hold that $\bar{c}_2[\hat{t}]$ resides on the segment $[\bar{r}_2[t], \bar{r}_1[t]]$ and that $d(\bar{r}_1[\hat{t}], \bar{r}_2[\hat{t}]) \leq d(\bar{r}_1[t_2], \bar{r}_2[t_2]) = D$. Now, since $d(\bar{r}_1[t_2], \bar{c}_1[t_2]) = d(\bar{r}_1[t_2], \bar{r}_2[t_2])/2 = D/2$, and $\bar{r}_2[\hat{t}]$ and $\bar{c}_2[\hat{t}]$ are to the left of $\bar{r}_1[\hat{t}]$ they are restricted to the segment left of $\bar{r}_1[\hat{t}]$ and to the right of $\bar{r}_2[t_2]$ of length $D/2 \leq L/2$.
3. $\bar{r}_1[\hat{t}] \in [\bar{r}_2[\hat{t}], \bar{c}_2[\hat{t}]]$. Since $d(\bar{r}_2[\hat{t}], \bar{c}_2[\hat{t}]) \leq L/2$ we are done.   □

Theorem 2. *In the fully asynchronous model, for two robots in d dimensional space, Algorithm* Go_to_COG *converges.*

Proof. Apply Lemma 8 to each dimension separately. □

References

1. Parker, L., Touzet, C.: Multi-robot learning in a cooperative observation task. In: Distributed Autonomous Robotic Systems 4. (2000) 391–401
2. Jung, D., Cheng, G., Zelinsky, A.: Experiments in realising cooperation between autonomous mobile robots. In: Proc. Int. Symp. on Experimental Robotics. (1997)
3. Parker, L., Touzet, C., Fernandez, F.: Techniques for learning in multi-robot teams. In Balch, T., Parker, L., eds.: Robot Teams: From Diversity to Polymorphism. A. K. Peters (2001)
4. Kawauchi, Y., Inaba, M., Fukuda, T.: A principle of decision making of cellular robotic system (CEBOT). In: Proc. IEEE Conf. on Robotics and Automation. (1993) 833–838
5. Beni, G., Hackwood, S.: Coherent swarm motion under distributed control. In: Proc. DARS'92. (1992) 39–52
6. Murata, S., Kurokawa, H., Kokaji, S.: Self-assembling machine. In: Proc. IEEE Conf. on Robotics and Automation. (1994) 441–448
7. Mataric, M.: Interaction and Intelligent Behavior. PhD thesis, MIT (1994)
8. Parker, L.: Designing control laws for cooperative agent teams. In: Proc. IEEE Conf. on Robotics and Automation. (1993) 582–587
9. Parker, L.: On the design of behavior-based multi-robot teams. J. of Advanced Robotics **10** (1996)
10. Balch, T., Arkin, R.: Behavior-based formation control for multi-robot teams. IEEE Trans. on Robotics and Automation **14** (1998)
11. Wagner, I., Bruckstein, A.: From ants to a(ge)nts. Annals of Mathematics and Artificial Intelligence **31, special issue on ant-robotics** (1996) 1–5
12. Cao, Y., Fukunaga, A., Kahng, A., Meng, F.: Cooperative mobile robots: Antecedents and directions. In: Proc. Int. Conf. of Intel. Robots and Sys. (1995) 226–234
13. Cao, Y., Fukunaga, A., Kahng, A.: Cooperative mobile robotics: Antecedents and directions. Autonomous Robots **4** (1997) 7–23
14. Ando, H., Suzuki, I., Yamashita, M.: Formation and agreement problems for synchronous mobile robots with limited visibility. In: Proc. IEEE Symp. of Intelligent Control. (1995) 453–460
15. Cieliebak, M., Flocchini, P., Prencipe, G., Santoro, N.: Solving the robots gathering problem. In: Proc. 30th Int. Colloq. on Automata, Languages and Programming. (2003) 1181–1196
16. Cieliebak, M., Prencipe, G.: Gathering autonomous mobile robots. In: Proc. 9th Int. Colloq. on Structural Information and Communication Complexity. (2002) 57–72
17. Prencipe, G.: CORDA: Distributed coordination of a set of atonomous mobile robots. In: Proc. 4th European Research Seminar on Advances in Distributed Systems. (2001) 185–190
18. Sugihara, K., Suzuki, I.: Distributed algorithms for formation of geometric patterns with many mobile robots. Journal of Robotic Systems **13** (1996) 127–139

19. Suzuki, I., Yamashita, M.: Distributed anonymous mobile robots - formation and agreement problems. In: Proc. 3rd Colloq. on Structural Information and Communication Complexity. (1996) 313–330
20. Suzuki, I., Yamashita, M.: Distributed anonymous mobile robots: Formation of geometric patterns. SIAM J. on Computing **28** (1999) 1347–1363
21. Flocchini, P., Prencipe, G., Santoro, N., Widmayer, P.: Hard tasks for weak robots: The role of common knowledge in pattern formation by autonomous mobile robots. In: Proc. 10th Int. Symp. on Algorithms and Computation. (1999) 93–102
22. Prencipe, G.: Distributed Coordination of a Set of Atonomous Mobile Robots. PhD thesis, Universita Degli Studi Di Pisa (2002)
23. Flocchini, P., Prencipe, G., Santoro, N., Widmayer, P.: Gathering of autonomous mobile robots with limited visibility. In: Proc. 18th Symp. on Theoretical Aspects of Computer Science. (2001) 247–258
24. Ando, H., Oasa, Y., Suzuki, I., Yamashita, M.: A distributed memoryless point convergence algorithm for mobile robots with limited visibility. IEEE Trans. Robotics and Automation **15** (1999) 818–828
25. Cohen, R., Peleg, D.: Robot convergence via center-of-gravity algorithms. Technical Report MSC 04-2, Weizmann Institue of Science, Rehovot, Israel (2004)
26. Suzuki, I., Yamashita, M.: Agreement on a common x-y coordinate system by a group of mobile robots. In: Proc. Dagstuhl Seminar on Modeling and Planning for Sensor-Based Intelligent Robots. (1996)
27. Goldstein, H.: Classical Mechanics. 2nd edn. Addison-Wesley, Reading, MA, USA (1980) 672 pages.

F-Chord: Improved Uniform Routing on Chord[*]
(Extended Abstract)

Gennaro Cordasco, Luisa Gargano, Mikael Hammar,
Alberto Negro, and Vittorio Scarano

Dipartimento di Informatica ed Applicazioni
Università di Salerno,
84081 Baronissi (SA), Italy
{cordasco,lg,hammar,alberto,vitsca}@dia.unisa.it

Abstract. We propose a family of novel schemes based on Chord retaining all positive aspects that made Chord a popular topology for routing in P2P networks. The schemes, based on the Fibonacci number system, allow to improve on the maximum/average number of hops for lookups and the routing table size per node.

1 Introduction

In this paper, we propose a family of new routing schemes that reduce the routing table size, and the maximum/average number of hops for lookup requests in Chord–like systems [15] without introducing any other protocol overhead. The improvement is obtained with no harm to the simplicity and ease of programming that are some of the many good characteristics that made Chord a popular choice.

The basis of Chord can be seen as a ring of N identifiers labelled from 0 to $N-1$. The edges, representing the overlay network, go from identifier x to identifier[1] $x + 2^i$, for each $x \in \{0, \ldots, N-1\}$ and $i < \log N$. The degree and the diameter are $\log N$, the average path length is $(\log N)/2$. Routing is greedy, never overshooting the destination.

Because of low diameter and average path length, Chord offers fast lookup algorithms. By having also low degree, it provides efficient join/leave of nodes since the cost depends on the diameter and the degree and is, in fact, upper bounded by their product. Chord is scalable: with $n \leq N$ nodes present in the network the same performance (in terms of n rather than N) can be obtained w.h.p.. Efficient routing in Chord is easy (a greedy algorithm is optimal) due to the fact that Chord is *uniform*: x is connected to y iff $x + z$ is connected to $y + z$. Since we are restricting ourselves to uniform routing schemes, we can use

---

[*] Work partially supported by EU RTN project ARACNE and by Italian FIRB project WebMinds.

[1] Throughout the paper, arithmetics on node identities is always mod N where N is the number of identifiers. Similarly, all the logarithms are base 2, unless differently specified.

R. Králóvič and O. Sýkora (Eds.): SIROCCO 2004, LNCS 3104, pp. 89–98, 2004.
© Springer-Verlag Berlin Heidelberg 2004

the term *jump of size s* to indicate the existence in the overlay network of an edge from x to $x + s$ for any identifier $x = 0, \ldots, N - 1$ (e.g. Chord has jumps 2^i, for $i < \log N$).

Uniformity is a crucial requirement, since it makes any system a good candidate for real implementations: besides simplicity in the implementation it also offers an optimal greedy routing algorithm without node congestion [16]. On the other hand, it is known that, as long as uniformity is required, the $O(\log N)$ values for the number of jumps and diameter cannot be asymptotically improved. Chord's values of the three main parameters (degree vs. diameter and average path length) can be improved if one removes the uniformity request [4, 7–10, 12].

Because of its practicality and its known bounds, it assumes a certain relevance to improve the performance of Chord while retaining simplicity and scalability. Moreover, due to the above, it is interesting and practically useful to improve known results for uniform systems, even only by constant factors. Our objective is to show improved bounds on all the important parameters of a P2P system that affect lookup time and join/leave cost, i.e., degree, diameter and average path length. To this aim we will propose and analyze a novel family of uniform routing schemes.

The lookup process in Chord can be seen as a binary search on an interval of N elements. A natural question to ask is whether the lookup can be realized with a more efficient search technique that can be translated into a uniform overlay network. Efficiency is measured in terms of degree and maximum/average path length. We notice that the problem poses several restrictions on the search model since we are assuming that all nodes are alike and, therefore, only queries taken from a globally given set can be used at any step.

In this paper, we consider search techniques in the above model imposed by uniform routing algorithms (that is, when a fixed set of jumps is available). To the best of our knowledge, while the problem is related to several search problems investigated in the literature, no useful results on problems totally fitting the above model and goals are known (see [3] and references therein quoted). A previous work in the P2P context in this direction is contained in [16]. Our starting point is Fibonacci search [2].

1.1   Our Results

Let $Fib(i)$ be the i-th Fibonacci number. We prove that any uniform algorithm that uses up to δ jumps and has diameter d can reach at most $N(\delta, d)$ consecutive identifiers where $N(\delta, d) \leq Fib(\delta + d + 1)$. This gives us a tradeoff of $1.44 \log N$ on the sum of the degree and the diameter in any P2P network using uniform routing on N identifiers. We then show a family of routing scheme F-Chord(α) that, besides improving all Chord parameters, also reaches equality for any choice of the parameters in the inequality above with $|\delta - d| \leq 1$.

Our analysis has been carried out by considering the N-size identifier space. We use a standard technique (see [15] for technical details) to manage the situation when $n < N$ nodes are in the network. In this case, in fact, we assume that the n nodes are uniformly distributed at random on the N identifiers, therefore we can consider a ring of n "chunks" each with N/n identifiers.

Since in each chunk there are at most $O(\log n)$ nodes w.h.p., any deterministic result on the diameter in terms of N can be easily translated in the same result in terms of n, with high probability. The same argument can be used for the degree. In fact, the distance between two nodes is at least N/n^2 w.h.p.. If $\delta(N)$ is the degree when all the N identifiers are used, then the number of finger pointers that are useful (and consequently stored) with n nodes is at most $2\delta(N)$, w.h.p..

1.2   Related Work

The Chord system was introduced in [15] to allow efficient lookup in a Distributed Hash Table (DHT). By using logarithmic size routing tables in each node, Chord allows to find in logarithmic n number of routing hops the node of a P2P system that is responsible for a given key. Adding or removing a node is accomplished at a cost of $O(\log^2 N)$ messages. A thorough study of uniform system is given in [16].

In general, Chord can be improved at the expenses of uniformity [7, 9, 10]. Recently, some non-greedy routing algorithms were proposed, that use De-Bruijn based DHT [4, 11], whose goal is to reach an optimal trade-off between degree and path length and, in particular, allow routing in $\Omega(\log N / \log\log N)$ with logarithmic degree.

One can also improve the results by eliminating the deterministic requirement. In fact, it is possible (see [8, 12]) to route greedily in $\Theta(\log N / \log\log N)$ with logarithmic degree by using randomization and the so called *neighbor-of-neighbor (NoN) approach*: a node uses, at each step, its neighbor's neighbors to make greedy decisions. It is not difficult to see that these techniques can be easily adapted to our scheme thus obtaining similar results.

However, in this paper we focus on *deterministic* and *uniform* routing schemes. Among other advantages, they offer an optimal greedy routing strategy that provides simplicity, fault tolerance (as long as some node has edges toward destination, the routing succeeds) and locality (messages flow only on the portion of ring between source and destination), as noted in [12].

1.3   Organization of the Paper

The rest of the paper is organized as follows. First of all, we provide the proof of the lower bound in Section 3. Then, in Section 4 we introduce the F-Chord(α) family and in Section 5 we prove some of its properties with regards to the degree, diameter, average number of hops, and congestion. Finally, in Section 6 we conclude with some remarks and give some open problems.

2   Fibonacci Numbers

We, briefly, recall here some basic facts on Fibonacci numbers which will be used in the sequel (see [6]). Let $Fib(i)$ denote the i-th Fibonacci number. They are defined as $Fib(0) = 0, Fib(1) = 1$ and, for each $i>1$,

$$Fib(i) = Fib(i-1) + Fib(i-2).$$

For each index i, it holds

$$Fib(i) = [\phi^i/\sqrt{5}],$$

where $\phi = 1.618\ldots$ is the golden ratio and $[\ ]$ represents the nearest integer function. Furthermore,

$$\sum_{i=0}^{p} Fib(i) \cdot Fib(p-i) = \frac{1}{5}\{p[Fib(p+1) + Fib(p-1)] - Fib(p)\}.$$

3   The Lower Bound

In this section, we furnish a tradeoff of $1.44 \log N$ on the sum of the degree and the diameter in any P2P network using uniform routing on N identifiers. Namely, we prove the following theorem.

Theorem 1 *Let $N(\delta, d)$ denote the maximum number of consecutive identifiers obtainable trough a uniform algorithm using up to δ jumps (i.e. degree δ) and diameter d. For any $\delta \geq 0$, and $d \geq 0$, it holds that*

$$N(\delta, d) \leq Fib(\delta + d + 1) \tag{1}$$

Remark: *It can be shown that inequality (1) is strict whenever $|\delta - d| > 1$. On the other hand, we will exhibit a routing scheme that reaches equality for any choice of the parameters with $|\delta - d| \leq 1$.*

Proof. We proceed by induction on the sum $\delta + d$. Trivially, $N(0,0) = N(1,0) = N(0,1) = 1 = Fib(1) = Fib(2)$. Assume that (1) holds for any δ and d with $\delta + d < x$. We will show that for any degree y and diameter z with $y + z = x$ it holds that $N(y, z) \leq Fib(x+1)$. We distinguish three cases on the number of times the first (i.e. the biggest) jump is repeated. We recall that the assumption of a uniform algorithm implies that jumps can be used to build paths in a greedy manner with respect to their size, that is, on each path jumps of the same size can be assumed to be consecutive.

Case 1: the first jump appears at most once on each path. In this case, each path, either starting with the given jump or not, will never use this jump again. This implies that (cfr. Figure 1(a))

$$N_1(y, z) \leq N(y-1, z) + N(y-1, z-1)$$
$$\leq Fib(y+z) + Fib(y+z-1) = Fib(y+z+1) = Fib(x+1).$$

Case 2: the first jump appears at most twice on each path. Since we have two equal jumps, the size of the jump cannot exceed the maximum number of nodes reachable using degree $y - 1$ and diameter $z - 1$ remaining after the first jump has been used (cfr. Figure 1 (b)). The part remaining after the second jump cannot exceed the maximum number of nodes reachable using the remaining degree $y - 1$ and diameter $z - 2$. Hence

$$N_2(y, z) \leq 2N(y - 1, z - 1) + N(y - 1, z - 2)$$
$$\leq 2Fib(y + z - 1) + Fib(y + z - 2) = Fib(x + 1).$$

Case 3: the first jump appears at most ℓ times on each path for some $2 < \ell \leq z$. As in Case 2, we can deduce that (cfr. Figure 1 (c)) $N_3(y, z) \leq \ell N(y - 1, z - \ell + 1) + N(y - 1, z - \ell)$, which in turn can be shown to give

$$N_3(y, z) < Fib(y + z + 1) = Fib(x + 1).$$

From the above inequalities, we get

$$N(y, z) \leq \max\{N_1(y, z), N_2(y, z), N_3(y, z)\} \leq Fib(x + 1).$$

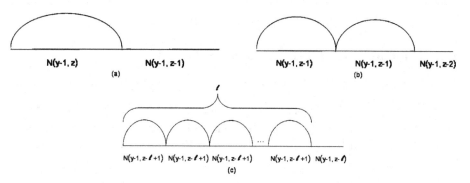

Fig. 1. The three cases in the proof of Theorem 1.

4   F-Chord: A Fibonacci-Based Chord-Like System

The purpose of this section is to introduce our schemes that are based on the Fibonacci number system applied to a Chord-like network, namely a ring of N identifiers labelled from 0 to $N - 1$.

The idea of using a different base (other than 2) for representing IDs (and consequently to route among them) is not new. It was already pointed out in [15], in fact, that any base $b < 2$ can be used for a Chord-like routing that trades off routing table size (i.e. degree) with number of hops (i.e. diameter). As a result, the jumps would be b^i, for each $i < \log_b N$ and, consequently, Chord's results are slightly changed: the degree is raised to $\log_b N$ while the diameter is lowered to $\log_{b/(b-1)} N$. By using a Chord-based scheme with base ϕ, one could, therefore, improve on the diameter by paying off a corresponding increase in the degree.

Nevertheless, by exploiting the many properties of Fibonacci numbers, we can define and analyze a family of uniform routing algorithms that offers improved performances on the various parameters – number of jumps, diameter, average path length, and edge congestion. Each member of the family can be obtained by tuning the parameter α, with $1/2 \le \alpha \le 1$. In particular, for $\alpha = 1/2$ the corresponding algorithm meets the bound in Theorem 1.

In this section, we define the families by first defining the generic Chord extension to Fibonacci numbers mentioned above. Successively we present the two families of routing schemes F-Chord.

4.1   The Fibonacci Routing Scheme Fib-Chord

For sake of simplicity, we first introduce the Fibonacci routing scheme Fib-Chord. Let $N \in (Fib(m - 1), Fib(m)]$. The scheme uses $m - 2$ jumps of size $Fib(i)$, for $i = 2, \ldots, m - 1$. Recall that $m \approx \log_\phi N = 1.44042 \log N$.

The results of next section will imply, in particular, that, by using Fib-Chord over a set of N identifiers, the degree (i.e., the number of jumps) is $1.44042 \log N - 2$, the diameter (i.e., the number of hops) is $0.72021 \log N$, and the average path length is $0.39812 \log N$.

4.2   The Family of Routing Schemes F-Chord

Intuitively, we obtain the two families of routing schemes by taking Fib-Chord and pruning them, either starting from smaller size jumps or from larger size jumps. The pruning is realized by eliminating a certain quantity (that is related to the parameter α) of the jumps with odd indices. More formally, our family is indeed composed by two subfamilies named F_a-Chord and F_b-Chord defined below.

Definition 1. Let $Fib(m - 1) < N \le Fib(m)$ and $\alpha \in [1/2, 1]$.
a) The F_a-Chord(α) scheme uses the $\lceil \alpha(m - 2) \rceil$ jumps

$$Fib(2i), \text{ for } i = 1, \ldots, \lfloor (1 - \alpha)(m - 2) \rfloor$$

and

$$Fib(i), \text{ for } i = 2\lfloor (1 - \alpha)(m - 2) \rfloor + 2, \ldots, m - 1.$$

b) The F_b-Chord(α) scheme uses the $\lceil \alpha(m - 2) \rceil$ jumps

$$Fib(i), \text{ for } i = 2, \ldots, m - 2\lfloor (1 - \alpha)(m - 2) \rfloor$$

and

$$Fib(2i), \text{ for } i = \left\lceil \frac{m - 2\lfloor (1 - \alpha)(m - 2) \rfloor}{2} \right\rceil + 1, \ldots, \lfloor (m - 1)/2 \rfloor.$$

We will use the name F-Chord(α) whenever we want to indicate any of the two schemes F_a-Chord(α) and F_b-Chord(α). Notice here that Fib-Chord $=$ F-Chord(1) $(=F_a$-Chord$(1)=F_b$-Chord$(1))$; moreover, F_a-Chord$(1/2) = F_b$-Chord$(1/2)$.

5   Properties of F-Chord

In this section we investigate and bound the main parameters of the F-Chord systems.

For the rest of this section, we always assume that $Fib(m-1) < N \leq Fib(m)$ and $\alpha \in [1/2, 1]$.

5.1   Diameter

Theorem 2   *For any value of α, the diameter of F-Chord(α) is* $\lfloor m/2 \rfloor \approx 0.72021 \log N$.

Proof. Since we use the greedy strategy it is enough to consider the case $\alpha = 1/2$, i.e., when there are only jumps of even Fibonacci index.

We can prove, by induction, that after performing i jumps, the search interval is reduced to $Fib(m - 2i + 1)$. In fact, if $i = 1$ then we are considering a jump of size greater or equal to $Fib(m-2)$, and since $N \leq Fib(m)$ the search interval is bounded by $Fib(m) - Fib(m-2) = Fib(m-1)$. Consider the situation after t jumps. If we took a jump in the previous step, the size of the search interval is limited by $Fib(m - 2t + 1)$. Hence, $Fib(m - 2t + 1) - Fib(m - 2t) = Fib(m - 2t - 1)$. If we did not take a jump in the previous step, the search interval is limited by $Fib(m - 2t)$. Performing the jump restricts the search interval to $Fib(m - 2t) - Fib(m - 2t - 2) = Fib(m - 2t - 1)$. For $i = \lfloor m/2 \rfloor$ the interval is thus reduced to $Fib(2) = 1$ and the search ends.

5.2   Average Path Length

We can precisely evaluate the average path length of F-Chord(α). Let us, first, denote by $L_1(i, m)$ (resp. $L_{1/2}(i, m)$) the load for an edge of size $Fib(i)$ on all intervals of size $Fib(m)$ when we use F-Chord(1) (resp. the F-Chord(1/2)).

The following properties can be easily obtained:

1. $L_1(m, m) = 0$;
2. $L_1(m - 1, m) = Fib(m - 2)$;
3. $L_1(i, m) = L_1(i, m - 1) + L_1(i, m - 2), i < m - 2$.

Indeed, either the destination is one of the first $Fib(m-1)$ nodes or among the last $Fib(m-2)$ nodes in the interval. To go from the first part to the second part we issue a jump of size $Fib(m - 1) \neq Fib(i)$. That jump is hence not counted in $L_1(i, m)$.

Moreover, for $L_{1/2}$ it can be proved that:

4. $L_{1/2}(m - 1, m) = Fib(m - 2)$;
5. $L_{1/2}(m - 2, m) = Fib(m - 1) + Fib(m - 3)$;
6. $L_{1/2}(i, m) = L_{1/2}(i, m - 1) + L_{1/2}(i, m - 2)$ for $i < m - 2$.

We can prove the following lemma.

Lemma 1. *By using properties **1**–**6**, it holds that:*

1) $L_1(i,m) = Fib(i-1)Fib(m-i)$ *for* $1 \le i \le m$.
2) $L_{1/2}(2i,m) = Fib(2i-1)Fib(m-2i) + Fib(2i+1)Fib(m-2i-1)$ *for* $2 \le 2i \le m-1$.

Theorem 3 *The average path length of the F-Chord(α) scheme ,is bounded by*

$$0.39812 \log N + (1-\alpha)0.24805 \log N + 1.$$

Proof. We sketch the proof for F_a-Chord(α), computation for F_b-Chord(α) is quite similar. For sake of simplicity, we assume that $N = Fib(m)$. The sum of the total link load in the network, denoted by $S_\alpha(m)$, equals the sum of the lengths of each path. It is easy to show that:

$$S_\alpha(m) = \sum_{i=1}^{m-2} Fib(i)Fib(m-i-1) + \sum_{i=1}^{\lfloor(1-\alpha)(m-2)\rfloor} Fib(2i-1)Fib(m-2i-1)$$

We have

$$S_\alpha(m) = \sum_{i=1}^{m-2} Fib(i)Fib(m-i-1) + \sum_{i=1}^{\lfloor(1-\alpha)(m-2)\rfloor} Fib(2i-1)Fib(m-2i-1)$$

$$= \frac{1}{5}[(m-1)(Fib(m)+Fib(m-2))-Fib(m-1)] +$$

$$+ \sum_{i=1}^{\lfloor(1-\alpha)(m-2)\rfloor} Fib(2i-1)Fib(m-2i-1)$$

$$= \frac{1}{5}[(m-1)(Fib(m)+Fib(m-2))-Fib(m-1)]$$

$$+ \sum_{i=3}^{\lfloor(1-\alpha)(m-2)\rfloor} Fib(2i-1)Fib(m-2i-1) + Fib(m-3) + 2Fib(m-5)$$

Hence, one can find the desired value of the average path length by observing that:

$$\frac{S_\alpha(m)}{Fib(m)} < \frac{1}{5}\left[(m-1)\left(1+\frac{1}{\phi^2}\right) - \frac{1}{\phi}\right] + \frac{(1-\alpha)(m-2)Fib(5)Fib(m-7)}{Fib(m)} + 1$$

The following are immediate corollaries of the above Theorems 2 and 3.

Corollary 1 *The F-Chord(1) scheme (=Fib-Chord) has degree $1.44042 \log N - 2$, diameter equal to $0.72021 \log N$, and average path length equal to $0.39812 \log N$.*

Corollary 2 *The F-Chord(1/2) scheme has degree and diameter both equal to $0.72021 \log N$ and the average path length is $0.52215 \log N$.*

We notice that F-Chord(1/2) meets the bound in Theorem 1.

Corollary 3 *For each $< \alpha \in [0.58929, 0.69424]$, the F_a-Chord(α) and F_b-Chord (α) schemes improve on Chord in all parameters (number of jumps, diameter, and average path length).*

5.3   Edge Congestion

We remind the reader that, by using a uniform scheme, there is no node congestion. Therefore, here we focus on edge load, that is defined as the number of times it is used by all routes from every node to every other node.

Analysis of $L_1(i, m)$ and $L_{1/2}(i, m)$ shows that $L_{1/2}(i, m)$ is decreasing with $i > 1$, whereas $L_1(i, m)$ decreases until it reaches its minimum at $i = \lceil m/2 \rceil$, after which it is increasing. Hence, the maximum and the minimum loads are given by

$$max\{L_1(i, m), L_{1/2}(i, m)\} = L_{1/2}(2, m) = Fib(m - 1) + Fib(m - 3)$$
$$min\{L_1(i, m), L_{1/2}(i, m)\} = L_1(\lceil m/2 \rceil, m) = Fib(\lceil m/2 \rceil - 1)Fib(\lceil m/2 \rceil),$$

and the ratio between the two is bounded by 9/4, which can be viewed as the worst possible congestion. On the other hand, in the literature congestion is sometimes defined as the maximum load divided by the average load. With this definition the congestion g_α for the F-Chord(α) scheme lies between $g_{1/2} \leq g_\alpha \leq g_1$, and the boundaries are swiftly calculated to $g_{1/2} = 1.18034$ and $g_1 = 1.38197$.

5.4   A Comment on Scalability

In the case of F-Chord(1) and F-Chord(1/2), with $n \leq N$ nodes in the network, results can be easily rewritten in terms of n with high probability, as already noticed in [15, 16]. When choosing a parameter $\alpha \in (1/2, 1)$, the presence of $n < N$ nodes will automatically tune (due to the fact that the shortest jumps will point to the same node): to $\alpha_a(n) > \alpha$ if one is using F_a-Chord(α); to $\alpha_b(n) < \alpha$ if one is using F_b-Chord(α). Thus it is possible to choose which member of the family is most suitable for the requirements of the application. If a lower average path length is desired one can choose F_a-Chord(α) (here, fixing α assures an upper bound of $\lceil \alpha(m-2) \rceil$ on the node degree for large values of n) while if a low degree is the main goal one can choose F_b-Chord(α) (in this case fixing α assures an upper bound on the average path length as in Theorem 3).

6   Conclusions and Open Problems

We have described a family of simple algorithms that (1) improves uniform routing on Chord and (2) is of practical interest. In fact, the designer can choose which member of the family is most suitable for the requirements of the application. For example, in a distributed file-system application one may want to prefer lower average path length over worst case (and thus choose F-Chord(1)), while in an application where fast delivery is paramount, one would choose a faster worst case over the average (i.e. F-Chord(1/2)) with a whole range of intermediate choices by using appropriate values of α.

Since any greedy routing requires $\Omega(\log N)$ hops when the degree is logarithmic [8], we believe that it is meaningful to improve the multiplicative constants in front of the $\log n$ since the results obtained by deterministic and uniform algorithms must be compared, practically, to the more theoretically appealing

$O(\log N/\log\log N)$ that, for some values of n could have performances comparable to the deterministic and uniform algorithms (that are much much simpler to realize and deploy).

It eluded us to find an optimal average path length routing algorithm (once the node degree anf the uniformity requirements are fixed). The search schemes used in F-Chord is close to optimal but not for all values of N.

References

1. P. Druschel and A.Rowstron, Pastry: Scalable, distribute object location and routing for large-scale peer-to-peer systems. in Proc. of the 18th IFIP/ACM Inter. Conf. on Distr. Systems Platforms (Middleware 2001), Nov. 2001.
2. Ding-Zhu Du, Frank K. Hwang, Combinatorial Group testing and its applications. World Scientific, 2000.
3. S. Kapoor, E.M.Reingold, Optimum Lopsided Binary Trees. Journal of ACM, Vol. 36, No.3, July 1989, pp. 573–590.
4. M. F. Kaashoek and D. R. Krager, Koorde: A simple degree-optimal distributed hash table. in IPTPS, Feb. 2003.
5. P. Ganesan, G. S. Manku, Optimal Routing in Chord. to appear in Proc. of SODA '04, 2004.
6. R. Graham, O. Patashnik, D.E. Knuth, Concrete Mathematics. Addison-Wesley, 1994.
7. A. Kumar, S. Merugu, J. Xu and X. Yu: Ulysses, A Robust, Low-Diameter, Low-Latency Peer-to-peer Network. to appear in the Proc. of IEEE Inter. Conf. on Network Protocols (ICNP 2003), Atlanta, Nov. 2003.
8. G. S. Manku, M. Naor, U. Wieder, Know thy Neighbor's Neighbor: The Power of Lookahead in Randomized P2P Networks. to appear in Proc. of STOC 2004.
9. D. Malkhi, M. Naor and Ratajczak, Viceroy: A Scalable and Dynamic Emulation of the Butterfly. in Proceedings of the 21st ACM Symposium on Principles of Distributed Computing (PODC 2002), Aug. 2002.
10. M. Naor and U. Wieder, A Simple Fault Tolerant Distributed Hash Table. in IPTPS, Feb. 2003.
11. M. Naor and U. Wieder, Novel architectures for p2p applications: the continous-discrete approach. in Proc. of 15th ACM Symp. on Parallel Algorithms and Architectures (SPAA), 2003.
12. M. Naor, U. Wieder, Know thy Neighbor's Neighbor: Better Routing for Skip-Graphs and Small Worlds. to appear in Proc. of IPTPS 2004.
13. S. Ratnasamy, S. Shenker, and I. Stoica, Routing Algorithms for DHTs: Some Open Questions. in IPTPS, 2002.
14. S. Ratnasamy, P. Francis, M. Handley, R. Karp, and S.Shenker, A scalable content-addressable network. in Proc. ACM SIGCOMM, Aug. 2001.
15. I. Stoica, R. Morris, D. Liben-Nowell, D. R. Karger, M. F. Kaashoek, F. Dabek, H. Balakrishnan, Chord: A Scalable Peer-to-peer Lookup Protocol for Internet Applications. in IEEE/ACM Trans. on Networking, 2003.
16. J. Xu, On the Fundamental Tradeoffs between Routing Table Size and Network Diameter in Peer-to-Peer Networks. in the Proc. of IEEE INFOCOM 2003, May 2003.
17. B. Y. Zhao, J. Kubiatowicz, and A. Joseph, Tapestry: An infrastructure for fault-tolerant wide-area location and routing. Tech. Rep. UCB/CSD-01-1141, University of California at Berkeley, Computer Science Department, 2001.

Swapping a Failing Edge of a Shortest Paths Tree by Minimizing the Average Stretch Factor[*]

Aleksej Di Salvo[1] and Guido Proietti[1,2]

[1] Dipartimento di Informatica, Università di L'Aquila,
Via Vetoio, 67010 L'Aquila, Italy
aleksejdis@tin.it, proietti@di.univaq.it
[2] Istituto di Analisi dei Sistemi ed Informatica "Antonio Ruberti", CNR,
Viale Manzoni 30, 00185 Roma, Italy

Abstract. We consider a 2-edge connected, undirected graph $G = (V, E)$, with n nodes and m non-negatively real weighted edges, and a single source shortest paths tree (SPT) T of G rooted at an arbitrary node r. If an edge in T is temporarily removed, it makes sense to reconnect the nodes disconnected from the root by adding a single non-tree edge, called a *swap edge*, instead of rebuilding a new optimal SPT from scratch. In the past, several optimality criteria have been considered to select a best possible swap edge. In this paper we focus on the most prominent one, that is the minimization of the average distance between the root and the disconnected nodes. To this respect, we present an $O(m \log^2 n)$ time and linear space algorithm to find a best swap edge for every edge of T, thus improving for $m = o\left(n^2 / \log^2 n\right)$ the previously known $O(n^2)$ time and space complexity algorithm.

Keywords: Network Survivability, Single Source Shortest Paths Tree, Swap Algorithms.

1 Introduction

One of the most common operations in computer networks is the broadcasting of a message from a source node to every other node of the network. When the broadcasting procedure represents an important part of the network activity, it makes sense to use a topology that enables the information to reach all the nodes in the fastest possible way, that is a *single source shortest paths tree* (SPT) rooted in the source node. The SPT, as any tree-based network topology, may present some problems regarding link malfunctioning, however: the smaller is the number of links, the higher is the average traffic for each link and, consequently, the higher is the risk of a link overloading. Furthermore, the failure of a single link may cause the disconnection of a wide part of the network.

Two different approaches can be followed to solve the problem of a link failure: either rebuilding a new SPT from scratch, or using a single non-tree

---

[*] This work has been partially supported by the Research Project GRID.IT, funded by the Italian Ministry of Education, University and Research.

R. Královič and O. Sýkora (Eds.): SIROCCO 2004, LNCS 3104, pp. 99–110, 2004.

edge (called a *swap edge*) to replace the failing link and reconnect the network, thus obtaining the so-called *swap tree*. The former guarantees the construction of a most efficient network on the graph without the failing link, but it is very expensive both in terms of set-up costs and of time complexity for computing a new SPT [8]. Indeed, a new SPT may be completely different from the existing one, and therefore the updating of a large amount of nodes may be necessary. Furthermore, the most efficient known algorithm for computing an SPT has an $O(m+n\log n)$ time complexity, and therefore the precomputation of a new SPT for every possible link failure in the network requires an $O(mn + n^2\log n)$ time complexity. Thus, if the failing link is supposed to be quickly restored, and in any other case in which changes on the network structure are expensive, it may be preferable to use a swap edge, thus minimizing the number of nodes to be updated, rather than rebuilding a new SPT.

Several different functions have been described in [5] to characterize a possible best swap edge. In particular, among all the defined functions, the average distance from the root represents the most suitable choice to find a swap edge that matches the criterium on which the original SPT is based: reaching every node from the root by following a shortest possible path. It is worth noticing that by adopting such swap strategy, the average *stretch factor* of the swap tree w.r.t. a new SPT (i.e., the average ratio between the distance from the root to a disconnected node in the swap tree and in a new SPT) is bounded by 3 [5]. Moreover, experimental results show that the tree obtained from the swap is functionally very close to a new SPT computed from scratch [6]. Therefore, swapping is not only faster than rebuilding, but it even gives a good network in terms of path lengths.

In this paper we focus exactly on the problem of finding, for every edge of the SPT, a swap edge that minimizes the average distance from the root to every node in the disconnected area of the network. An $O(n^2)$ time and space complexity algorithm is known for this problem [5], while an $O(m + n)$ time and space one is described in [7] for the simpler problem of finding a best swap edge for a single failing link. In this paper we improve the former bound by presenting an $O(m\log^2 n)$ time and linear space complexity algorithm, running in the standard RAM model. In this way, we obtain a better space complexity for every instance of the problem, and we reduce the time complexity for graphs in which $m = o\left(\frac{n^2}{\log^2 n}\right)$.

The paper is organized as follows: in Section 2 we give some basic definitions that will be used throughout the paper; in Section 3 we present our algorithm; in Section 4 we show few results that will form the basis for the time complexity analysis of our algorithm; finally, in Section 5, we will formally analyze our algorithm.

2   Basic Definitions

Let $G = (V, E)$ be a graph in which V is the set of nodes and E is the set of edges; every edge $e = (u, v)$ has associated a non-negative real *weight* $w(e)$, also denoted as $w(u,v)$. A *simple path* is a sequence of nodes $p = (u_1, u_2, \ldots u_k) = u_1 \rightarrow u_k$ such that $(u_i, u_{i+1}) \in E$, $i \in [1, k-1]$, and $u_i \neq u_j$ for every $i, j \in [1, k], i \neq j$; if $u_1 = u_k$ the path is called a *cycle*. Let $w(p) = \sum_{i=1}^{k-1} w(u_i, u_{i+1})$ be the *length* of the path p. The *distance* $d(u, v)$ between the nodes u, v is defined as the length of a shortest path connecting the nodes. A *connected graph* is a graph in which every couple u, v of nodes has a path $u \rightarrow v$ linking them; a graph is said to be *2-edge connected* if the deletion of any edge leaves the graph connected.

A *tree* is a connected graph without any cycle. If we root a tree T at an arbitrary node r, then a node x is said to be an *ancestor* of a node y in T if the path from r to y in T contains x; in this case, y is a *descendant* of x. A *single source shortest paths tree* (SPT) of G rooted in r is a tree $T = (V, E_T)$ made up by shortest paths linking r with every other node of the graph. Let $T_v = (V_v, E_v)$ be the subtree of T rooted at v. In the rest of the paper, we will denote by $|T_v|$ the number of nodes in T_v, and by $w(T_v) = \sum_{x \in T_v} d(v, x)$. The *nearest common ancestor* (NCA) of a given couple of nodes x, y is the lowest node of the tree that is an ancestor of both x and y (for details see [3]); we will indicate it by $nca(x, y)$.

Let $e = (u, v)$ be any tree edge, with u parent of v in T. We indicate with $C_e = \{(x, y) \in E \setminus E_T | (x \in V \setminus V_v) \wedge (y \in V_v)\}$ the set of *swap edges* for e, i.e., the edges that may be used to replace e for maintaining the tree connected. In a 2-edge connected graph, we have that $C_e \neq \emptyset$, $\forall e \in E_T$. Given a non-tree edge f, let $life(f) = \{e \in E_T | f \in C_e\}$, i.e., the set of tree edges for which f is a swap edge; we will indicate $life(\alpha) \cap life(\beta)$ by writing $life(\alpha, \beta)$. For any $f \in C_e$, let $T_{e/f}$ and $d_f(x, y)$ be the tree obtained by swapping e with f and the distance between nodes x, y in $T_{e/f}$, respectively, and let $F(e, f) = \sum_{x \in T_v} d_f(r, x)$. Let $\phi(e, f) = F(e, f)/|T_v|$ denote the average distance between r and nodes in T_v. Notice that $|T_v|$ does not depend on the swap edge, thus finding a swap edge minimizing ϕ is equivalent to finding a swap edge minimizing F. Hence, a *best swap edge* f^* for a tree edge e is defined as

$$f^* = \operatorname{argmin}\{F(e, f) | f \in C_e\}.$$

In the sequel, we will study the problem of finding, for every edge of the SPT, a best swap edge minimizing the function F, and then minimizing the function ϕ.

3   High-Level Description of the Algorithm

Let us first recall the definition of *lower envelope* of a set of functions:

Definition 1. *Let $\mathcal{F} = \{f_1(x), \ldots, f_k(x)\}$ be a set of functions, where $f_i : x \in D_i \subseteq \mathbb{R} \mapsto \mathbb{R}$. The function*

$$L_{\mathcal{F}} : x \in D_{\mathcal{F}} = \bigcup_{i=1}^{k} D_i \mapsto \min\{f_i(x) \mid i \in [1, k] \wedge x \in D_i\} \in \mathbb{R}$$

is named the lower envelope *of $\mathcal{F}$.*

Let us consider the nodes of the tree as ordered in any arbitrary post-order. For every tree edge $e = (u, v)$, and for every non-tree edge α, let the *swap function* associated with α be defined as follows:

$$f_\alpha(v) = \begin{cases} F(e, \alpha) & \text{if } \alpha \in C_{e=(u,v)}; \\ \text{undefined} & \text{otherwise.} \end{cases}$$

If we arrange the nodes on the x axis and the values of the f_α functions on the y axis, then solving our problem reduces to finding the lower envelope of $\mathcal{F} = \{f_\alpha(v) \mid \alpha \text{ is a non-tree edge}\}$.

In [4] it was presented an efficient comparison-based algorithm to compute the lower envelope of a set of continuous real functions. Such algorithm relies on the concept of *intersection* between functions, and can be extended to our case by introducing the notion of *inversion* between pairs of swap functions. Basically, an inversion between two swap functions f_α, f_β with $life(\alpha, \beta) \neq \emptyset$, is defined by a couple of tree edges $e = (u, v), e' = (v, z) \in life(\alpha, \beta)$ such that u is the parent of v and v is the parent of z in T, and for which the following holds: either $f_\alpha(z) \leq f_\beta(z)$ and $f_\alpha(v) > f_\beta(v)$, or $f_\alpha(z) \geq f_\beta(z)$ and $f_\alpha(v) < f_\beta(v)$. Node v is said to be an *inversion node* for α, β.

The algorithm provided in [4] relies on the efficient answering to a set of three queries that, for our problem, can be rephrased as follows:

Q_1: Given a swap function f_α, return the minimum and the maximum node (with respect to the selected post-order numbering) for which f_α is defined;

Q_2: Given a couple of swap functions and a node $v \in V$, return their minimum inversion node greater than v (w.r.t. the current post-order), if any;

Q_3: Given a couple of swap functions f_α, f_β and a node v belonging to the path induced by $life(\alpha, \beta)$, compare $f_\alpha(v)$ and $f_\beta(v)$.

In the following section, we show how to answer efficiently to these queries, and we provide a bound on the number of inversions of a couple of functions f_α, f_β.

4   Answering the Queries

Given two swap edges $\alpha = (y_\alpha, x_\alpha), \beta = (y_\beta, x_\beta)$, observe that either $life(\alpha, \beta)$ is empty, or it consists of a set of edges which form a path in T. In the latter case, such a path consists either of: (i) a subpath of a root-leaf path in T, or (ii) two edge-disjoint subpaths of T which start from the node $nca(y_\alpha, x_\alpha) = nca(y_\beta, x_\beta)$ and proceed downwards. Notice that case (ii) can be easily reduced to case (i) (it suffices to transform the swap edge α into edges $\alpha' = (y_\alpha, nca(y_\alpha, x_\alpha))$ with $w(\alpha') = w(\alpha) + d(r, x_\alpha)$, and $\alpha'' = (x_\alpha, nca(y_\alpha, x_\alpha))$ with $w(\alpha'') = w(\alpha) + d(r, y_\alpha)$, and similarly for β). Hence, in the following we will assume, for the sake of simplicity, that $life(\alpha, \beta) = \{e_0, \dots, e_p\}$, where e_i precedes e_{i+1} along a root-leaf path of T, $i = 0, \dots, p - 1$.

Trivially, query Q_1 can be answered in constant time, since the nodes we are searching for coincide with the endnodes of the given swap edge. Concerning the query Q_2, performing it efficiently is harder. In the following, given two swap edges α, β, if the lower endpoint of $life(\alpha, \beta)$ coincides with any endvertex of α or β, then we say the two edges are *related, unrelated* otherwise. Next, we show separately how to manage related and unrelated swap edges.

4.1   Related Swap Edges

We start by proving two lemmas concerned with related edges:

Lemma 1. *Let* $\alpha = (y_\alpha, x_\alpha), \beta = (y_\beta, x_\beta)$ *be related swap edges such that* $life(\alpha, \beta) = \{e_0, \ldots, e_p = (v_{p-1}, v_p)\}$, *where* $x_\beta = v_p$ *is an ancestor of* x_α. *If there exists an edge* $e_k \in life(\alpha, \beta)$ *such that* $F(e_k, \beta) \leq F(e_k, \alpha)$, *then* $d_\beta(r, x_\beta) \leq d_\alpha(r, x_\beta)$.

Proof. In the following, we adopt the following notation (see Figure 1). Let $\{e_{p+1}, \ldots, e_{p+q}\}$ be the set of edges in $life(\alpha)$ which lie below e_p. Let v_i be the endvertex of e_i farthest from the root, and for $i = 0, \ldots, p + q - 1$, let $S_i = T_{v_i} - T_{v_{i+1}}$, i.e., the subtree of T_{v_i} induced by the removal of all the nodes in $T_{v_{i+1}}$. Finally, we set $S_{p+q} = T_{v_{p+q}}$. Then

$$F(e_k, \beta) = \sum_{i=k}^{p+q} w(S_i) + \sum_{i=k}^{p} \left(d_\beta(r, x_\beta) + \sum_{\ell=i+1}^{p} w(e_\ell) \right) |S_i| +$$

$$+ \sum_{i=p+1}^{p+q} \left(d_\beta(r, x_\beta) + \sum_{\ell=p+1}^{i} w(e_\ell) \right) |S_i|;$$

$$F(e_k, \alpha) = \sum_{i=k}^{p+q} w(S_i) + \sum_{i=k}^{p} \left(d_\alpha(r, x_\alpha) + \sum_{\ell=p+1}^{p+q} w(e_\ell) + \sum_{\ell=i+1}^{p} w(e_\ell) \right) |S_i| +$$

$$+ \sum_{i=p+1}^{p+q} \left(d_\alpha(r, x_\alpha) + \sum_{\ell=p+q}^{i} w(e_\ell) \right) |S_i|.$$

Recalling that, by definition, $d_\alpha(r, x_\beta) = d_\alpha(r, x_\alpha) + \sum_{\ell=p+1}^{p+q} w(e_\ell)$, and assuming by contradiction that $d_\beta(r, x_\beta) > d_\alpha(r, x_\beta)$, we have

$$F(e_k, \beta) > \sum_{i=k}^{p+q} w(S_i) + \sum_{i=k}^{p} \left(d_\alpha(r, x_\alpha) + \sum_{\ell=p+1}^{p+q} w(e_\ell) + \sum_{\ell=i+1}^{p} w(e_\ell) \right) |S_i| +$$

$$+ \sum_{i=p+1}^{p+q} \left(d_\alpha(r, x_\alpha) + \sum_{\ell=p+1}^{p+q} w(e_\ell) + \sum_{\ell=p+1}^{i} w(e_\ell) \right) |S_i| \geq F(e_k, \alpha),$$

thus contradicting the assumptions.   □

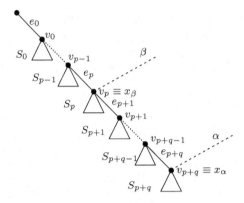

Fig. 1. Non-tree edges α, β, where the endnode x_α of α is a descendant of the endnode x_β of β; edges α, β are both swap edges for the tree edges $\{e_0, \ldots, e_p\}$.

The following result derives from Lemma 1:

Lemma 2. *Under the same assumptions of the previous lemma, we have that* $F(e_j, \beta) \leq F(e_j, \alpha), \forall j = 0, \ldots, k - 1.$

Proof. We have

$$F(e_j, \beta) = F(e_k, \beta) + \sum_{i=j}^{k-1} w(S_i) + \sum_{i=j}^{k-1} \left(d_\beta(r, x_\beta) + \sum_{\ell=i+1}^{p} w(e_\ell) \right) |S_i|;$$

$$F(e_j, \alpha) = F(e_k, \alpha) + \sum_{i=j}^{k-1} w(S_i) + \sum_{i=j}^{k-1} \left(d_\alpha(r, x_\beta) + \sum_{\ell=i+1}^{p} w(e_\ell) \right) |S_i|.$$

From the assumptions, we have that $F(e_k, \beta) \leq F(e_k, \alpha)$. Moreover, from Lemma 1 we know that $d_\beta(r, x_\beta) \leq d_\alpha(r, x_\beta)$, thus obtaining $F(e_j, \beta) \leq F(e_j, \alpha)$.   □

This result introduces an important monotonicity property: if an edge β gives a better result than a lower one α in swapping an edge e_k, then β will always be better than α for every edge above e_k.

The next lemma focuses on the case in which the lower swap edge α gives initially a better result than the higher one β, and provides a condition to establish for which failing edge, if any, β will became preferable w.r.t. α while climbing up the tree.

Lemma 3. *Let* $\alpha = (y_\alpha, x_\alpha), \beta = (y_\beta, x_\beta)$ *be related swap edges such that* $life(\alpha, \beta) = \{e_0, \ldots, e_p = (v_{p-1}, v_p)\}$, *where* $x_\beta = v_p$ *is an ancestor of* x_α. *Let* $\rho = x_\beta \to x_\alpha$ *and let* $F(e_p, \alpha) \leq F(e_p, \beta)$. *Then, there exists* $k \in [0, p)$ *such that* $F(e_k, \beta) \leq F(e_k, \alpha)$ *if and only if*

$$\sum_{i=k}^{p-1} |S_i| > \frac{F(e_p, \beta) - F(e_p, \alpha)}{d_\alpha(r, x_\alpha) + w(\rho) - d_\beta(r, x_\beta)}. \tag{1}$$

Proof. We write $F(e_k, \beta)$ and $F(e_k, \alpha)$ by using the same technique used above:

$$F(e_k, \beta) = F(e_p, \beta) + \sum_{i=k}^{p-1} w(S_i) + \sum_{i=k}^{p-1} \left(d_\beta(r, x_\beta) + \sum_{\ell=i+1}^{p} w(e_\ell) \right) |S_i|; \qquad (2)$$

$$F(e_k, \alpha) = F(e_p, \alpha) + \sum_{i=k}^{p-1} w(S_i) + \sum_{i=k}^{p-1} \left(d_\alpha(r, x_\alpha) + w(\rho) + \sum_{\ell=i+1}^{p} w(e_\ell) \right) |S_i|. \qquad (3)$$

From (3) and (2), the condition $F(e_k, \beta) \le F(e_k, \alpha)$ can be rewritten as:

$$F(e_p, \beta) + d_\beta(r, x_\beta) \sum_{i=k}^{p-1} |S_i| \le F(e_p, \alpha) + \left(d_\alpha(r, x_\alpha) + w(\rho) \right) \sum_{i=k}^{p-1} |S_i|,$$

which is equivalent to (1), thus concluding the proof. □

4.2   Unrelated Swap Edges

Next, we prove two results concerned with unrelated swap edges.

Lemma 4. *Let* $\alpha = (y_\alpha, x_\alpha), \beta = (y_\beta, x_\beta)$ *be unrelated swap edges such that* $life(\alpha, \beta) = \{e_0, \ldots, e_p = (v_{p-1}, v_p)\}$, *where both* x_α *and* x_β *descend from* v_p *in* T. *If* $F(e_p, \alpha) \le F(e_p, \beta)$ *and* $F(e_k, \beta) \le F(e_k, \alpha)$ *for some* $k \in [0, p)$, *then* $F(e_j, \beta) \le F(e_j, \alpha), \forall j = 0, \ldots, k - 1$.

Proof. Let x_a, x_b be the children of v_p that are ancestors of x_α, x_β in T, respectively; let $a = (v_p, x_a)$, $b = (v_p, x_b)$, and let $\rho = x_a \rightarrow x_\alpha, \mu = x_b \rightarrow x_\beta$ (see Figure 2). We start by writing the values of $F(e_k, \alpha)$ and $F(e_k, \beta)$:

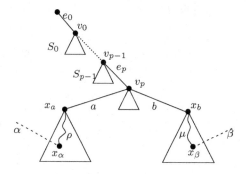

Fig. 2. Unrelated swap edges α, β and the edges $\{e_0, \ldots, e_p\} = life(\alpha, \beta)$ that they both can swap (splines denote paths).

$$F(e_k, \alpha) = F(e_p, \alpha) + \sum_{i=k}^{p-1} w(S_i)$$

$$+ \sum_{i=k}^{p-1} \left(d_\alpha(r, x_\alpha) + w(\rho) + w(a) + \sum_{\ell=i+1}^{p} w(e_\ell) \right) |S_i|;$$

$$F(e_k, \beta) = F(e_p, \beta) + \sum_{i=k}^{p-1} w(S_i)$$

$$+ \sum_{i=k}^{p-1} \left(d_\beta(r, x_\beta) + w(\mu) + w(b) + \sum_{\ell=i+1}^{p} w(e_\ell) \right) |S_i|.$$

From the assumptions $F(e_k, \beta) \leq F(e_k, \alpha)$ and $F(e_p, \alpha) \leq F(e_p, \beta)$, we have that the above two equations imply

$$w(b) + d_\beta(r, x_\beta) + w(\mu) \leq w(a) + d_\alpha(r, x_\alpha) + w(\rho). \tag{4}$$

Then, the cost of swapping a generic edge e_j, $j \in [0, k)$, with α and β is:

$$F(e_j, \alpha) = F(e_k, \alpha) + \sum_{i=j}^{k-1} \left(d_\alpha(r, x_\alpha) + w(\rho) + w(a) + \sum_{\ell=i+1}^{k} w(e_\ell) \right) |S_i|$$

$$+ \sum_{i=j}^{k-1} w(S_i);$$

$$F(e_j, \beta) = F(e_k, \beta) + \sum_{i=j}^{k-1} \left(d_\beta(r, x_\beta) + w(\mu) + w(b) + \sum_{\ell=i+1}^{k} w(e_\ell) \right) |S_i|$$

$$+ \sum_{i=j}^{k-1} w(S_i).$$

We know from the assumptions that $F(e_k, \beta) \leq F(e_k, \alpha)$, thus we can conclude the proof by using (4).   □

The next result deals with the same situation of Lemma 4, giving us a condition to identify, if any, an edge in which β gives a better result than α.

Lemma 5. *Let* $\alpha = (y_\alpha, x_\alpha), \beta = (y_\beta, x_\beta)$ *be unrelated swap edges such that* $life(\alpha, \beta) = \{e_0, \ldots, e_p = (v_{p-1}, v_p)\}$, *where both* x_α *and* x_β *descend from* v_p *in* T. *If* $F(e_p, \alpha) \leq F(e_p, \beta)$, *then there exists an edge* $e_k \in life(\alpha, \beta)$ *such that* $F(e_k, \beta) < F(e_k, \alpha)$ *if and only if*

$$\sum_{i=k}^{p-1} |S_i| > \frac{F(e_p, \beta) - F(e_p, \alpha)}{d_\alpha(r, x_\alpha) + w(\rho) + w(a) - d_\beta(r, x_\beta) - w(\mu) - w(b)}. \tag{5}$$

Proof. We can rewrite the condition $F(e_k, \beta) < F(e_k, \alpha)$ in the following way, by using a technique very similar to that used above:

$$F(e_p, \beta) + \left(d_\beta(r, x_\beta) + w(\mu) + w(b)\right) \sum_{i=k}^{p-1} |S_i| <$$

$$F(e_p, \alpha) + \left(d_\alpha(r, x_\alpha) + w(\rho) + w(a)\right) \sum_{i=k}^{p-1} |S_i|$$

which is equivalent to condition (5), thus concluding the proof.   □

4.3   Answering to Queries Q_2 and Q_3

The results of Lemmas 1-5 can be summarized by the following:

Proposition 1. *Let* α, β *be swap edges such that* $life(\alpha, \beta) = \{e_0, \dots, e_p\}$. *If* $F(e_p, \alpha) \leq F(e_p, \beta)$, *then one of the following two conditions must hold:*

1. $F(e_i, \alpha) \leq F(e_i, \beta) \; \forall i = 0, \dots, p-1$;
2. $\exists j \in [0, p) | \left(\forall i \in (j, p], \; F(e_i, \alpha) \leq F(e_i, \beta)\right) \wedge \left(\forall i \in [0, j], \; F(e_i, \alpha) > F(e_i, \beta)\right)$.   □

Hence, the following two cases are possible:

1. If the edges are related, then from Lemma 3 the inversion node v_j, if it exists, must satisfy the following condition:

$$|T_{v_j}| \geq |T_{v_p}| + \frac{F(e_p, \beta) - F(e_p, \alpha)}{d_\alpha(r, v_j) + w(\rho) - d_\beta(r, v_j)}. \tag{6}$$

2. Otherwise, if the edges are unrelated, then from Lemma 5 the following must hold to ensure the existence of the inversion node v_j:

$$|T_{v_j}| \geq |T_{v_p}| + \frac{F(e_p, \beta) - F(e_p, \alpha)}{d_\alpha(r, x_\alpha) + w(a) + w(\rho) - w(b) - d_\beta(r, x_\beta) - w(\mu)}. \tag{7}$$

Thus, finding the inversion node reduces to searching for the lowest node with a certain number of descendants on a given path. Concerning the evaluation of conditions (6) and (7), in [7] it has been proved the following: Let $e = (u, v)$ be a tree edge and $f = (x, y) \in C_e$, and let $w(T, x)$ $(w(T, v)$, respectively) denote the sum of the lengths of all the paths in T starting from x $(v$, respectively) and leading to every node in T; then, the function $F(e, f)$ can be written in this way:

$$F(e, f) = w(T, x) - w(T, v) + w(T_v) - d(v, x) \cdot (|T| - |T_v|) + |T_v| \cdot (d(r, y) + w(f)).$$

Terms $|T_v|$ and $d(r, y)$ can be easily obtained in constant time, after a linear time preprocessing of T. Moreover, in [7] it is also proved that it is possible to compute $w(T, x)$ for every node $x \in V$ in $O(n)$ time. From this, it follows that we can compute $F(e_p, \alpha)$ and $F(e_p, \beta)$ in constant time. Since all other terms in (6) and (7) are available in $O(1)$ time after linear time preprocessing, it follows that these two conditions can be evaluated in $O(1)$ time.

The above analysis shows that to establish the query time for Q_2, it remains to bound the number of times conditions (6) and (7) need to be tested to find their inversion node. We can prove the following result:

Lemma 6. (Query time for Q_2) *Let $\alpha = (y_\alpha, x_\alpha), \beta = (y_\beta, x_\beta)$ be swap edges such that $life(\alpha, \beta) = \{e_0, \ldots, e_p = (v_{p-1}, v_p)\}$, where both x_α and x_β descend from v_p in T. Then, it is possible to compute their inversion node, if any, in $O(\log n)$ time and linear space, after linear time preprocessing.*

Proof. Fixed any node v_j in $life(\alpha, \beta)$, we can evaluate conditions (6) and (7) in $O(1)$ time. Hence, we have to search for the lowest node in $life(\alpha, \beta)$ with the needed number of descendants.

First of all, we compute the endpoints of $life(\alpha, \beta)$, which can be easily expressed in terms of NCA queries, by noticing that the lower endpoint is given by $nca(x_\alpha, x_\beta)$, while the higher one is the lowest node among $nca(x_\alpha, y_\alpha)$ and $nca(x_\beta, y_\beta)$. In this way, we also realize whether α, β are related or not. In [3] it is showed a technique to find the NCA of a pair of nodes in constant time, after linear time preprocessing. Thus, it is possible to find the endnodes of $life(\alpha, \beta)$ in $O(1)$ time, after linear time preprocessing. Afterwards, we execute a binary search over the path $life(\alpha, \beta)$ in the following way: we first jump up to node v_0 and we check if it satisfies the appropriate condition; if not, there is no inversion node, and we are done. Otherwise, we have to search in the path $v_0 \to v_p$. To this aim, we jump up to the node $v_{p/2}$, and we verify if it satisfies the appropriate condition: if yes, we continue to search in $v_{p/2} \to v_p$, otherwise we look at $v_0 \to v_{p/2}$. By proceeding in this way, we can find the inversion node in $O(\log n)$ steps. In [2] it is illustrated a technique to jump up an arbitrary number of nodes in a tree in $O(1)$ time, after linear time preprocessing, hence we can find the inversion node in $O(\log n)$ time and linear space, after linear time preprocessing. □

On the other hand, concerning the query time for Q_3, we can prove the following:

Lemma 7. (Query time for Q_3) *Let f_α, f_β be the swap functions of the swap edges α, β, and let $life(\alpha, \beta) = \{e_0, \ldots, e_p\}$. Then, it is possible to compare $f_\alpha(v_i)$ and $f_\beta(v_i)$, $i \in [0, p]$, in $O(\log n)$ time, after linear time preprocessing.*

Proof. We evaluate the functions in v_p, and we compute the inversion node v_j of f_α and f_β, if any, in $O(\log n)$ time, as shown in Lemma 6. W.l.o.g., let $f_\alpha(v_p) \le f_\beta(v_p)$. From Proposition 1, we know that f_α, f_β invert in at most one point, hence $f_\alpha(v_i) \le f_\beta(v_i) \iff i \le j$. □

5   Analysis of the Algorithm

To analyze our algorithm, we start by recalling some important results about the relationships between *Davenport-Schinzel sequences* and the *lower envelope sequence* of a set of functions (for details see [1] and [4]):

Definition 2. *Let* $k, s \in \mathbb{N}^+$. *A sequence* $U = (u_1, \ldots u_p)$ *of integers* $u_i \in [1, k]$ *is a Davenport-Schinzel sequence of order* s *with* k *elements, say* $DS(k, s)$, *if* $u_i \neq u_{i+1}, \forall i < p$, *and there exist no* $s + 2$ *indexes* $1 \leq i_1 \leq \cdots \leq i_{s+2} \leq p$ *such that*

$$u_{i_1} = u_{i_3} = u_{i_5} = \cdots = u_{i_{s+1}} = a, \ u_{i_2} = u_{i_4} = u_{i_6} = \cdots =_{i_{s+2}} = b, \ a \neq b.$$

The complexity of a $DS(k, s)$ *is defined as* $\lambda_s(k) = \max\{|U| \,|\, U \in DS(k, s)\}$.

Definition 3. *Let* $\mathcal{F} = \{f_1(x), \ldots, f_k(x)\}$ *be a set of functions, and let* $L_{\mathcal{F}}(x)$ *denote the lower envelope of* $\mathcal{F}$. *Let* p *be the minimum number of intervals* $I_1, \ldots, I_p$ *in which* $D_{\mathcal{F}}$ *(the domain of* $L_{\mathcal{F}}(x)$*) can be partitioned and such that*

$$\forall j \in [1, p] \, \exists i_j \,|\, L_{\mathcal{F}}(x) = f_{i_j}(x), \ \forall x \in I_j.$$

Then, the lower envelope sequence *of* $\mathcal{F}$ *is the sequence* $U_{\mathcal{F}} = (i_1, \ldots, i_p)$.

We are now ready to prove the main result:

Theorem 1. *Given a 2-edge connected, undirected graph* $G = (V, E)$, *with* n *nodes and* m *non-negatively real weighted edges, and given a SPT* T *of* G, *the problem of finding, for every tree edge, a best swap edge which minimizes the average distance between the root of* T *and the disconnected nodes, can be solved on a standard RAM model in* $O(m \log^2 n)$ *time and linear space complexity.*

Proof. Let $\mathcal{F}$ be a set of k functions intersecting pairwise in at most s points. Then, the lower envelope sequence of $\mathcal{F}$ is a $DS(k, s+2)$ [1]. Moreover, the lower envelope of $\mathcal{F}$ can be computed by performing $O(\lambda_{s+1}(k) \log k)$ comparisons and by using linear space [4].

In our case, $\mathcal{F} = \{f_\alpha(v) \,|\, \alpha$ is a non-tree edge$\}$. Since we have $k = m - n + 1$ swap functions, and given that from Proposition 1 every pair of swap functions has at most $s = 1$ inversions, it turns out that to compute the lower envelope of $\mathcal{F}$ we need $O(\lambda_2(m) \log m)$ comparisons. Since $\lambda_2(m) = 2m + 1$ [1], this means we need $O(m \log m)$ comparisons. In every comparison, we have to answer any of the queries described in the previous section, and so from Lemmas 6 and 7, we know that every query can be answered in $O(\log n)$ time after linear time preprocessing. It follows that the algorithm has $O(m \log^2 n)$ time complexity. Both the algorithm and the precomputations use linear space, and then the space complexity is linear. □

References

1. P.K. Agarwal and M. Shariri, Davenport-Schinzel sequences and their geometric applications, Cambridge University Press, New York, 1995.
2. M.A. Bender and M. Farach-Colton, The level ancestor problem simplified. Proc. of the 4th Latin American Symp. on Theoretical Informatics (LATIN 2002), Vol. 2286 of Lecture Notes in Computer Science, Springer-Verlag, 508–515.
3. D. Harel and R.E. Tarjan, Fast algorithms for finding nearest common ancestors. SIAM Journal on Computing, 13(2) (1984) 338–355.
4. J. Hershberger, Finding the upper envelope of n line segments in $O(n \log n)$ time. Information Processing Letters, 33 (1989) 169–174.
5. E. Nardelli, G. Proietti, and P. Widmayer, Swapping a failing edge of a single source shortest paths tree is good and fast. Algorithmica, 36(4) (2003) 361–374.
6. G. Proietti, Dynamic maintenance versus swapping: an experimental study on shortest paths trees. Proc. 3rd Workshop on Algorithm Engineering (WAE 2000), Vol. 1982 of Lecture Notes in Computer Science, Springer-Verlag, 207–217.
7. G. Proietti, Computing a single swap edge in a shortest paths tree by minimizing the average distance from the root, manuscript available at http : //www.di.univaq.it/~proietti/paper.ps.
8. H. Ito, K. Iwama, Y. Okabe, and T. Yoshihiro, Polynomial-time computable backup tables for shortest-path routing. Proc. of the 10th Int. Colloquium on Structural Information and Communication Complexity (SIROCCO 2003), Vol. 17 of Proceedings in Informatics, Carleton Scientific, 163–177.

Improved Bounds for Optimal Black Hole Search with a Network Map

Stefan Dobrev[1], Paola Flocchini[1], and Nicola Santoro[2]

[1] SITE, University of Ottawa
{sdobrev,flocchin}@site.uottawa.ca
[2] School of Computer Science, Carleton University
santoro@scs.carleton.ca

Abstract. A *black hole* is a harmful host that destroys incoming agents without leaving any observable trace of such a destruction. The *black hole search* problem is to unambiguously determine the location of the black hole. A team of agents, provided with a network map and executing the same protocol, solves the problem if at least one agent survives and, within finite time, knows the location of the black hole.

It is known that a team must have at least *two* agents. Interestingly, two agents with a map of the network can locate the black hole with $O(n)$ moves in many highly regular networks; however the protocols used apply only to a narrow class of networks. On the other hand, any universal solution protocol must use $\Omega(n \log n)$ moves in the worst case, regardless of the size of the team.

A universal solution protocol has been recently presented that uses a team of just *two* agents with a map of the network, and locates a black hole in at most $O(n \log n)$ moves. Thus, this protocol has both *optimal size* and *worst-case-optimal cost*. We show that this result, far from closing the research quest, can be significantly improved.

In this paper we present a *universal* protocol that allows a team of *two* agents with a network map to locate the black hole using at most $O(n + d \log d)$ moves, where d is the diameter of the network. This means that, without losing its universality and without violating the worst-case $\Omega(n \log n)$ lower bound, this algorithm allows two agents to locate a black hole with $\Theta(n)$ cost in a very large class of (possibly unstructured) networks.

Keywords: Distributed Algorithms, Distributed Computing, Mobile Agents, Harmful Host, Undetectable Failure, Size and Cost Optimal Protocols

1 Introduction

In networked systems that support autonomous *mobile agents*, a main concern is how to develop efficient agent-based *system protocols*; that is, to design protocols that will allow a team of identical simple agents to cooperatively perform (possibly complex) system tasks. Example of basic tasks are *wakeup*, *traversal*, *rendez-vous*, *election*. The coordination of the agents necessary to perform these

R. Královič and O. Sýkora (Eds.): SIROCCO 2004, LNCS 3104, pp. 111–122, 2004.
© Springer-Verlag Berlin Heidelberg 2004

tasks is not necessarily simple or easy to achieve. In fact, the computational problems related to these operations are definitely non trivial, and a great deal of theoretical research is devoted to the study of conditions for the solvability of these problems and to the discovery of efficient algorithmic solutions; e.g., see [1–8, 13].

At an abstract level, these environments can be described as a collection $\mathcal{E}$ of autonomous mobile *agents* (or *robots*) located in a graph G. The agents have computing capabilities and bounded storage, execute the same protocol, and can move from node to neighboring node. They are *asynchronous*, in the sense that every action they perform (computing, moving, etc.) takes a finite but otherwise unpredictable amount of time. Each node of the network, also called *host*, provides a storage area called *whiteboard* for incoming agents to communicate and compute, and its access is held in fair mutual exclusion. The research concern is on determining what tasks can be performed by such entities, under what conditions, and at what cost.

At a practical level, in these environments, *security* is the most pressing concern, and possibly the most difficult to address. Actually, even the most basic security issues, in spite of their practical urgency and of the amount of effort, must still be effectively addressed (for a survey, see [14]).

Among the severe security threats faced in these environments, a particularly troublesome one is a *harmful host*; that is, the presence at a network site of harmful stationary processes. This threat is acute not only in unregulated non-cooperative settings, such as Internet, but also in environments with regulated access and where agents cooperate towards common goals. In fact, a local (hardware or software) failure might render a host harmful. The problem posed by the presence of a harmful host has been intensively studied from a software engineering point of view (e.g., see [12, 15, 16]), and recently also from an algorithmic prospective [9–11].

Obviously, the first step in any solution to such a problem must be to *identify*, if possible, the harmful host; i.e., to determine and report its location. Following this phase, a "rescue" activity would conceivably be initiated to deal with the destructive process resident there. Depending on the nature of the danger, the task to identify the harmful host might be difficult, if not impossible, to perform.

A particularly harmful host is a *black hole*: a host that *disposes* of visiting agents upon their arrival, leaving *no observable trace* of such a destruction. The task is to unambiguously determine and report the location of the black hole, and will be called *black hole search*. The searching agents start from the same safe site and follow the same set of rules; the task is successfully completed if, within finite time, at least one agent survives and knows the location of the black hole. Note that this type of highly harmful host is not rare; for example, in asynchronous networks, the undetectable crash failure of a site will transform that site into a black hole.

Black hole search is a non trivial problem, its difficulty aggravated by the combination of absence of any trace of destruction (outside the black hole) and asynchrony of the agents. It has been investigated focusing on identifying conditions for its solvability and determining the smallest number of agents needed for its solution [9–11]. Some conditions are very simple; for example, the graph G must be biconnected, the team must consist of at least *two* agents, and n must be known. A complete characterization has been provided for *ring* networks [10]. Our interest is in *universal* (or *generic*) solution protocols, i.e. protocols that can be used in every biconnected network.

The efficiency of a solution protocol is measured first and foremost by the *number of agents* used by the solution. This value, called *size*, depends on many factors, including the topology of the network, the amount of *a priori* information the agents have about the network, etc. In particular, in an arbitrary network, if the topology of the network is known, *two agents* suffice ! [11]. Indeed, this surprising result can be achieved by several protocols. The second efficiency measure is the *number of moves*, called *cost*, performed by the agents. Clearly the research interest is in the design of size-optimal universal solutions (i.e., using a team of just two agents) that are also cost-efficient.

Sometimes the network has special properties that can be exploited to obtain a lower cost network-specific protocol. For example, two agents can locate a black hole with only $O(n)$ moves in a variety of highly structured interconnection networks such as *hypercubes*, square *tori* and *meshes*, *wrapped butterflies*, *star graphs* [9]. On the other hand, there are networks where $\Omega(n \log n)$ moves are required *regardless of the number of agents*; such is for example the case of ring networks [10]. The lower bound for rings implies an $\Omega(n \log n)$ lower bound on the worst case cost complexity of any *universal* protocol.

Indeed, a universal solution protocol has been recently presented that has both *optimal size* and *worst-case-optimal cost*. In fact, it uses a team of just *two* agents with a map of the network, and locates a black hole in at most $O(n \log n)$ moves [11]. Surprisingly, this result does not close the research quest.

In this paper, we show that it is possible to considerably improve the bound on cost without increasing the team size. In fact, we present a *universal* protocol, *Explore and Bypass*, that allows a team of *two* agents with a map of the network to locate a black hole with cost $O(n + d \log d)$, where d denotes the diameter of the network.

This means that, without losing its universality and without violating the worst-case $\Omega(n \log n)$ lower bound, this algorithm allows two agents to locate a black hole with $\Theta(n)$ cost in a very large class of (possibly unstructured) networks: those where $d = O(n/\log n)$.

Importantly, there are many networks with $O(n/log n)$ diameter in which the existing protocols [9, 11] fail to achieve the $O(n)$ bound. A simple example of such a network is the *wheel*, a ring with a central node connected to all ring nodes, where the central node is very slow: those protocols will require $O(n \log n)$ moves.

2   Definitions and Terminology

2.1   Framework

Let $G = (V, E)$ be a simple biconnected graph; let $n = |V|$ be the size of G and d be its diameter. At each node x, there is a distinct label (called port number) associated to each of its incident links (or ports); let $\lambda_x(x, z)$ denote the label associated at x to the link $(x, z) \in E$, and λ_x denote the overall injective mapping at x. The set $\lambda = \{\lambda_x | x \in V\}$ of those mappings is called a *labelling* and we shall denote by (G, λ) the resulting edge-labelled graph. The nodes of G can be *anonymous* (i.e., without unique names).

Operating in (G, λ) is a set $\mathcal{A}$ of distinct autonomous mobile agents. The agents can move from node to neighbouring node in G, have computing capabilities and bounded computational storage, obey the same set of behavioral rules (the *protocol*). The agents are *asynchronous* in the sense that every action they perform (computing, moving, etc) takes a finite but otherwise unpredictable amount of time. Initially, the agents are in the same node h, called *home base*. Each agent has a map of the labelled graph (G, λ) where the location of the home base is indicated.

Each node has a bounded amount of storage, called *whiteboard*; $O(\log n)$ bits suffice for all our algorithms. Agents communicate by reading from and writing on the whiteboards; access to a whiteboard is gained fairly in mutual exclusion.

2.2   Black Hole Search

A *black hole* is a node where resides a stationary process that destroys any agent arriving at that node; no observable trace of such a destruction will be evident to the other agents. The location of the black hole is unknown to the agents. The *Black Hole Search* problem is to find the location of the black hole. More precisely, the problem is solved if at least one agent survives, and all surviving agents know the location of the black hole.

The main measure of complexity of a solution protocol $\mathcal{P}$ is the number of agents used to locate the black hole, called the *size* of $\mathcal{P}$. Clearly, at least two agents are needed to locate the black hole. On the other hand, two agents with a map of the network suffice to locate the black hole.

The other measure of complexity of a protocol $\mathcal{P}$ is the total number of moves performed by the agents, called the *cost* of $\mathcal{P}$.

2.3   Exploration

Let T be a spanning-tree of G rooted in the home base of the agents. For each node v we define T_v to be the subtree of T rooted at v. We will slightly abuse the notation and use T_v to mean both the subtree rooted at v and the set of the nodes of this subtree. For a given T_v we define the *components* of T_v to be the connected components of the graph induced in G by the nodes of $T_v \setminus \{v\}$. Clearly, each component of T_v is a union of one or more of its subtrees (plus the connecting edges). As usual, $|S|$ means the size of the set S (we will almost exclusively talk about sets of nodes, sometimes about set of edges). For a given node v the *level* of v is defined as its distance (in T) from the root. Given two nodes v and w, $\pi_{v,w}$ denotes the unique directed path from v to w in T. For a given path π we denote by π^{-1} the reversal of this path. Given a node v and a set $S = \{e_1, e_2, \ldots, e_k\}$ of edges of T, we denote by $C^v \setminus S$ the connected component containing v obtained from T by removing the edges in S.

We say that a node is *unexplored* if it has not been visited by an agent. A node v is *explored* if all the nodes of T_v have been visited. A node v is *safe* if it has been visited, but there are *unexplored* nodes in T_v.

Similarly, each *port* can be classified as: (1) *unexplored* – no agent has yet arrived/departed via that port; (2) *dangerous* – an agent has left via this port, but none has arrived from there; (3) *used* – an agent arrived via that port and (4) *explored* – the port is *used* and all nodes in the corresponding subtree has been explored. Obviously, a *used* port does not lead to black hole; on the other hand, both *unexplored* and *dangerous* ports might lead to it.

To ensure that at least one agent survives, we will not allow an agent to leave through a *dangerous* port. To prevent the execution from stalling, we will require any *dangerous* port not leading to the black hole to be made *used* as soon as possible. This is accomplished using the following technique called *Cautious Walk* [9–11]:

Cautious Walk

Whenever an agent a leaves a node u through an *unexplored* port p (transforming it into *dangerous*) leading to a node v, upon its arrival and before proceeding somewhere else, a returns to u (transforming that port into *used*). Node u will be called the *last safe* node of a until a is back in u to make the port *used*.

If agent b reaches the last safe node u of a (i.e., a has not returned yet from v), b will perceive a as being *blocked on* (u, v) (or simply, *blocked at* v); this perception will persist until b becomes aware that a has indeed returned from v.

3   Algorithm *Explore and Bypass*

The proposed algorithm *Explore and Bypass* ($\mathcal{E}\&\mathcal{B}$) is a rather complex protocol that uses several quite different techniques and strategies; it allows a team of just *two* agents with a map of the network to solve the Black Hole Location using at most $O(n + d \log d)$ moves.

In the following we will first describe the overall strategy. The structure of the main modules and the algorithmic techniques they employ are described next. The full description can be found in the appendix, as the handling of numerous cases does not fit in the limited space available. The overall structure and the major modules of Algorithm $\mathcal{E}\&\mathcal{B}$ are captured in Fig. 1.

In assessing the overall *cost*, our goal will be to charge the moves caused by each module to the nodes explored during the execution of that module. As most of the activities consist of traversals of unexplored nodes (and possibly a second traversal of newly explored nodes), this works quite well. The activities that cannot be easily accounted for in this way are counted as *overhead* and the cost is charged to a higher level module.

3.1   Overall Strategy

In Algorithm $\mathcal{E}\&\mathcal{B}$, the two agents, a and b, cooperatively explore the network to locate the black hole. The exploration is achieved by the two agents performing

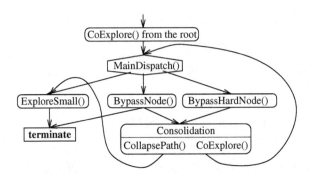

Fig. 1. The overall structure of the Algorithm $\mathcal{E}\&\mathcal{B}$.

mainly a *cooperative depth-first* traversal of a spanning-tree T of the network, using the other links as bypasses if needed. Special traversal techniques are employed in some particular cases to avoid an unnecessary increase in the overall cost of the protocol. The tree T used is a *breadth-first* spanning-tree rooted in the home base of the agents; this choice is made for efficiency reasons that will become apparent later.

The algorithm starts with both agents cooperatively exploring the tree T (module CoExplore) without passing through non-T edges. Eventually, no more nodes can be explored without using non-T edges: an agent a arrives at a node u and discovers that the only non-explored edge is a *dangerous* edge (u, v) (thus, b is blocked at v). When this occurs, there are several possible actions that a may take.

The decision on which action a will take (procedure MainDispatch()) depends on the structure of the set U of still unexplored nodes: (1) If U is small ($|U| \leq d$), a will execute procedure ExploreSmall(). (2) If the largest unexplored subtree T_m hanging from v is not too big ($|U \setminus T_m| \geq d$), a will execute procedure BypassNode(); otherwise (3) a will execute procedure BypassHardNode(). Agent a informs b about the decided action by leaving a message in the whiteboard at u.

If agent b returns to u, it will discover a's decision, and coordinate its activity with a. The goal is to re-establish a situation similar to the initial one, where both agents cooperatively explore a single (sub)tree (and eventually one of them perceives the other as blocked on a link). Although the goal might not be achieved, the agents will always return to a configuration when MainDispatch() can be applied again:

Property UVW: An agent (say, b) is blocked on (u, v), and the other agent (say, a) is in u. Either the only *unexplored* nodes are the nodes of T_v, or there exists a neighbour w of v such that the *unexplored* nodes are the nodes of $T_v \setminus T_w$ (if v is a child of u and w is a child of v) or $T_v \setminus T_u$ (if u is a child of v and v is a child of w)).

Summarizing, the two agents perform a cooperative depth-first traversal of a breadth-first spanning-tree of the network, until an agent finds the other blocked on a link and all unexplored nodes are on the other side of that link. In this case, the agent will either execute ExploreSmall() (a procedure that leads the protocol to terminate without any other call to any other procedure), or will bypass the blocked agent using either BypassNode() or BypassHardNode(); every time the latter cases occur, the level (i.e., the distance from the root) in T of the node being bypassed will become larger, so eventually the ExploreSmall() action will be chosen.

3.2   Cooperative Exploration of a Tree

Procedure CoExplore() is the principal exploration procedure. The algorithm starts with both agents executing CoExplore(x) from the home base x. It is a *cooperative depth-first* traversal in which each agent avoids the part being explored by the other (for as long as possible).

When executing CoExplore(Node u), the agent, say a, determines if there is an unexplored subtree T_v of u. If so, it (recursively) explores T_v using cautious walk. That is, it will: mark the port from u to v as *dangerous*; go to v; (if it survives) mark the port from v to u as *safe*, return to u, mark the port from u to v as *safe*, and check for messages – if any, execute the corresponding code, otherwise return to v, recursively execute CoExplore(v), and finally mark the port from u to v as *explored*. This process is repeated as long as there is an unexplored subtree of u.

There are now two cases. If all incident edges are *explored*, the tree T_u does not contain the black hole; in this case, the recursive call ends. Otherwise, there is only one non-explored (i.e., *safe* or *dangerous*) link left, the one leading to the subtree where the other agent is still working. Let (u, z) be such a link. If (u, z) is *safe*, a goes to z to join the other agent in exploring T_z, executing CoExplore(z). If (u, z) is *dangerous*, the co-exploration cannot continue, and the agent executes MainDispatch(u, z, nil) to decide which action to take.

The cost of CoExplore() is linear in the number of nodes explored, as it is essentially the traversal of a tree.

3.3   Exploring a Small Forest

When an agent finds the other blocked at a node, the agent executes Procedure ExploreSmall() if there are at most d *unexplored* nodes left; i.e. $|U| \leq d$. The main idea is that the unexplored part is so small that it can be efficiently handled by executing the existing $O(n \log n)$ algorithm of [11] on U. However, that algorithm requires that the graph is biconnected, while U might not be. The main task of ExploreSmall() is therefore to explore U in such a way that, eventually, the part U' of U still left unexplored satisfies the following property:

Property Small: There is a path π of length $O(d)$ in $G \setminus U'$ such that adding π to U' results in a biconnected graph U''.

The description of how to obtain U' from U using $O(n)$ moves is rather technical and can be found in the appendix. When only U' is left unexplored, the path π is identified and the algorithm from [11] is applied to U''.

Since $|U| \leq d$, $U' \subseteq U$ and $|\pi| \in O(d)$, also $|U''| \in O(d)$ and the complexity of applying this algorithm to U'' is $O(d \log d)$. In other words, ExploreSmall() requires at most $O(n + d \log d)$ moves in total by the two agents.

3.4   Exploring a Forest of Small Components

If agent a finds agent b blocked at a node v, a must be able to bypass v to continue the exploration. BypassNode() is the branch taken in MainDispatch() if the largest connected component of $U \setminus \{v\}$ is not too big (i.e. there are at least d other nodes in U).

As a result of BypassNode(), either v is identified as the black hole, or all unexplored nodes will be located in a single subtree hanging from v; in the latter case, either PropertyUVW holds or ExploreSmall() is called. We will now describe its modules in some details.

Main Action – Procedures BypassNode() and ReleasedB(). When agent a finds agent b blocked on the link (u, v), a must be able to bypass v to continue the exploration. BypassNode() is the branch taken in MainDispatch() if the largest connected component of $U \setminus \{v\}$ is not too big (i.e. there are at least d other nodes in U); if/when b returns to u, it executes procedure ReleasedB().

The main idea comes from observing that 1) each unexplored component can be reached from u by a safe bypass path β (otherwise v would disconnect G) and 2) $|\beta| \in O(d)$ (from the structure of the explored and unexplored parts).

The trees (i.e., the connected components) in the forest $U \setminus \{v\}$ must be reached by a to be explored. To control the amount of moves to reach each of those trees, the exploration of the trees will be done in order and with a different technique, depending on a the size of the tree. More precisely:

1. Agent a first explores the largest components (i.e., the ones containing a subtree of size at least d); the cost of reaching any such tree is $O(d)$, hence it can be "charged" to the cost of exploring that tree (increasing the multiplicative constant but not the order of magnitude of that cost).
2. After all these components have been explored (and assuming agent b is still blocked) a proceeds to explore the smaller ones according to the following strategy:
   a traverses the whole *explored* part in one sweep and whenever it finds a link leading to an unexplored component C, a explores C, then returns and continues with the traversal. If b is still blocked, the only unexplored node left is v that is identified as the black hole.

What makes the situation complex is the fact that b could become unblocked at any time; the algorithm must be able to handle correctly and efficiently every configuration this fact might cause. This goal is achieved by carefully exploring the components using a special technique (procedure ExploreComponent) described in the next section.

The activity of agent b, when (if) it becomes unblocked and finds the message <BypassNode>, is prescribed by procedure ReleasedB(): when (if) it is released,

b follows the trace of a, until it reaches the last safe node visited by a; b then leaves there a message, telling a to finish its component, and goes on exploring the remaining unexplored subtrees of v. If a finishes its part, we are exactly in the situation where the agents continue the co-exploration of T_v as if no bypassing had happened. However, if b finishes first while a is still working on its component, the structure of the remaining unexplored part can be quite complex. In this case, the unexplored part must be consolidated so that PropertyUVW holds and MainDispatch() can be applied. This is accomplished by procedure CollapsePath() described later.

ExploreComponent. To simplify the consolidation (without increasing the complexity), the components are explored in a special manner: Consider the *component graph* H where nodes correspond to the subtrees of component C, and edges correspond to connections (in G) between subtrees. The component graph H is "visited" using a depth-first traversal, where "visiting a node" in H means fully exploring the corresponding subtree before going to the next one.

Such an approach guarantees the following property to hold:

Property P1: At any moment there is only one partially explored subtree $T_{v'}$ of component C.

This property will allow b, when unblocked, to tell a to finish only the current subtree $T_{v'}$, not the whole component C.

Consider now the partially explored subtree $T_{v'}$. Let y be the node in which a entered $T_{v'}$ and let $\pi_{yv'}$ be the path from y to the root v'. After agent a enters $T_{v'}$, it performs a DF-traversal of $T_{v'}$ starting from y and satisfying the following restriction: If an edge (h, k) belongs to $\pi_{yv'}$, it will be the last edge explored from h. Such an approach guarantees that the unexplored part of $T_{v'}$ has the following structure:

Property P2: Let w be the last explored node on the path from y to the root v', and let w' be the next node (if $w \neq v'$) on this path. The path from w' to v', together with all subtrees hanging from this path, is unexplored. If a is on the link from w to w', then all subtrees hanging from w are explored; otherwise, a subtree of w contains a and is partially explored, while the other subtrees are either fully explored, or unexplored.

Notice that, if b has done all its work and a is still in $T_{v'}$, property P2 tells us that the unexplored part consists just of a path and the trees hanging from this path. In order to reach a configuration where PropertyUVW is finally satisfied (and, thus MainDispatch() can be called again), this path is explored using the procedure CollapsePath() described in the next section.

As each subtree is traversed at most twice, the complexity of ExploreComponent() is linear in the number of explored nodes.

CollapsePath. Procedure CollapsePath() is called when all subtrees of v except one have been explored. The remaining unexplored nodes form a path π_e with unexplored subtrees hanging from this path. The agents divide the remaining unexplored path π_e into two disjoint parts with about equal number of nodes (the nodes of the hanging unexplored subtrees are counted as well), and each agent goes on exploring its part. Note that there is a safe bypass path β of length $O(d)$ connecting the opposite ends of π_e : if π_e is in the first subtree T' of component C, we use as bypass β the path that agent a had used to reach C; otherwise, β passes through the (already explored) subtree T'' from which the entry point of T' was reached (note that T'' is connected to the already explored node v).

The agent that finishes first uses β to reach the other one and split the workload again. This is repeated until there is a single unexplored node in the path or the number of unexplored nodes becomes less then d. In the first case, PropertyUVW is satisfied, in the second case ExploreSmall() is directly executed.

It may happen (because the agents are also exploring the subtrees hanging from the path) that the whole path is explored. In such a case, both agents end up co-exploring a single subtree hanging from this path and eventually one of them will become blocked, creating a configuration corresponding to the first case of PropertyUVW.

Each round of halving the unexplored path involves traversing a bypass of length $O(d)$. However, as the number of unexplored nodes is about halved at each round, and CollapsePath() terminates when there are less then d nodes left, the number of rounds with less than d newly explored nodes is $O(1)$. In other words, the overhead of of CollapsePath() is only $O(d)$.

3.5   Exploring a Forest with a Large Component

When agent a must bypass agent b blocked at a node v, if $U \setminus \{v\}$ has one overwhelmingly large subtree T_m, then BypassNode() may be inefficient: It can happen that an overhead of $O(d)$ is incurred, but the one remaining partially explored subtree of v is T_m itself but only $o(d)$ new nodes have been explored. Since in the worst case, this can happen $O(d)$ times, the total overhead would be an unacceptable $O(d^2)$.

To avoid such an overhead, we must make sure there is enough progress; in this way, while the overhead is not avoided, it has no chance to accumulate. This is done by procedure BypassHardNode().

Let x be the node dividing $T|_U = U \cap T$ (the tree induced in T by the nodes of U) into components of size at most $|U|/2$. BypassHardNode() uses the path $\pi_{v,x}$ from v to x as the pivot, i.e. after BypassHardNode() there is either a single unexplored subtree of $T|_U$ hanging from the fully explored $\pi_{v,x}$, or there is one unexplored node $w \in \pi_{v,x}$ and the subtrees hanging from it are unexplored.

In this way, we are ensured that the number of unexplored nodes is at least halved and MainDispatch() can be applied again. Moreover, the number of newly explored nodes is at least $d/2$, i.e. there are enough of them to distribute the $O(d)$ overhead among them.

There is a major problem with this approach: while all components of $U \setminus \{v\}$ are reachable from the already explored nodes, the same is not true for the components of $U \setminus \pi_{v,x}$. We must make sure that at least some components (or the path itself) are reachable.

To achieve that, the idea of BypassHardNode() is to proceed along the path from v to x, taking as long a jump ahead as possible: in other words, a goes to the furthest reachable subtree $T_{w'}$ hanging from $\pi_{v,x}$ (w is the parent if w') and explores from there (first $T_{w'}$, then the path from w to x and the hanging subtrees), while b (after/if becoming unblocked) will take care of the path (and hanging subtrees) between v and w. If a finishes first, the remaining unexplored path between v and w will be collapsed and the number of unexplored nodes is at least halved. If b finishes first, there are several possible cases but two principal outcomes:

(1) The unexplored nodes are limited to the subtree $T_{w'}$. The corresponding unexplored path is collapsed (procedure CollapsPath() is called).

(2) We are essentially in the same situation as we started, but we have moved along the line from v to x to at least w. The process (bypassing) is repeated until x is reached and/or the unexplored nodes are limited to a subtree (and possibly its root) hanging from the path (thus, Collapsepath() can be called to terminate).

In case 2) a new bypass path will be needed to reach a smaller area of unexplored nodes. The crucial observation is that the bypass does not go below w, otherwise T'_w would not have been the furthest reachable subtree. This means that the bypass of iteration $i > 1$ uses only nodes explored in the previous iteration.

As each bypass path (except in the last iteration, when it is used in CollapsePath()) is used only a constant number of times, the overhead of iteration $i > 1$ can be charged to iteration $i - 1$. We are left with the overhead $O(d)$ for the first iteration and for the possible CollapsePath() in the last iteration. Since BypassHardNode() has explored at least $d/2$ nodes, this overhead can be distributed among them.

4 Analysis

Due to space constraints, we omit correctness and complexity analysis of Algorithm $\mathcal{E}\&\mathcal{B}$. Full proofs can be found in the technical report.

References

1. S. Alpern. The Rendezvous search problem. SIAM J. of Control and Optimization 33, 673 - 683, 1995.
2. E. Arkin, M. Bender, S. Fekete, and J. Mitchell. The freeze-tag problem: how to wake up a swarm of robots. In 13th ACM-SIAM Symposium on Discrete Algorithms (SODA '02), pages 568–577, 2002.
3. B. Awerbuch, M. Betke, and M. Singh. Piecemeal graph learning by a mobile robot. Information and Computation 152, 155–172, 1999.
4. L. Barrière, P. Flocchini, P. Fraigniaud, and N. Santoro. Capture of an intruder by mobile agents. In 14th ACM-SIAM Symp. on Parallel Algorithms and Architectures (SPAA '02), 200-209, 2002.

5. L. Barrière, P. Flocchini, P. Fraigniaud, and N. Santoro. Election and rendezvous in fully anonymous systems with sense of direction. In 10th Colloquium on Structural Information and Communication complexity (SIROCCO '03), 17-32, 2003.

6. X. Deng and C. H. Papadimitriou. Exploring an unknown graph. J. of Graph Theory, 32:265-297, 1999.

7. A. Dessmark, P. Fraigniaud and A. Pelc. Deterministic rendezvous in graphs. In 11th European Symposium on Algorithms (ESA'03) 2003.

8. K. Diks, P. Fraigniaud, E. Kranakis, and A. Pelc. Tree exploration with little memory. In 13th ACM-SIAM Symposium on Discrete Algorithms (SODA '02), 588–597, 2002.

9. S. Dobrev, P. Flocchini, R. Král'ovič, G. Prencipe, P. Ružička, and N. Santoro. Optimal search for a black hole in common interconnection networks. In Proc. of Symposium on Principles of Distributed Systems (OPODIS'02), 2002.

10. S. Dobrev, P. Flocchini, G. Prencipe, and N. Santoro. Mobile agents searching for a black hole in an anonymous ring. In Proc. of 15th Int. Symp. on Distributed Computing (DISC'01), 166-179, 2001.

11. S. Dobrev, P. Flocchini, G. Prencipe, and N. Santoro. Searching for a black hole in arbitrary networks: Optimal mobile agent protocols. In Proc. of 21st ACM Symposium on Principles of Distributed Computing (PODC'02), 153-162, 2002.

12. F. Hohl. A Model of attacks of malicious hosts against mobile agents. In ECOOP Workshop on Distributed Object Security and 4th Workshop on Mobile Object Systems, LNCS 1603, 105-120, 1998.

13. E. Kranakis, D. Krizanc, N. Santoro, and C. Sawchuk. Mobile agent rendezvous in a ring. In 23rd International Conference on Distributed Computing Systems (ICDCS'03), 2003.

14. R. Oppliger. Security issues related to mobile code and agent-based systems. Computer Communications 22, 12, 1165-1170, 1999.

15. T. Sander and C. F. Tschudin. Protecting mobile agents against malicious hosts. In Conf on Mobile Agent Security, LNCS 1419, 44–60, 1998.

16. J. Vitek and G. Castagna. Mobile computations and hostile hosts. In D. Tsichritzis, editor, Mobile Objects, 241-261. University of Geneva, 1999.

Sparse Additive Spanners
for Bounded Tree-Length Graphs*

Yon Dourisboure and Cyril Gavoille

LaBRI, Université Bordeaux 1
{yon.dourisboure,gavoille}@labri.fr

Abstract. This paper concerns construction of additive stretched spanners with few edges for n-vertex graphs having a tree-decomposition into bags of diameter at most δ, i.e., the tree-length δ graphs. For such graphs we construct additive 2δ-spanners with $O(\delta n \log n)$ edges, and additive 4δ-spanners with $O(\delta n)$ edges. This provides new upper bounds for chordal graphs for which $\delta = 1$. We also show a lower bound, and prove that there are graphs of tree-length δ for which every multiplicative δ-spanner (and thus every additive $(\delta - 1)$-spanner) requires $\Omega(n^{1+1/\Theta(\delta)})$ edges.

Keywords: additive spanner, tree-decomposition, tree-length, chordality

1 Introduction

Let G be a connected graph with n vertices. A subgraph H of G is an (s, r)-*spanner* if $d_H(u, v) \leqslant s \cdot d_G(u, v) + r$ for all pair of vertices u, v of G. An $(s, 0)$-spanner is also termed *multiplicative s-spanner*, and an $(1, r)$-spanner is termed *additive r-spanner*. An (s, r)-spanner is also an $(s + r, 0)$-spanner (in particular, an additive r-spanner is a multiplicative $(r + 1)$-spanner), but the reverse is false in general.

The main objective is to construct for a graph an (s, r)-spanner with few edges. There are many applications of spanners, for example, the complexity of a lot of distributed algorithms depends on the number of messages, itself depending on the number of edges [23, 21]. Sparse spanners occur also in the efficiency, of compact routing schemes [24]. Unfortunately, given an arbitrary graph G and three integers s, r and m, determine whether G admits an (s, r)-spanner with m or fewer edges, is NP-complete [22], even if we restrict $r = 0$ (see also [6, 5, 7, 11, 18] for complexity issue). Best known results on (s, r)-spanners for general graphs are summarized in Table 1.

An interesting question still left open is to know whether every graph has an additive $(2k - 2)$-spanner with $O(n^{1+1/k})$ edges, for $k > 2$. In the affirmative, this would generalize the result of [12, 16] ($k = 2$) and implies also the observation

* Supported by the European Research Training Network COMBSTRU-HPRN-CT-2002-00278.

Table 1.

(s, r)-spanner	edges	reference
$(1, 2)$	$\Theta(n^{3/2})$	[12, 16][a]
$(2k - 1, 0)$	$O(n^{1+1/k})$	[2, 27, 3][b]
$(k - 1, 2k - 4)$	$O(kn^{1+1/k})$	[16], $k \geqslant 4$ even
$(k - 2 + \epsilon, 2k - 2 - \epsilon)$	$O(\epsilon^{-1}kn^{1+1/k})$	[16], $k \geqslant 3$ odd, $\epsilon > 0$
$(k - 1 + \epsilon, 2k - 4 - \epsilon)$	$O(\epsilon^{-1}kn^{1+1/k})$	[16], $k \geqslant 4$ even, $\epsilon > 0$
$(1 + \epsilon, \beta(\epsilon, k))$	$O(\beta(\epsilon, k)n^{1+1/k})$	[16][c], $k \geqslant 2$

[a] The bound of [12] was $O(n^{3/2} \log^{O(1)} n)$ edges but with a better running time, $O(n^{5/2})$ v.s. $O(n^2)$.

[b] This observation due to [2] is based on the classical result (see [1]) that every graph with at least $\frac{1}{2}n^{1+1/k}$ edges has a cycle of length at most $2k$, for every $k \geqslant 1$. [27] and [3] gave respectively an $O(kmn^{1/k})$ and $O(km)$ time algorithm for the construction of such spanner.

[c] Actually $\beta(\epsilon, k) = k^{\max\{\log \log k - \log \epsilon, \log 6\}}$ for $2 \leqslant k \leqslant \log n$ and fixed $0 < \epsilon < 1$.

of [2]. Another interesting question is to known whether the $O(n^{1+1/k})$ edge bound for multiplicative $(2k - 1)$-spanner is tight or not. This bound directly relies to an 1963 Erdös Conjecture [17] on the existence of graphs with $\Omega(n^{1+1/k})$ edges and girth at least $2k + 2$. This has been proved only for $k = 1, 2, 3$ and $k = 5$.

Better bounds can be achieved if we restrict spanners to be trees, or if particular classes of graphs are considered: planar graphs [18], and more structured graphs (e.g., see [20] and [26] for a survey). Among them, the class of *chordal* graphs is of particular interests [6, 22, 9]. A graph is *k-chordal* if its induced cycles are of length at most k. Chordal graphs coincide with 3-chordal. Here below are summarized the best constructions for k-chordal graphs.

chordal	(s, r)-spanner	edges	reference
3	$(2, 0)$	$\Theta(n^{3/2})$	[22]
3	$(3, 0)$	$O(n \log n)$	[22]
3	$(1, 3)$	$O(n \log n)$	[9]
k	$(1, k + 1)$	$\Theta(n)$	[9], $k \geqslant 3$

Tree-decomposition is a rich concept introduced by Robertson and Seymour [25] and is widely used to solve various graph problems. In particular efficient algorithms exist for graphs having a tree-decomposition into subgraphs (or *bags*) of bounded size, i.e., for bounded *tree-width* graphs.

The *tree-length* of a graph G is the smallest integer δ for which G admits a tree-decomposition into bags of diameter at most δ. It has been formally introduced in [15], and extensively studied in [13]. Chordal graphs are exactly the graphs of tree-length 1, since a graph is chordal if and only if it has a tree-decomposition in cliques (cf. [10]). AT-free graphs, permutation graphs, and distance-hereditary graphs are of tree-length 2. More generally, [19] showed that k-chordal graphs have tree-length at most $k/2$. However, there are graphs with bounded tree-length and unbounded chordality[1], like the wheel. So, bounded tree-length graphs is a larger class than bounded chordality graphs.

For several problems involving distance computation, like the design of approximate distance labeling schemes [19] or of near-optimal routing schemes [13], tree-length δ graphs are a natural generalization of chordal graphs, and their tree-decomposition induced can be successfully used. In this paper we highlight a new property of bounded tree-length graphs: the design of sparse additive spanners. The following table summarizes the bounds we have obtained on the minimum number of edges of additive spanner.

tree-length	(s,r)-spanner	edges
δ	$(1, 2\delta)$	$O(\delta n \log n)^a$
δ	$(1, 4\delta)$	$O(\delta n)$
δ	$(\delta, 0)$	$\Omega(n^{1+\epsilon})$, $\epsilon \geqslant 1/\lceil \delta/2 \rceil$
		for^b $\delta = 1, 2, 3, 4, 5, 6, 9, 10$

^{a} In a forthcoming paper [14], we show that the number of edges can be reduced to $O(\delta n + n \log n)$.

^{b} The last line relies to the Erdös's Conjecture, but in any case $\epsilon \geqslant 1/\Theta(\delta)$ (cf. Table of Corollary 1).

Thus our first result provides an additive 2-spanner with $O(n \log n)$ edges for chordal graphs (for $\delta = 1$), improving [9] and also implying [22].

In this paper, we also compare our algorithm to the Chepoi-Dragan-Yan's algorithm (CDY) used successfully for k-chordal graphs [9]. For small chordality k, our algorithm produces an additive $2k$-spanner (or $(2k - 2)$-spanner of odd k) with $O(n)$ edges (recall that $\delta \leqslant k/2$), whereas CDY's algorithm constructs an additive $(k + 1)$-spanner with $O(n)$ edges, which is better for $k \geqslant 4$. However, we show in Section 3 that the CDY's algorithm cannot be used for bounded tree-length graphs of large chordality. More precisely, we construct a worst-case graph of tree-length 3 and chordality $\Omega(n^{1/3})$ for which the CDY's algorithm produces an $\Omega(n^{1/3})$-spanner with $O(n)$ edges. A generic algorithm would certainly combine both algorithms.

The lower bound shows that every additive $o(\delta)$-spanner requires $\Omega(n^{1+\epsilon})$ edges. However, combined with our two upper bounds, this naturally leads to the question of whether there exists, for every tree-length δ graph, an additive $O(\delta)$-spanner with $O(n \log n)$ or even with $O(n)$ edges.

We now recall the definition of *tree-decomposition* introduced by Robertson and Seymour in their work on graph minors [25]. A tree-decomposition of a graph G is a tree T whose vertices, called *bags*, are subsets of $V(G)$ such that:

1. $\bigcup_{X \in V(T)} X = V(G)$;
2. for all $\{u, v\} \in E(G)$, there exists $X \in V(T)$ such that $u, v \in X$; and
3. for all $X, Y, Z \in V(T)$, if Y is on the path from X to Z in T then $X \cap Z \subseteq Y$.

The *length* of tree-decomposition T of a graph G is $\max_{X \in V(T)} \max_{u,v \in X} d_G(u, v)$, and the *tree-length of G* is the minimum, over all tree-decompositions T of G, of the length of T.

---

[1] The *chordality* is the smallest k such that the graph is k-chordal.

A well-known invariant related to tree-decompositions of a graph G is the *tree-width*, defined as minimum of $\max_{X \in V(T)} |X| - 1$ over all tree-decompositions T of G. We stress that the tree-width of a graph is not related to its tree-length. For instance cliques have unbounded tree-width and tree-length 1, whereas cycles have tree-width 2 and unbounded tree-length.

The paper is organized as follows. In Section 2, we present the first algorithm (Line 1 of the previous table). Section 3 presents the second algorithms (Line 2) and the CDY's algorithm, both based on a *layering-tree* of the graph. We conclude in Section 4 with the lower bound.

2   Additive 2δ-Spanner with $O(\delta n \log n)$ Edges

Theorem 1. *Every n-vertex graph of tree-length δ has an additive 2δ-spanner with $O(\delta n \log n)$ edges.*

The remaining of this section concerns the proof of Theorem 1. For this purpose we need several ingredients, and of two basic properties.

It is well known that every tree T has a vertex u, called *median*, such that each connected component of $T \setminus \{u\}$ has at most $\frac{1}{2}|V(T)|$ vertices. A *hierarchical tree* of T is then a rooted tree H defined as follows: the root of H is the median of T, u, and its children are the roots of the hierarchical trees of the connected components of $T \setminus \{u\}$. Observe that T and H share the same vertex set, and that the depth of H is at most[2] $\log |V(T)|$.

A tree-decomposition is *reduced* if any bag is contained in no other bags. A leaf of such decomposition contains necessarily a vertex contained in none other bags. Thus by induction the number of bags of a reduced tree-decomposition does not exceed $\max \{n - 1, 1\}$ for an n-vertex connected graph (cf. [4]).

From now, G is a (connected) graph with n vertices and of tree-length δ. T denotes a tree-decomposition of G of length δ supposed reduced, and H denotes a hierarchical tree of T. So, the depth of H is at most $\log n$. We denote by $\mathrm{NCA}_H(U, V)$ the *nearest common ancestor* of U and V in H.

Property 1. Let U, V be two vertices of T and let Q be the path in T from U to V, and let $Z = \mathrm{NCA}_H(U, V)$. Then, $Z \in Q$, and Z is an ancestor in H of all the vertices of Q.

Proof. By construction, the subtree induced by Z and its descendants in H is a connected component of T, say A. Thus, Z, U, V are in A, but U and V are in two different components of $T \setminus \{Z\}$. Thus in T, the path Q from U to V is wholly contained in A and intersects Z. So, $Z \in Q$ and Z is ancestor of all vertices of Q in H. $\square$

We assume that T is rooted. For every vertex u of G, the *bag* of u, denoted by $\mathcal{B}(u)$, is a bag X of T of minimum depth such $u \in X$. Observe that, once T has been fixed, $\mathcal{B}(u)$ is unique for each u.

---

[2] All the logs are in base two.

Property 2. For every edge $\{u, v\}$ of G, either

1. $\mathcal{B}(u)$ is an ancestor of $\dot{\mathcal{B}}(v)$ in T and $u \in \mathcal{B}(v)$, or
2. $\mathcal{B}(v)$ is an ancestor of $\mathcal{B}(u)$ in T and $v \in \mathcal{B}(u)$.

Proof. Let X be any bag of T with a vertex x. Since $x \in X \cup \mathcal{B}(x)$, by Rule 3 of tree-decomposition's definition and by minimality of the depth of $\mathcal{B}(x)$, we have that X is a descendant of $\mathcal{B}(x)$. Now, by Rule 2 there is a bag, say Y, containing u and v. It follows that Y is a descendant of $\mathcal{B}(u)$ and of $\mathcal{B}(v)$, i.e., either $\mathcal{B}(u)$ is an ancestor of $\mathcal{B}(v)$ or the reverse. If $\mathcal{B}(u)$ is an ancestor of $\mathcal{B}(v)$, then $\mathcal{B}(v)$ is on the path from $\mathcal{B}(u)$ to Y in T. By Rule 3, $u \in \mathcal{B}(v)$. Similarly if $\mathcal{B}(v)$ is an ancestor of $\mathcal{B}(u)$, then $v \in \mathcal{B}(u)$. □

We associate with every bag X of T a shortest path spanning tree of G rooted at an arbitrary vertex $r_X \in X$. A vertex u of G is *good* w.r.t. S_X if $\mathcal{B}(u)$ is a descendant of X in H. Otherwise u is *bad* w.r.t. S_X. Clearly a vertex u can be good for some trees and bad for others, but since H is of depth at most $\log n$, u is good for at most $\log n$ trees.

For every tree S_X, we define S'_X, the tree obtained from S_X by removing recursively each bad leaf w.r.t. S_X. The spanner of G claimed by Theorem 1 is simply the graph defined by $G' := \bigcup_{X \in V(T)} S'_X$.

Lemma 1. G' *is an additive 2δ-spanner of G.*

Proof. Let u, v be two vertices of G, and let $Z = \mathrm{NCA}_H(\mathcal{B}(u), \mathcal{B}(v))$. By definition, u, v are good in S_Z, thus u, v are both in S'_Z. By Property 1, Z belongs to the path from $\mathcal{B}(u)$ to $\mathcal{B}(v)$ in T, thus there is a vertex $z \in Z$ such that $d_G(u, v) = d_G(u, z) + d_G(z, v)$. Since S'_Z is a shortest path rooted at r_Z, we have $d_{G'}(u, v) \leqslant d_{S'_Z}(u, v) \leqslant d_G(u, r_Z) + d_G(r_Z, v)$. Note that $d_G(z, r_Z) \leqslant \delta$, and by the triangle inequality, we have:

$$
\begin{aligned}
d_{G'}(u, v) \;\leqslant\; d_{S'_Z}(u, v) &\leqslant d_G(u, z) + d_G(z, r_Z) + d_G(r_Z, z) + d_G(z, v) \\
&\leqslant d_G(u, z) + d_G(z, v) + 2d_G(z, r_Z) \\
&\leqslant d_G(u, v) + 2\delta \;.
\end{aligned}
$$

□

The difficult part is to upper bound the number of edges.

Lemma 2. G' *has $O(\delta n \log n)$ edges.*

Proof. For every vertex u of G and every bag X of T, we define $\mathrm{cost}(u, X) = 1$ if $u \in S'_X$, and $\mathrm{cost}(u, X) = 0$ if $u \notin S'_X$. Clearly, the number of vertices of S'_X is $\sum_{u \in V(G)} \mathrm{cost}(u, X)$, and so the total number of edges of G' is

$$
|E(G')| \;\leqslant\; \sum_{X \in T} \left(\left(\sum_{u \in V(G)} \mathrm{cost}(u, X) \right) - 1 \right) \;<\; \sum_{X \in T} \sum_{u \in V(G)} \mathrm{cost}(u, X)
$$

The problem to estimate this upper bound is that neither $\sum_u \mathrm{cost}(u, X)$, nor $\sum_X \mathrm{cost}(u, X)$ are easy to calculate. To overcome this problem, we assume that each vertex u has a "charge" for each bag X, denoted by $\mathrm{charge}(u, X)$, that can be exchanged with its neighbors under the following condition. Initially, $\mathrm{charge}(u, X) = \mathrm{cost}(u, X)$, for all u and X. Then, while there is a vertex u and a bag X such that:

1. u is bad in S'_X, and
2. $\mathrm{charge}(u, X) > 0$,

$\mathrm{charge}(u, X)$ is decremented and $\mathrm{charge}(v, X)$ incremented, where v is one of the children of u in S'_X (observe that there is no leaf u in S'_X that is bad in S'_X). Such a procedure converges, and we denote by $\mathrm{charge}^*(u, X)$ the final charge of u w.r.t. X.

Clearly, $\sum_X \sum_u \mathrm{cost}(u, X) = \sum_X \sum_u \mathrm{charge}^*(u, X)$. On the other hand we have:

$$\sum_X \sum_u \mathrm{charge}^*(u, X) = \sum_u \sum_X \mathrm{charge}^*(u, X) \leqslant n \cdot \max_u \left\{ \sum_X \mathrm{charge}^*(u, X) \right\}$$

From the above procedure, if u is bad in S'_X then $\mathrm{charge}^*(u, X) = 0$. Say in other words, if u is good in t trees, then the sum $\sum_X \mathrm{charge}^*(u, X)$ has at most t non-null terms. We have seen that a vertex u is good in at most $\log n$ trees. Thus, $\sum_X \mathrm{charge}^*(u, X) \leqslant \max_X \{\mathrm{charge}^*(u, X)\} \cdot \log n$. To summarize,

$$|E(G')| < \sum_X \sum_u \mathrm{charge}^*(u, X) \leqslant n \log n \cdot \max_{u, X} \{\mathrm{charge}^*(u, X)\}$$

It remains to show that $\mathrm{charge}^*(u, X) = O(\delta)$, for all u and X. We assume that u is good in S'_X (otherwise we have seen that $\mathrm{charge}^*(u, X) = 0$). Observe that, in S'_X, u can receive the charge of v only if: 1) v is bad, 2) v is an ancestor of u, and 3) there is no good vertex on the path from u to v in S'_X. So, let v be the nearest ancestor of u that is good in S'_X (v is set to the root if such vertex does not exist). The charge received by u (w.r.t. X) is thus at most $d_{S'_X}(u, v)$, and thus its total charge is $\mathrm{charge}^*(u, X) \leqslant 1 + d_{S'_X}(u, v)$.

Let $P = x_0, x_1, \ldots, x_p$ be the path in S'_X from $u = x_0$ to $v = x_p$. P is a shortest path in G and $p = d_G(u, v) \geqslant 1$. Our aim is to show that $p \leqslant 3\delta$. Let Q be the path in T from $\mathcal{B}(u)$ to $\mathcal{B}(v)$.

Claim. For every $0 < i < p$, $\mathcal{B}(x_i) \notin Q$.

Let $Z = \mathrm{NCA}_H(\mathcal{B}(u), \mathcal{B}(v))$. By construction of P, every $x_i \in P \setminus \{x_0, x_p\}$ is bad, i.e., $\mathcal{B}(x_i)$ is not a descendant of Z. By Property 1, all the bags of Q are descendant of Z, so $\mathcal{B}(x_i) \notin Q$, for every $0 < i < p$, as claimed.

Let j be the largest index such that $x_j \in \mathcal{B}(u)$. Obviously, $j \leqslant \delta$ because x_0, x_j belongs to the same bag, $\mathcal{B}(x_0)$. W.l.o.g. $p > j$, since otherwise $p \leqslant \delta$ and we are done. Let $Y = \mathrm{NCA}_T(\mathcal{B}(u), \mathcal{B}(v))$.

Claim. $x_j \in Y$.

If $\mathcal{B}(u) = Y$, then we are done. Assume, $\mathcal{B}(u) \neq Y$. If $j = 0$, then we get a contradiction applying Property 2 on the edge $\{u, x_1\}$. Indeed, either $\mathcal{B}(x_1)$

is an ancestor in T of $\mathcal{B}(u)$ and $x_1 \in \mathcal{B}(u)$, contradicting $j = 0$, or $\mathcal{B}(u)$ is an ancestor of $\mathcal{B}(x_1)$. In this latter case, since the path P intersects Y and $\mathcal{B}(u) \neq Y$, the sub-path $x_1, \ldots, x_p$ must intersect $\mathcal{B}(u)$ in some x_i with $i \geq 1$, contradicting again $j = 0$. So we have $0 < j < p$, and thus $\mathcal{B}(x_j) \notin Q$. Now, $\mathcal{B}(x_j)$ must be an ancestor in T of $\mathcal{B}(u)$, because $x_j \in \mathcal{B}(u)$. Since $\mathcal{B}(x_j) \notin Q$, it follows that $\mathcal{B}(x_j)$ is an ancestor in T of all the ancestors of $\mathcal{B}(u)$ in Q. In particular x_j belongs to all the ancestors of $\mathcal{B}(u)$ in Q, so $x_j \in Y$ as claimed.

Let k be the largest index such that $x_k \in Y$. Since $d_G(x_j, x_k) \leq \delta$, we have $k \leq 2\delta$. W.l.o.g. $k < p$, since otherwise $p \leq 2\delta$ and we are done. Since $p > k$, the vertex x_{k+1} exists. Let $Y' = \mathrm{NCA}_T(\mathcal{B}(x_{k+1}), \mathcal{B}(x_p))$.

Claim. $Y' \in Q$, and $x_k \in Y'$.

First observe that for every $t > k$, $\mathcal{B}(x_t)$ is a descendant of Y. Otherwise, since the sub-path $x_t, \ldots, x_p$ must intersect Y, Y would contain some x_i with $i > k$, contradicting the definition of k. In particular, $\mathcal{B}(x_{k+1})$ is a descendant of Y. Since $\mathcal{B}(x_{k+1})$ is descendant of Y', it follows that Y' is a descendant of Y, and thus Y' is on the path in T from Y to $\mathcal{B}(x_p)$. I.e., $Y' \in Q$ as claimed. Let us apply Property 2 on the edge $\{x_k, x_{k+1}\}$. $\mathcal{B}(x_k)$ is ancestor of Y ($x_k \in Y$), and thus $\mathcal{B}(x_k)$ is ancestor of $\mathcal{B}(x_{k+1})$, and thus $x_k \in \mathcal{B}(x_{k+1})$. Thus x_k belongs to all the bags of the path in T from Y to $\mathcal{B}(x_{k+1})$. In particular, $x_k \in Y'$ completing the proof of the claim.

The sub-path $x_{k+1}, \ldots, x_p$ must intersect Y', say in some vertex x_ℓ. Since $\ell > k$, $\mathcal{B}(x_\ell)$ is a descendant of Y. It is also an ancestor of Y'. Thus, $\mathcal{B}(x_\ell)$ belongs to the path in T from Y to Y', i.e., $\mathcal{B}(x_\ell) \in Q$. Since $\mathcal{B}(x_i) \notin Q$ for $i < p$, it turns out that $\ell = p$. We conclude with the fact that $d_G(x_k, x_\ell) \leq \delta$, thus $p = \ell \leq k + \delta \leq 3\delta$.

Therefore, $|E(G')| < (3\delta + 1)n \log n$, completing the proof of Lemma 2. □

Theorem 1 directly follows from lemmas 1 and 2.

3   Additive Spanner with a Linear Number of Edges

We present in this section an algorithm to construct for any tree-length δ graph an additive 4δ-spanner with $O(\delta n)$ edges. Since for every graph the tree-length is at most half the chordality, it follows that this spanner is as well an additive $2k$-spanner (or $(2k-2)$-spanner of odd k) with $O(kn)$ edges if the graph is k-chordal. This latter construction is far from the optimal. Indeed, Chepoi et al [9] have presented an algorithm which computes, for any graph G of chordality k, an additive $(k+1)$-spanner with $O(n)$ edges.

The CDY's algorithm, and our algorithm as well, is based on a *Layering-tree*, technique we present in Section 3.1 with the CDY's algorithm. Our variant is detailed in Section 3.2. In Section 3.3, we compare both algorithms and show that, there are tree-length 3 graph for which the CDY's algorithm returns an additive $\Omega(n^{1/3})$-spanner whereas our algorithm guarantees an additive 12-spanner[3] with $O(n)$ edges.

---

[3] Actually, on the counter-example, it produces an additive 3-spanner.

3.1   Layering-Tree and the CDY's Algorithm

Let G be a graph with a distinguished vertex s. For every integer $i \geqslant 0$, we define $L^i := \{u \in V(G) \mid d_G(s, u) = i\}$. A *layering partition* of G is a partition of each set L^i into $L^i_1, \ldots, L^i_{p_i}$ such that $u, v \in L^i_j$ if and only if there exists a path from u to v using only intermediate vertices w such that $d_G(s, w) \geqslant i$.

Let H be the graph whose vertex set is the collection of all the parts L^i_j. In H, two vertices L^i_j and $L^{i'}_{j'}$ are adjacent if and only if there exists $u \in L^i_j$ and $v \in L^{i'}_{j'}$ such that u and v are adjacent in G (see Fig. 1 for an example). The vertex s is called the *source* of H.

Lemma 3. [8] *The graph H, called* layering-tree *of G, is a tree and is computable in linear time.*

In each part W of H, a special vertex r_W is distinguished[4] (grayed vertices on Fig. 1(a)).

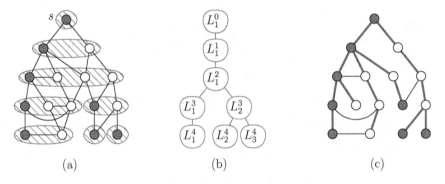

$$(a) \qquad\qquad (b) \qquad\qquad (c)$$

Fig. 1. (a) A 5-chordal graph, (b) a layering-tree of it, and (c) the CDY's spanner.

The CDY's spanner of G is composed of the following edges (see Fig. 1(c)):

1. all the edges of a shortest path spanning tree S of G rooted at s, and
2. for every vertex u of a part W of H, the edge $\{x, y\}$ (if it exists) such that x is the nearest ancestor of u in S with a neighbor y ancestor of r_W.

3.2   Additive $O(\delta)$-Spanner with $O(\delta n)$ Edges

Theorem 2. *Every n-vertex graph of tree-length δ has an additive 4δ-spanner with at most $(2\delta + 1)(n - 1)$ edges.*

The construction is also based on the layering-tree H of G, presented in Section 3.1, and on the following result:

---

[4] In the original CDY's algorithm r_W is given by a fixed policy we do not detail here. Here, we consider that any vertex of W can be chosen, it is simpler but more powerful.

Lemma 4. *Let H be a layering-tree of G. For every part W of H, there is a vertex r of G, called center of W, such that for all $u, v \in W$, $d_G(u, v) \leqslant 3\delta$ and $d_G(u, r) \leqslant 2\delta$. Moreover, for every δ, these bounds are best possible.*

Proof. Let T be a tree-decomposition of G of length δ, W.l.o.g. T is supposed rooted at a bag containing s, the source of H. Let W be a part of H at distance i of s. Let X be the bag of T that is the nearest common ancestor of all the of bags containing a vertex of W, and let $d_X = \max_{u, v \in X} d_G(u, v)$ be its diameter. Let us prove that for every $u \in W$, $d_G(u, X) \leqslant \delta$. In this way, we will prove that:

- $\forall u, v \in W$, $d_G(u, v) \leqslant d_G(u, X) + d_X + d_G(v, X) \leqslant 3\delta$;
- $\forall u \in W$ and $\forall r \in X$, $d_G(u, r) \leqslant d_G(u, X) + d_X \leqslant 2\delta$.

Let u be an arbitrary vertex of W. Consider a vertex $v \in W$ such that there are two bags, U and V, such that: $u \in U$, $v \in V$, and $X = \mathrm{NCA}_T(U, V)$ (we check that v, U, V exist). Let P be a shortest paths from s to u, P intersects X. Let x be the closest from s vertex in $P \cap X$. Since u, v are both in W, there exists a path Q from u to v using only intermediate vertices w such that $d_G(s, w) \geqslant i$. Q intersects X at a vertex r (see Fig. 2(a)).

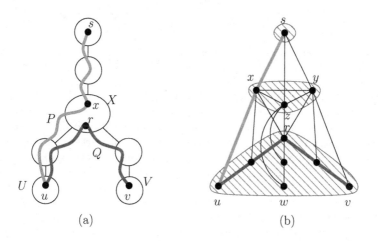

(a)   (b)

Fig. 2. A part of H is of diameter at most 3δ and of radius at most 2δ.

Note that $d_G(s, u) = i = d_G(s, x) + d_G(x, u)$ and $d_G(s, r) \leqslant d_G(s, x) + \delta$. So, $d_G(s, r) \leqslant i + \delta - d_G(x, u)$. If $d_G(x, u) \geqslant \delta + 1$ then $d_G(s, r) \leqslant i - 1$: a contradiction since $r \in Q$. So, $d_G(u, X) \leqslant d_G(u, x) \leqslant \delta$ as claimed.

These bounds are best possible for each $\delta \geqslant 1$. For $\delta = 1$, the graph depicted on Fig. 2(b) is chordal, u, v, w belong to the same part and $d_G(u, v) = d_G(u, w) = d_G(v, w) = 3$. By replacing each edge by a path of length δ, the tree-length of this subdivision increases to δ, u, v, w still belong to the same part and are at distance 3δ. We check that a center c for W can be chosen arbitrarily among

$\{x, y, z, r, s\}$ and attains a radius 2δ. Moreover, if $c \notin \{x, y, z, r, s\}$, one can prove that either $d_G(u, c) > 2\delta$ or $d_G(v, c) > 2\delta$ or $d_G(w, c) > 2\delta$. Thus the radius of the part containing u, v, w is exactly 2δ.   □

The spanner satisfying Theorem 2 is simply the graph defined by $G' := S \cup \bigcup_{W \in V(H)} S_W$, where S is a shortest path tree spanning G, H a layering-tree of G, and S_W a shortest path tree spanning W and rooted at a center of W.

Lemma 5. *G' is an additive 4δ-spanner of G.*

Proof. Let u, v be two vertices of G, and let A, B be the two parts of H containing respectively u and v. Let us show that every path from u to v must intersect the part $W = \text{NCA}_H(A, B)$.

This clearly holds if $W = A$ or $W = B$. If $W \neq A$ and $W \neq B$, (so in particular $A \neq B$), then, by definition of H, every path intermediate vertex of a path from u to v must intersect an ancestor of A and of B. So, by induction, it must intersect the nearest common ancestor of A and of B, W.

Let $u', v' \in W$ be the ancestors in S of u and v. We observe that $d_G(u, u') = d_H(A, W)$. Since G' contains a shortest path spanning tree of G rooted at s, it follows that $d_H(A, W) = d_{G'}(u, u')$, and finally, $d_G(u, u') = d_{G'}(u, u')$. Similarly, $d_G(v, v') = d_{G'}(v, v')$, and thus $d_{G'}(u, u') + d_{G'}(v, v') \leqslant d_G(u, v)$.

Using the tree S_W contained in G' and rooted at the center of W, and by Lemma 4, we have $d_{G'}(u', v') \leqslant 4\delta$. Therefore, we obtain:

$$d_{G'}(u, v) \leqslant d_{G'}(u, u') + d_{G'}(u', v') + d_{G'}(v', v) \leqslant d_G(u, v) + 4\delta .$$

□

Lemma 6. *G' has at most $(2\delta + 1)(n - 1)$ edges.*

Proof. S has $n - 1$ edges. Every tree S_W has at most $|W|$ leaves and so at most $2\delta|W|$ edges, except when W is the root of H. In this latter case, S_W contains no edges. The parts of H are disjoint, so the number of edges of G' is at most $n - 1 + 2\delta(n - 1) = (2\delta + 1)(n - 1)$.   □

3.3   Counter-Example

Theorem 3. *There is a graph of tree-length 3 and with $n + o(n)$ vertices for which the CDY's algorithm, for any choice of the special vertices, constructs an additive $\Omega(n^{1/3})$-spanner with $O(n)$ edges.*

The remaining of this section is devoted to the proof of Theorem 3.

The *Cartesian product* of two graphs A and B is the graph denoted by $A \times B$ such that $V(A \times B) = \{(x, y) \mid x \in V(A), y \in V(B)\}$, and $E(A \times B) = \{((x, x'), (y, y')) \mid (x = x' \text{ and } (y, y') \in E(B)) \text{ or } (y = y' \text{ and } (x, x') \in E(A))\}$. E.g., the mesh is the Cartesian product of two paths. Let K_t and P_t denote respectively the complete graph and the path with t vertices.

We set $D_p = K_p \times K_p$. The counter-example, denoted by G_0, is the graph $D_t \times P_{t-1}$, so composed of $t - 1$ copies $D_t^1, \ldots, D_t^{t-1}$ of D_t, with an extra vertex

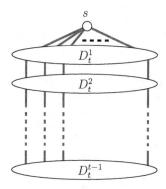

Fig. 3. Counter-example G_0.

s connected to all the vertices of D_t^1 (see Fig. 3). To every vertex u of G_0, $u \neq s$, we denote by $P(u)$ the copy of the path P_{t-1} containing u. Hereafter, we set $t := \lceil n^{1/3} \rceil$, so that G_0 has $t^2(t-1) + 1 = n + O(n^{2/3})$ vertices.

A subgraph H of a graph G is *isometric* if $d_H(x,y) = d_G(x,y)$, for all $x, y \in V(H)$. It is a natural generalization of induced subgraph (any isometric subgraph is clearly an induced subgraph). We have:

Lemma 7. [15] *The tree-length of any isometric subgraph of G is no more than the tree-length of G.*

Let u, v, w be three vertices of some D_t^i inducing a path of length two (since D_t is of diameter two, such vertices exists). We check that the graph induced by the vertices of the three paths $P(u), P(v), P(w)$ is a isometric mesh of G_0. This mesh has $t - 1$ rows and 3 columns.

Lemma 8. G_0 *has chordality at least* $2t = \Omega(n^{1/3})$, *and tree-length 3 for* $t \geq 5$.

Proof. In a $(t-1) \times 3$ mesh, the perimeter is an induced cycle of the mesh of length $2t$. Since this mesh is an isometric subgraph of G_0, it follows that G_0 is of chordality at least $2t$.

It is proved in [15] that the tree-length of the mesh with p rows and q columns is $\min \{p, q\}$ if $p \neq q$ or p is even, and is $p-1$ otherwise. In particular, the $(t-1) \times 3$ mesh has tree-length 3 if $t \geq 5$. By Lemma 7, G_0 has tree-length at least 3 for $t \geq 5$.

We obtain a tree-decomposition of G_0 of length 3 by considering a path $X_0, X_1, \ldots X_{t-2}$ where $X_0 = \{s\} \cup V(D_t^1)$, and $X_i = V(D_t^i) \cup V(D_t^{i+1})$ for $i \geq 1$. $\qquad \square$

A *dominating set* of a graph G is a set of vertices R such that for every vertex u of G either $u \in R$ or u is adjacent to a vertex of R.

Lemma 9. *If R is a dominating set of D_t, then $|R| \geq t$.*

Proof. The graph D_t is the union of two disjoint sets of cliques $\mathcal{K}_1$ and $\mathcal{K}_2$, each one composed of t disjoint copies of K_t, so that every edge belongs either to a clique of $\mathcal{K}_1$ or of $\mathcal{K}_2$. Every clique of $\mathcal{K}_1$ intersects each clique of $\mathcal{K}_2$ and vice et versa. Assume $|R| < t$. By the Pigeon Hole Principle, there is a clique $A \in \mathcal{K}_1$ with no vertices of R. Similarly, there is a clique $B \in \mathcal{K}_2$ with no vertices of R. The cliques A and B share exactly one vertex, say u (otherwise there would exists an edge that belongs to a clique of $\mathcal{K}_1$ and to a clique of $\mathcal{K}_2$). All the incident edges of u belongs either to A or to B. It follows that u is not adjacent to any vertex of R: a contradiction. □

Proof (of Theorem 3). Let G' be the spanner obtained by CDY's algorithm applied on the source s of G_0. The parts of H are the set $L^0 = \{s\}$, and $L^i = V(D_t^i)$ for $i \geqslant 1$. The spanning tree S rooted at s used in G' contains exactly the edges incident to s and the edges of the paths P_{t-1}. No edge of any D_t^i is contained in S.

We assume that a special vertex r_i has been arbitrarily selected for each part L^i of H, and let $R = \{r_1, \ldots, r_{t-1}\}$.

Let $u \in L^i$ be a vertex of G_0, $i \neq 0$. Observe that if u and r_i are not adjacent in G_0, then there is no edge in G_0 (and thus in G') between $P(u)$ and $P(r_i)$. Let R' be the projection of R on D_t^{t-1}: $R' = \{u \in D_t^{t-1} \mid V(P(u)) \cap R \neq \varnothing\}$. $|R'| = |R| = t - 1$, so R' is not a dominating set of D_t^{t-1} (Lemma 9).

Let v be a vertex of D_t^{t-1} with no neighbors in R', and let $r' \in R'$. From the above observation, in G', there is no edge between $P(v)$ and $P(r')$. During the second phase of the CDY's algorithm, only edges incident to r_i are added, for all $i \geqslant 1$. It follows that every vertex of $P(v)$ has no incident edges in G', excepted those of $P(v)$. So, $d_{G'}(v, r') \geqslant 2(t-1) + 2 = \Omega(n^{1/3})$ whereas $d_{G_0}(v, r') = 2$. G' is an additive $\Omega(n^{1/3})$-spanner, as claimed. □

4   Lower Bound

Let $m(n, g)$ be the maximum number of edges contained in a graph with n vertices and of girth at least g. It is clear that there exist at least an n-vertex graph for which every additive $(g-3)$-spanner (or multiplicative $(g-2)$-spanner) needs $m(n, g)$ edges. Indeed, any graph G of girth g and of $m(n, g)$ edges has no proper additive $(g-3)$-spanner: removing any edge $\{u, v\}$ of G implies that $d_H(u, v) \geqslant g - 1 = d_G(u, v) + g - 2$ and thus that H is not an additive $(g-3)$-spanner of G.

Theorem 4. *For each $\delta \geqslant 1$, there exists a graph of $n + 3\delta - 2$ vertices and of tree-length δ for which every multiplicative δ-spanner (and thus every additive $(\delta - 1)$-spanner) needs $m(n, \delta + 2) + 3\delta - 1$ edges.*

Proof. Consider a graph G with n vertices, a girth at least $\delta + 2$, and with $m(n, \delta + 2)$ edges. The diameter of G is at most δ. Indeed, otherwise G has two vertices, say u and v, at distance $\delta + 1$. So augmenting G by the edge u, v would provide a graph with n vertices, a girth at least $\delta + 2$, and with $m(n, \delta + 2) + 1$

edges: a contradiction with the definition of $m(n, \delta + 2)$. So G is of tree-length at most its diameter, that is $\leqslant \delta$.

Now we construct a graph G^* obtained from G by selecting an edge of G, say $\{u, v\}$, and by adding a path of length $3\delta - 1$, so that G^* contains a cycle C of length 3δ. The graph G^* has $n + 3\delta - 2$ vertices, a girth at least $\delta + 2$, and $m(n, \delta + 2) + 3\delta - 1$ edges. Again, G^* does not contain any proper multiplicative δ-spanner.

The tree-length of G^* is exactly δ observing that the tree-length of a graph composed of two subgraphs, say G and C, sharing a vertex or an edge is the maximum between the tree-length of G and the tree-length of C (because G and C are isometric subgraphs, and the common vertex or edge can be used to combined both optimal tree-decompositions). As shown in [15], the tree-length of a cycle of length $k = 3\delta$ is $\lceil k/3 \rceil = \delta$. □

An Erdös Conjecture [17] claims existence of n-vertex graphs with $\Omega(n^{1+1/k})$ edges and of girth at least $2k + 2$. This has been proved only for $k = 1, 2, 3$ and $k = 5$. It is known however that there are graphs of girth at least $2k + 2$ with $\Omega(n^{1+1/(2k)})$ edges. From Theorem 4, we have:

Corollary 1. *For every constant $\delta \geqslant 1$, there are graphs with $O(n)$ vertices and tree-length δ for which every multiplicative δ-spanner requires $\Omega(n^{1+\epsilon})$ edges, where $\epsilon \geqslant 1/\lceil \delta/2 \rceil$ for $\delta \leqslant 6$. Moreover, for every δ, $\epsilon \geqslant 1/\Theta(\delta)$, where the best current lower bound on ϵ is given by the table below.*

Proof. For each fixed integer $k \geqslant 1$, let $f(k)$ be the largest real such that there exists an n-vertex graph of girth at least $2k + 2$ and with $\Omega(n^{1+f(k)})$ edges. We have $m(n, 2k + 2) = \Omega(n^{1+f(k)})$.

Consider the worst-case graph G_δ given by Theorem 4. It has at most $5n/2$ vertices (recall that $\delta < n/2$ because the chordality of a graph is at most $n - 1$), and at least $m(n, \delta + 2)$. Note that $m(n, \delta + 2) \leqslant m(n, \delta + 1)$. So, G_δ has at least $m(n, 2 \lceil \delta/2 \rceil + 2) = \Omega(n^{1+f(\lceil \delta/2 \rceil)})$ edges.

It is known that $f(k) = 1/k$ for all $k \geqslant 1$, if the Erdös's Conjecture holds. The following table summarizes the best known results on $f(k)$. Complete references can be found in [27].

$k = \lceil \delta/2 \rceil$	$f(k)$
$1, 2, 3, 5$	$= 1/k$
4	$\geqslant 1/(k + 1)$
$6, 7$	$\geqslant 1/(k + 2)$
$k = 2r,\ r \geqslant 4$	$\geqslant 1/(3k/2 - 1)$
$k = 2r - 1,\ r \geqslant 5$	$\geqslant 1/(3k/2 - 3/2)$

□

Acknowledgements

We would like to thank F. Dragan for fruitful discussions about Lemma 4.

References

1. Noga Alon, Shlomo Hoory, and Nathan Linial. The Moore bound for irregular graphs. Graph and Combinatorics, 18(1):53–57, 2002.
2. Ingo Althöfer, Gautam Das, David Dobkin, Deborah Joseph, and José Soares. On sparse spanners of weighted graphs. Discrete & Computational Geometry, 9(1):81–100, 1993.
3. Surender Baswana and Sandeep Sen. A simple linear time algorithm for computing a $(2k - 1)$-spanner of $O(n^{1+1/k})$ size in weighted graphs. In 30th International Colloquium on Automata, Languages and Programming (ICALP), volume 2719 of LNCS, pages 384–396. Springer, July 2003.
4. Hans Leo Bodlaender. A linear time algorithm for finding tree-decompositions of small treewidth. SIAM Journal on Computing, 25:1305–1317, 1996.
5. Ulrik Brandes and Dagmar Handke. NP-completeness results for minimum planar spanners. Discrete Mathematics & Theoretical Computer Science, 3(1):1–10, 1998.
6. Andreas Brandstädt, Feodor F. Dragan, H.-O. Le, and Van Bang Le. Tree spanners on chordal graphs: Complexity and algorithms. Theoretical Computer Science, 310:329–354, 2004.
7. Leizhen Cai and Derek G. Corneil. Tree spanners. SIAM Journal on Discrete Mathematics, 8(3):359–387, 1995.
8. Victor D. Chepoi and Feodor F. Dragan. Distance approximating trees in graphs. In Elsevier, editor, 6th International Conference on Graph Theory (ICGT). Electronical Notes in Discrete Mathematics, August 2000.
9. Victor D. Chepoi, Feodor F. Dragan, and Chenyu Yan. Additive spanners for k-chordal graphs. In 5th Italian Conference on Algorithms and Complexity (CIAC), volume 2653 of LNCS, pages 96–107. Springer, May 2003.
10. Reinhard Diestel. Graph Theory (second edition), volume 173 of Graduate Texts in Mathematics. Springer, February 2000.
11. Yevgeniy Dodis and Sanjeev Khanna. Designing networks with bounded pairwise distance. In 30th Annual ACM Symposium on Theory of Computing (STOC), pages 750–759, 1999.
12. Dorit Dor, Shay Halperin, and Uri Zwick. All-pairs almost shortest paths. SIAM Journal on Computing, 29:1740–1759, 2000.
13. Yon Dourisboure. Routage compact et longueur arborescente. PhD thesis, Université Bordeaux 1, Talence, France, December 2003.
14. Yon Dourisboure, Feodor F. Dragan, Cyril Gavoille, and Chenyu Yan. Improved spanners for bounded tree-length graphs, 2004. In preparation.
15. Yon Dourisboure and Cyril Gavoille. Tree-decomposition of graphs with small diameter bags. In 2nd European Conference on Combinatorics, Graph Theory and Applications (EUROCOMB), pages 100–104, September 2003.
16. Michael Elkin and David Peleg. $(1 + \epsilon, \beta)$-spanner constructions for general graphs. In 33rd Annual ACM Symposium on Theory of Computing (STOC), pages 173–182, July 2001.
17. Paul Erdös. Extremal problems in graph theory. In Publ. House Cszechoslovak Acad. Sci., Prague, pages 29–36, 1964.
18. Sándor P. Fekete and Jana Kremer. Tree spanners in planar graphs. Discrete Applied Mathematics, 108:85–103, 2001.
19. Cyril Gavoille, Michal Katz, Nir A. Katz, Christophe Paul, and David Peleg. Approximate distance labeling schemes. In 9th Annual European Symposium on Algorithms (ESA), volume 2161 of LNCS, pages 476–488. Springer, August 2001.

20. M.S. Madanlal, G. Venkatesan, and C. Pandu Rangan. Tree 3-spanners on interval, permutation and regular bipartite graphs. Information Processing Letters, 59(2):97–102, 1996.

21. David Peleg. Distributed Computing: A Locality-Sensitive Approach. SIAM Monographs on Discrete Mathematics and Applications, 2000.

22. David Peleg and Alejandro A. Schäffer. Graph spanners. Journal of Graph Theory, 13(1):99–116, 1989.

23. David Peleg and Jeffrey D. Ullman. An optimal synchornizer for the hypercube. SIAM Journal on Computing, 18:740–747, 1989.

24. David Peleg and Eli Upfal. A trade-off between space and efficiency for routing tables. Journal of the ACM, 36(3):510–530, July 1989.

25. Neil Robertson and Paul D. Seymour. Graph minors. II. Algorithmic aspects of tree-width. Journal of Algorithms, 7:309–322, 1986.

26. José Soares. Graphs spanners: A survey. Congressus Numerantium, 89:225–238, 1992.

27. Mikkel Thorup and Uri Zwick. Approximate distance oracles. In 33rd Annual ACM Symposium on Theory of Computing (STOC), pages 183–192, July 2001.

No-Hole $L(p, 0)$ Labelling
of Cycles, Grids and Hypercubes[*]

Guillaume Fertin[1], André Raspaud[2], and Ondrej Sýkora[3]

[1] LINA, FRE CNRS 2729
Université de Nantes, 2 rue de la Houssinière
BP 92208 44322 Nantes Cedex 3, France
fertin@lina.univ-nantes.fr
[2] LaBRI U.M.R. 5800, Université Bordeaux 1
351 Cours de la Libération, F33405 Talence Cedex, France
raspaud@labri.fr
[3] Department of Computer Science, Loughborough University
LE11 3TU, The United Kingdom
O.Sykora@lboro.ac.uk

Abstract. In this paper, we address a particular case of the general problem of λ labellings, concerning frequency assignment for telecommunication networks. In this model, stations within a given radius r must use frequencies that differ at least by a value p, while stations that are within a larger radius $r' > r$ must use frequencies that differ by at least another value q. The aim is to minimize the span of frequencies used in the network. This can be modelled by a graph labelling problem, called the $L(p, q)$ labelling, where one wants to label vertices of the graph G modelling the network by integers in the range $[0; M]$, while minimizing the value of M. M is then called the λ number of G, and is denoted by $\lambda_q^p(G)$.

Another parameter that sometimes needs to be optimized is the fact that all the possible frequencies (i.e., all the possible values in the span) are used. In this paper, we focus on this problem. More precisely, we want that: (1) all the frequencies are used and (2) condition (1) being satisfied, the span must be minimum. We call this the *no-hole* $L(p, q)$ labelling problem for G. Let $[0; M']$ be this new span and call the ν number of G the value M', and denote it by $\nu_q^p(G)$.

In this paper, we study a special case of no-hole $L(p, q)$ labelling, namely where $q = 0$. We also focus on some specific topologies: cycles, hypercubes, 2-dimensional grids and 2-dimensional tori. For each of the mentioned topologies cited above, we give bounds on the ν_0^p number and show optimality in some cases. The paper is concluded by giving new results concerning the (general, i.e. not necessarily no-hole) $L(p, q)$ labelling of hypercubes.

[*] This work was done while the two first authors were visiting the University of Loughborough, and was supported in part by the EPSRC grant GR/R37395/01 and by VEGA grant No. 2/3164/23.

1   Introduction

In this paper, we study the *frequency assignment problem*, that arises in wireless communication systems. We are interested here in minimizing the number of frequencies used in the framework where radio transmitters that are geographically close may interfere if they are assigned close frequencies. This problem has originally been introduced in [12] and later developed in [9], where it has been shown to be equivalent to a graph labelling problem, in which the nodes represent the transmitters, and any edge joins two transmitters that are sufficiently close to potentially interfere. The aim here is to label the nodes of the graph in such a way that:

- any two neighbours (transmitters that are very close) are assigned labels (frequencies) that differ by a parameter at least p ;
- any two vertices at distance 2 (transmitters that are close) are assigned labels (frequencies) that differ by a parameter at least q ;
- the greatest value for the labels is minimized.

It has been proved that under this model, we could assume the labels to be integers, starting at 0 [8]. In that case, the minimum range of frequencies that is necessary to assign to the vertices of a graph G is denoted $\lambda_q^p(G)$, and the problem itself is usually called the $L(p,q)$ labelling problem. The frequency assignment problem has been studied in many different specific topologies [8, 14, 17, 1, 3, 5, 13, 2, 15, 16]. The case $p = 2$ and $q = 1$ is the most widely studied (see for instance [6, 11, 10, 4]). Some variants of the model also exist, such as the following generalization where one gives k constraints on the k first distances (any two vertices at distance $1 \leq i \leq k$ in G must be assigned labels differing by at least δ_i). One of the issues also considered in the frequency assignment problem is the *no-hole labelling*, where one wants to use all the frequencies in the span. More precisely, we want that: (1) all the frequencies are used and (2) condition (1) being satisfied, the span must be minimum. Let $[0; M']$ be the span of frequencies that we obtain. We then call the ν number of G the value M', and we denote it by $\nu_q^p(G)$. We note that depending on the values of p, q and on the considered graph G, a no-hole labelling might not exist. In that case, we let $\nu_q^p(G) = \infty$. Hence, we clearly have $\nu_q^p(G) \geq \lambda_q^p(G)$ for any p, q and G.

In this paper, we study a special case of no-hole $L(p,q)$ labelling, namely where $q = 0$. We also focus on some specific topologies: cycles, d-dimensional hypercubes, 2-dimensional grids and 2-dimensional tori. For each of the mentioned topologies cited above, we give bounds on the ν_0^p number for any value of $p \geq 1$ and $d \geq 1$, and show optimality in some cases. We conclude the paper with new results concerning the (general, i.e. not necessarily no-hole) $L(p,q)$ labelling of the d-dimensional hypercube, H_d.

2   No-Hole $L(p,0)$ Labellings

Proposition 1 (General graphs). *For any $p \geq 1$ and any connected graph G, if a no-hole labelling of G exists, then:*

- $\nu_0^p(G) \geq 2p - 1$ if G is bipartite.
- $\nu_0^p(G) \geq 2p$ if G is not bipartite.

Proof. In any graph such that a no-hole $L(p,0)$ labelling exists, there must, by definition, exist at least one vertex of label 0. Let u be this vertex. Then all the neighbours of u must be labelled at least p. Since the labelling is no-hole, all the labels in the range $[0;p]$ must be used. This is true, in particular, for label $p-1$. Let v be a vertex whose label is $p-1$. Then v has all its neighbours labelled at least $2p-1$.

Now, let G be non bipartite, and suppose that $\nu_0^p(G) \leq 2p-1$. G has at least one odd cycle. Consider the vertices on this cycle, and let $i \geq 0$ be the minimum label among them, assigned to vertex v. If $i \geq p$, then the neighbours of v, not being of minimum label, must be assigned a label at least $2p$, a contradiction. Hence, $i \in [0;p-1]$. In that case, the two neighbours of v on the cycle, say w_1 and w_2, are assigned labels at least $p+i$, that is in the range $[p;2p-1]$. But the neighbours of w_1 and w_2 on the cycle must be assigned labels in the range $[0;p-1]$, etc. If we repeat this argument, we see that, when we will close the cycle, since it is odd, we will end up with a vertex z whose two neighbours, say x and y, are such that x is assigned a label in the range $[0;p-1]$, while y is assigned a label in the range $[p;2p-1]$. As $\nu_0^p(G) \leq 2p-1$, there is no possibility to label z in an $L(p,0)$ fashion, a contradiction.

Observation 1 *For any graph G of order n that admits a no-hole labelling, $n \geq \nu_0^p(G) + 1$*

Proof. Suppose a no-hole labelling for G exists. In order to be able to assign the vertices of G all labels in the range $[0; \nu_0^p(G)]$, we must have $n \geq \nu_0^p(G) + 1$.

Proposition 2 (Cycles). *For any $p \geq 1$ and any graph G:*

- $\nu_0^p(C_n) = 2p$ *for any odd $n \geq 2p + 1$*
- $\nu_0^p(C_n) = 2p - 1$ *for any even $n \geq 2p + 2$*

Proof. First, suppose that n is even. According to Proposition 1, we know that, if a no-hole $L(p,0)$ labelling for C_n exists, then $\nu_0^p(C_n) \geq 2p-1$. By Observation 1, we must have $n \geq 2p$. However, let v be the vertex which is assigned label p. In that case, both its neighbours must be assigned label 0, because only labels in the range $[0;2p-1]$ are allowed, and the gap between two neighbours is at least p. Hence, $n \geq 2p + 1$; but since n is even, we have $n \geq 2p + 2$. Suppose the vertices of C_n are numbered clockwise from 1 to n. We give the following labelling function c on C_n : (a) for any vertex v numbered $2i+1$ ($0 \leq i \leq p-1$), $c(v) = p+i$; (b) for any vertex v numbered $2i$ ($1 \leq i \leq p$), $c(v) = i-1$; (c) for any vertex v numbered $2i+1$, $i \geq p$, $c(v) = 2p-1$; (d) for any vertex v numbered $2i$ ($i \geq p+1$), $c(v) = 0$ (cf. for instance Figure 1 (right), where $n = 2p+2$).

Now, we show that this assignment is an $L(p,0)$ no-hole labelling of C_n, for any even $n \geq 2p+2$. First, consider any two neighbouring vertices j and $j+1$, $1 \leq j \leq 2p-1$. If j is even $j = 2i$, then $c(j) = i-1$, while $c(j+1) = p+i$, thus the gap of at least p is satisfied. If j is odd $j = 2i + 1$, then $c(j) = p + i$, while $c(j + 1) = i$, which also satisfies the $L(p,0)$ condition. Now consider two neighbours j and $j + 1$, with $2p + 1 \leq j \leq n - 1$. If j is even, then $c(j) = 0$ while $c(j + 1) = 2p - 1$, and if j is odd, then $c(j) = 2p - 1$ while $c(j + 1) = 0$. There still remain two cases to consider: (1) Vertices 1 and n and (2) vertices $2p$ and $2p + 1$. However, in case (1) we have $c(1) = p$ and $c(n) = 0$, while in case (2) we have $c(2p) = p - 1$ and $c(2p + 1) = 2p - 1$.

Consequently, the $L(p,0)$ condition is satisfied. There is no hole as from the definition of the labelling, one can see that all labels are used on vertices 0 to $2p$: odd vertices $2i + 1$, $0 \leq i \leq p - 1$ have labels from p to $2p - 1$, while even vertices $2i$, $1 \leq i \leq p$ are assigned labels from 0 to $p - 1$.

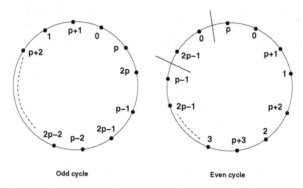

Odd cycle   Even cycle

Fig. 1. $L(p,0)$ no-hole labelling in cycles

Suppose that n is odd. By Proposition 1, we know that, if a no-hole $L(p,0)$ labelling for C_n exists, then $\nu_0^p(C_n) \geq 2p$. Now, clearly, if $n \leq 2p - 1$, then there is not enough vertices to use all the labels. Thus, we must have $n \geq 2p + 1$ in order that the no-hole labelling exists. Assume that $n \geq 2p + 1$, and the vertices of C_n are numbered clockwise from 1 to n. We define the following labelling function c to C_n: (a) for any vertex v numbered $2i + 1$ $(0 \leq i \leq p)$, $c(v) = p - i$; (b) for any vertex v numbered $2i$ $(1 \leq i \leq p)$, $c(v) = 2p - i + 1$; (c) for any vertex v numbered $2i + 1$, $i \geq p + 1$, $c(v) = 0$; (d) for any vertex v numbered $2i$ $(i \geq p + 1)$, $c(v) = p$ (cf. for instance Figure 1 (left), where $n = 2p + 1$).

The labelling is an $L(p,0)$ no-hole labelling of C_n, for any odd $n \geq 2p + 1$. First, consider any two neighbouring vertices j and $j + 1$, $1 \leq j \leq 2p$. If j is even $j = 2i$, then $c(j) = 2p - i + 1$, while $c(j + 1) = p - i$, thus the gap of at least p is satisfied. If j is odd $j = 2i + 1$, then $c(j) = p - i$, while $c(j + 1) = 2p - i$, which also fulfills the $L(p,0)$ condition. Now consider two neighbours j and $j + 1$, with $2p + 2 \leq j \leq n - 1$. If j is even, then $c(j) = p$ while $c(j + 1) = 0$, and if j is odd, then $c(j) = 0$ while $c(j+1) = p$. Now there remains some cases to consider: (1) $j = 2p + 1$ and $j + 1 = 2p + 2$ and (2) $j = n$ and $j + 1 = 0$ (that is, we "close" the cycle). But in both cases, we have $c(j) = 0$ and $c(j + 1) = p$. Thus, altogether, the $L(p,0)$ condition is satisfied. Now, by definition of the labelling, we can see that all the labels are used on vertices 0 to $2p$: vertices of the form

$2i + 1$, $0 \leq i \leq p$ are assigned labels from 0 to p, while vertices of the form $2i$, $1 \leq i \leq p$ are assigned labels from $p + 1$ to $2p$.

Proposition 3 (Hypercubes). *For any d–dimensional hypercube H_d such that $d \geq \frac{p+4}{2}$, $\nu_0^p(H_d) = 2p - 1$.*

Sketch of Proof. By Proposition 1, $\nu_0^p(H_d) \geq 2p - 1$. We will first show that $\nu_0^p(H_d) \leq 2p - 1$ (thus, proving the equality) for any $d \geq 2p - 1$; then, we will show that this result can be extended to any $d \geq \frac{p+4}{2}$. Suppose $d \geq 2p - 1$. The fact that $\nu_0^p(H_d) \leq 2p - 1$ is proved by homomorphism into the following graph G'_p: (a) the nodes of G'_p are the integers between 0 and $2p-1$ and (b) there is an edge between u and v in G'_p iff $|u - v| \geq p$. Clearly, G'_p represents the constraints on the $L(p,0)$ labelling, in the sense that any edge (u,v) of G'_p indicate that labels u and v can be assigned to neighboring nodes in H_d. We want to find an homomorphism $\mathcal{H}$ from H_d to G'_p, i.e. we want to find a mapping from $V(H_d)$ to $V(G')$, where every node v has an image $h(v)$ such that any edge (u,v) in H_d corresponds to an edge $(h(u), h(v))$ in G'_p. If we can do this, then we can find a labelling (more precisely, $c(v) = h(v)$ for any node v) that satisfies the $L(p,0)$ constraints. Furthermore as this labelling has to be no-hole, we also need that every node of G'_p is an image of at least one node of H_d. Let us partition the nodes of H_d into $d+1$ sets: for any $0 \leq i \leq d$, the set S_i corresponds to the nodes having i bits equal to 0 in its binary coordinates. By definition of the hypercube, for every $0 \leq i \leq d$, S_i is a stable set. In other words, all edges appear between different S_is. More precisely, all the edges of H_d appear between an S_i and an S_{i+1}. Let us define the homomorphism so that all nodes belonging to the same S_i have the same image by $\mathcal{H}$. Let h_i be the image by $\mathcal{H}$ of all the nodes of S_i. Then, for any $1 \leq i \leq d - 1$, h_i must be connected in G'_p to both h_{i+1} and h_{i-1}. Moreover, h_0 must be connected to h_1, and h_{d-1} must be connected to h_d. Hence, this induces a path starting at h_0, and ending at h_d, with edges (h_i, h_{i+1}) for any $0 \leq i \leq d - 1$. But we also want this labelling to be no-hole, hence this path must be hamiltonian. In other words, if we are able to find a hamiltonian path in G'_p, then there exists a homomorphism of H_d into G'_p. Clearly, since we have $d + 1$ sets S_i, and since each one has a unique image in G'_p, we must have $d + 1 \geq 2p$, that is $d \geq 2p - 1$.

Finally we need to show that G'_p contains a hamiltonian path ; it is as follows: $p, 0, p+1, 1, \ldots, i, p+i, i+1, p+i+1, \ldots, p-2, 2p-2, p-1, 2p-1$ (cf. Figure 2(left)).

Let v_j be any node of set S_j, $0 \leq j \leq d$. If $j = 2i$, we set $h(v_{2i}) = p + i$ for every $0 \leq i \leq p - 1$ and if $j = 2i + 1$, we set $h(v_{2i+1}) = i$ for every $0 \leq i \leq p - 1$. Finally, for any $j \geq 2p$, if j is of the form $2p + 2i$, we set $h(v_j) = 2p - 1$, and if j is of the form $2p + 2i + 1$, we set $h(v_j) = 0$.

We can show that the above result can be extended for any $d \geq p + 1$. This is obtained using the same kind of argument (that is, homomorphism into G'_p), but with a better mapping of the nodes (cf. Figure 2(right)).

The same goes to extending the result for any $d \geq \frac{p+4}{2}$, hence proving the proposition. Here again, the proof technique is the same as for the two previous cases. Roughly speaking, we consider H_d as 4 copies of H_{d-2}, connected between them by 2 perfect matchings. $\qquad \square$

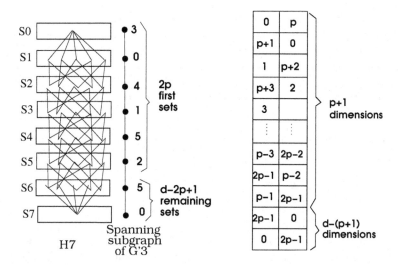

Fig. 2. (left) Homomorphism of H_d into G'_p, with $d = 7$ and $p = 3$; (right) Another homomorphism of H_d into G'_p, where nodes are represented by squares

Observation 2 *For any $p \geq 2$, if $\nu_0^p(H_d) = 2p - 1$, then $p \leq 2^{d-1} - d + 1$.*

Proof. Suppose that a no-hole $L(p,0)$ labelling of H_d exists, with $\nu_0^p(H_d) = 2p - 1$. Since H_d has 2^d vertices, we must have $2^d \geq 2p$. Since $p \geq 2$, labels $0, p - 1, p$ and $2p - 1$ are pairwise distinct. Moreover, by definition, there must exist a vertex labelled p, whose d neighbours must then be labelled 0. The same goes for any vertex labelled $p - 1$, whose d neighbours must then be labelled $2p - 1$. Hence, those four pairwise different labels are present on at least $2(d+1)$ vertices, leaving $2p - 4$ labels to be present on at most $2^d - 2(d + 1)$ vertices. Thus, we must have $2p - 4 \leq 2^d - 2(d + 1)$, which gives the result.

Proposition 4. *If $\nu_0^p(H_{d_0}) = 2p - 1$ for a given dimension d_0, then $\nu_0^p(H_{d'}) = 2p - 1$ for any $d' \geq d_0$.*

Proof. Consider H_{d_0}, for which we have a no-hole $L(p,0)$ labelling with $\nu_0^p(H_{d_0}) = 2p - 1$. We will show here a way to obtain a no-hole $L(p,0)$ labelling of H_{d_0+1} from the labelling of H_{d_0}. We recall that H_{d_0+1} is obtained from two copies of H_{d_0} joined by a perfect matching. Now consider a copy of H_{d_0}, having a no-hole $L(p,0)$ labelling with $\nu_0^p(H_{d_0}) = 2p - 1$: necessarily, there must exist a vertex in H_{d_0}, say x, whose label is p. Wlog (since hypercubes are vertex transitive), let the binary coordinates of x be as follows: $x = (0,0 \ldots 0)$. Since $\nu_0^p(H_{d_0}) = 2p-1$, all the neighbours of x must be labelled 0. Moreover, all the neighbours of x have exactly one bit equal to 1. By the same argument, we can see that all the vertices having 2 bits equal to 1 must be labelled in the range $[p; 2p - 1]$, while all the vertices having 3 bits equal to 1 must be labelled in the range $[0; p - 1]$. More generally, all the vertices having an even (resp. odd) number of bits equal to 1 are labelled in the range $[p; 2p-1]$ (resp. $[0; p - 1]$). Now, take a second copy of H_{d_0}, and label each vertex having an even (resp. odd) number of bits equal to 1 with label 0 (resp. $2p - 1$). This labelling is an $L(p, 0)$ labelling, but it is not no-hole. If we connect the corresponding vertices in both copies of H_{d_0}, the labelling we obtain remains $L(p, 0)$ (vertices labelled in the range $[p; 2p - 1]$ (resp. $[0; p - 1]$) in the first copy are connected to vertices labelled 0 (resp. $2p - 1$) in the second copy). Moreover, since it is no-hole in the first copy of H_{d_0}, it remains no-hole in H_{d_0+1}.

Proposition 5 (2-Dimensional grids $P_n \times P_m$). *For any p and $n \geq m \geq 1$ we have: $\nu_0^p(P_n \times P_m) = 2p - 1$, where*

1. *$n \cdot m - m + 1 \geq 2p$ if n is even and m is odd,*
2. *$n \cdot m - m \geq 2p$ otherwise.*

Sketch of Proof: Fill in the $P_n \times P_m$ grid (i.e. n rows and m columns) in the chessboard mode. Like in the chessboard where we have white and black alternating squares, we have in the "white" squares the labels from the range $[p; 2p-1]$ and in the "black" squares the labels $[0; p-1]$. Without loss of generality assume that the left upper square is white. Take the following labelling: put p in the left upper corner and subsequently put in the white squares from left to right and row by row the upper range labels: $p + 1, p + 2, ..., 2p - 1$. In the last row put in all white squares $2p - 1$. Further put 0 into all "black" squares of the first row of the grid. Starting with the left most square in the second row of the grid, we subsequently put into the "black" squares lower range labels from $[0; p - 1]$. The labelling is no-hole $L(p, 0)$ and $2p = m(n - 1) + 1$ if m is odd and n is even. Otherwise $2p = m(n - 1)$. □

Below is an example of 2-dimensional grids $G(5, 6)$ and $G(5, 5)$.

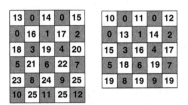

Fig. 3. No-hole $L(p, 0)$ labelling in 2D grids $G(n, m)$: (left) $m = 5, n = 6, p = 13$; (right) $m = n = 5, p = 10$

Proposition 6 (Consequence of Proposition 5). *For any $p, q \geq 0$ and $d \geq 4$ we have: $\nu_0^p(H_d) = 2p - 1$ when*

1. *$(\lfloor d/2 \rfloor + 1)\lceil d/2 \rceil + 2 \geq 2p$ if d is odd,*
2. *$(d/2 + 1)^2 - d/2 \geq 2p$ if d is even.*

Proof. Proof follows by combining Propositions 3 and 5.

Remark 1. Similar results to Propositions 5 and 6 can be obtained for the 3 and higher dimensional grids. The results will appear in the full version of this work. Below is an example of a no-hole labelling of a 3-dimensional grid (see Figure 4).

24	0	25	1
0	26	1	31
27	2	32	7
3	33	8	43

0	28	1	34
29	2	35	7
3	36	8	44
37	9	45	17

30	4	38	10
5	39	11	46
40	12	47	18
13	47	19	47

6	41	14	47
42	15	47	20
16	47	21	47
47	22	47	23

Fig. 4. A $L(24,0)$ no-hole labelling in $P_4 \times P_4 \times P_4$ with $\nu_0^{24} = 47$. Each of the 4 blocks represents a 2-Dimensional subgrid $P_4 \times P_4$, in which each square represents a node

By direct application of Proposition 1, we get the following result for the 2-dimensional toroidal meshes.

Proposition 7 (Tori). *For any $p \geq 1$ and $n, m \geq 3$:*

- $\nu_0^p(C_n \times C_m) \geq 2p - 1$ *if n and m are both even*
- $\nu_0^p(C_n \times C_m) \geq 2p$ *otherwise*

7	0	7	3
0	4	0	7
6	0	6	2
1	5	0	7

Fig. 5. $L(p,0)$ no-hole labelling in 2D-tori; an example for $C_4 \times C_4$ where $p = 4$

3 $L(p,q)$ Labellings of Hypercubes

We conclude this paper by giving new results concerning the (general, i.e. not necessarily no-hole) $L(p,q)$ labelling of H_d. Some results on this topic have been given in [7]. However, it is possible to improve them. This is the purpose of Proposition 8 below.

Proposition 8. *For any $p, q, \geq 0$ and $d \geq 1$:*

1. $\lambda_0^p(H_d) = p$
2. $(d-1)q \leq \lambda_q^0(H_d) \leq (2d-3)q$
3. $\lambda_q^p(H_d) \leq 2p + (2d-2)q - 1$.

Proof. (1) For any $d \geq 1$, H_d has at least two vertices. Consider a vertex v of H_d that is assigned label 0 in a $L(p,0)$ labelling of H_d. Such a vertex must exist, otherwise every label could be decreased by at least 1, leading to a better solution. v has at least one neighbour w, whose label must then be greater than or equal to p. Hence, $\nu_0^p(H_d) \geq p$. We can show that $\nu_0^p(H_d) \leq p$ by noticing that H_d is bipartite. Thus, if all the vertices of the first (resp. second) partition of H_d are labelled 0 (resp. p), the labelling we obtain is $L(p,0)$.

(2) Take any vertex u of H_d. It has d neighbors, all of them lying at distance 2 from each other. Hence, those d vertices must be assigned labels that differ by at least q. Since labels can begin at 0, it follows that the greatest label is greater than or equal to $(d-1)q$, showing that $\lambda_q^O(H_d) \geq (d-1)q$. The upper bound is obtained by the following labelling: suppose each node $v = (x_1, x_2 \ldots x_d)$ is defined by its (binary) coordinates in each of the d dimensions of H_d ; for any vertex $v = (x_1, x_2 \ldots x_d)$ of G_d, we define $c(v) = (\sum_{k=1}^{d-1} kqx_k) \bmod (2d-2)q$. Since $p = 0$, we only need to consider two vertices u and v lying at distance 2 in G_d, thus differing on two coordinates, say x_i and x_j, $1 \leq i \neq j \leq d$. We will consider two cases here: (a) $j = d$ and (b) $j \neq d$. In case (a), we have $|c(u) - c(v)| = iq \bmod (2d-2)q$, and since $1 \leq i \leq d-1$, $|c(u) - c(v)| \geq q$. In case (b), we either have $|c(u) - c(v)| = (i+j)q \bmod (2d-2)q$ or $|c(u) - c(v)| = (j-i)q \bmod (2d-2)q$, but since $1 \leq i \leq d-2$ and $i+1 \leq j \leq d-1$, $|c(u) - c(v)| \geq q$.

(3) Suppose each node $v = (x_1, x_2 \ldots x_d)$ is defined by its (binary) coordinates in each of the d dimensions of H_d. Every node $v = (x_1, x_2 \ldots x_d)$ is then assigned label $c(v) = \sum_{i=1}^d (p + (i-1)q)x_i \bmod (2p + (2d-2)q)$. Take two neighbors u and v in H_d, which thus differ on exactly one coordinate, say in x_j. Thus $|c(v) - c(u)| = p + (j-1)q \bmod (2p + (2d-2)q)$, that is $|c(v) - c(u)| = p + (j-1)q$. Since $1 \leq j \leq d$, we have that $|c(v) - c(u)| \geq p$. Now consider two nodes differing on two coordinates i and j, where, wlog, $i < j$ (since we are in H_d, necessarily $i \neq j$). We have two cases here: (a) $x_i = x_j$ or (b) $x_i \neq x_j$. In case (a), we obtain that $|c(v) - c(u)| = 2p + (i+j-2)q$, which is clearly greater than or equal to q. In case (b), $|c(v) - c(u)| = (j-i)q$, which is also greater than or equal to q since $j \neq i$.

Remark 2.

- Concerning the $L(0,q)$ labelling of H_d, we can show that $\lambda_q^0(H_3) = \lambda_q^0(H_4) = 3q$;
- We note that Proposition 8(3), when applied to the case $p = 2$ and $q = 1$, gives $\lambda_q^p(H_d) \leq 2d+1$, a value which coincides with the lower bound proved in [8].

4   Conclusion

In this paper, we have mainly considered the no-hole $L(p,0)$ labelling in different topologies, such as cycles, hypercubes and 2-dimensional tori. We also gave some bounds for the (not necessarily no-hole) $L(p,q)$ labelling of hypercubes, that improve the ones from [7].

Concerning no-hole $L(p,0)$ labellings, we have mainly based our study on the cases for which there exists a no-hole $L(p,0)$ labelling having the minimum number of labels (that is, minimum as stated in Proposition 1). Indeed, depending on the respective values of n (number of nodes of the considered network) and p, such a no-hole $L(p,0)$ labelling might not exist. Also, as remarked above, our work concerning 2-dimensional grids can be extended to any d-dimensional grids, $d \geq 3$ (this will appear in the full version of this work). We also note that a natural extension of this work is to study the no-hole $L(p,q)$ labelling of graphs, for any p,q.

References

1. A.A Bertossi, C.M. Pinotti, and R.B. Tan. Efficient use of radio spectrum in wireless networks with channel separation between close stations. In Proc. DIAL M for Mobility 2000 (Fourth International Workshop on Discrete Algorithms and Methods for Mobile Computing and Communications), 2000.
2. A.A Bertossi, C.M. Pinotti, and R.B. Tan. Channel assignment with separation for special classes of wireless networks : Grids and rings. In Proc. IPDPS'02 (International Parallel and Distributed Processing Symposium), pages 28–33. IEEE Computer Society, 2002.
3. H.L. Bodlaender, T. Kloks, R.B. Tan, and J. van Leeuwen. λ-coloring of graphs. In Proc. STACS 2000 : 17th Annual Symposium on Theoretical Aspect of Computer Science, volume 1770, pages 395–406. Lecture Notes Computer Science, Springer-Verlag Berlin, 2000.
4. T. Calamoneri and R. Petreschi. $L(2,1)$-labeling of planar graphs. In Proc. DIAL M for Mobility 2001 (Fifth International Workshop on Discrete Algorithms and Methods for Mobile Computing and Communications), pages 28–33, 2001.
5. G.J. Chang, W.-T. Ke, D. Kuo, D.D.-F. Liu, and R.K. Yeh. On $L(d,1)$-labelings of graphs. Discrete Mathematics, 220:57–66, 2002.
6. G.J. Chang and D. Kuo. The $L(2,1)$-labeling on graphs. SIAM J. Discrete Math., 9:309–316, 1996.
7. G. Fertin and R. Raspaud. $L(p,q)$ labeling of d-dimensional grids. Discrete Mathematics, 2003. Submitted.
8. J.R. Griggs and R.K. Yeh. Labeling graphs with a condition at distance two. SIAM J. Discrete Math., 5:586–595, 1992.
9. W.K. Hale. Frequency assignment : theory and applications. Proc. IEEE, 60:1497–1514, 1980.
10. P.K. Jha. Optimal $L(2,1)$-labeling of cartesian products of cycles with an application to independent domination. IEEE Trans. Circuits & Systems I: Fundamental Theory and Appl., 47:1531–1534, 2000.
11. P.K. Jha, A. Narayanan, P. Sood, K. Sundaram, and V. Sunder. On $L(2,1)$-labeling of the cartesian product of a cycle and a path. Ars Combin., 55:81–89, 2000.
12. B.H. Metzger. Spectrum management technique. Paper presented at the 38th National ORSA Meeting, Detroit, MI, 1970.
13. M. Molloy and M.R. Salavatipour. Frequency channel assignment on planar networks. In In Proc. 10th Annual European Symposium (ESA 2002), Rome, Italy, September 2002, volume 2461, pages 736–747. Lecture Notes Computer Science, Springer-Verlag Berlin, 2002.

14. D. Sakai. Labeling chordal graphs with a condition at distance two. SIAM J. Discrete Math., 7:133–140, 1994.
15. J. van den Heuvel. Radio channel assignment on 2-dimensional lattices. Annals of Combinatorics, 6:463–477, 2002.
16. J. van den Heuvel, R. A. Leese, and M. A. Shepherd. Graph labelling and radio channel assignment. Journal of Graph Theory, 29:263–284, 1998.
17. M. Whittlesey, J. Georges, and D.W. Mauro. On the λ-number of Q_n and related graphs. SIAM J. Discrete Math., 8:499–506, 1995.

Existence of Nash Equilibria
in Selfish Routing Problems

Alessandro Ferrante[1,2] and Mimmo Parente[1]

[1] Dipartimento di Informatica ed Applicazioni, Università degli Studi di Salerno
84081 Baronissi, Italy
{ferrante,parente}@dia.unisa.it
[2] Department of Computer Sciences, Purdue University
West Lafayette (INDIANA), USA
ferrante@cs.purdue.edu

Abstract. The problem of routing traffic through a congested network is studied. The framework is that introduced by Koutsoupias and Papadimitriou where the network is constituted by m parallel links, each having a finite capacity, and there are n selfish (noncooperative) agents wishing to route their traffic through one of these links: thus the problem sets naturally in the context of noncooperative games. Given the lack of coordination among the agents in large networks, much effort has been lavished in the framework of *mixed* Nash equilibria where the agent's routing choices are regulated by probability distributions, one for each agent, which let the system reach thus a stochastic steady state from which no agent is willing to unilaterally deviate. Recently Mavronicolas and Spirakis have investigated *fully mixed* equilibria, where agents have all non zero probabilities to route their traffics on the links. In this work we concentrate on constrained situations where some agents are forbidden (have probability zero) to route their traffic on some links: in this case we show that at most one Nash equilibrium may exist and we give necessary and sufficient conditions on its existence; the conditions relating the traffic load of the agents. We also study a dynamic behaviour of the network, establishing under which conditions the network is still in equilibrium when some of the constraints are removed. Although this paper covers only some specific subclasses of the general problem, the conditions found are all effective in the sense that given a set of yes/no routing constraints on each link for each agent, we provide the probability distributions that achieve the unique Nash equilibrium associated to the constraints (if it exists).

1 Introduction

We study the problem of routing traffic in a simple context proposed by Koutsoupias and Papadimitriou in [11]. We are given a network of m parallel links, each having a finite capacity (traffic rate) of carrying the traffic $w_1, \ldots, w_n$ of n agents. Each agent has an indivisible amount of traffic that wishes to route through a link and it has a delay given by the overall traffic routed by all the

R. Královič and O. Sýkora (Eds.): SIROCCO 2004, LNCS 3104, pp. 149–160, 2004.
© Springer-Verlag Berlin Heidelberg 2004

other agents on that link at the given link rate. In large networks, it is reasonable to assume a selfish behaviour of the agents, that is a lack of coordination[1] (see [11, 17]). Intuitively, each agent aims at minimizing the delay of *its own* traffic. Thus the problem sets naturally in a context of *noncooperative games* where the routes chosen by the users form a so called *mixed* Nash equilibrium, in the sense of classical game theory: a stochastic steady state of the game ([15, 16]). A probability distribution to route the traffic of an agent through the links, is called a *strategy* for that agent. Given a set of n strategies, the network converges to an equilibrium point (steady state) such that each agent has no convenience to modify unilaterally its routing strategy: in game theory this is called a Nash equilibrium (see [15, Chapter 3] for deep interpretations of mixed strategy Nash Equilibrium and [14]). Since Nash equilibria often lead to a performance degradation of the network, in [11] (where the authors introduce also the idea to calculate Nash Equilibrium as a solution of linear equation system, see also [12]) it has been proposed to measure this degradation in terms of *coordination ratio*: the ratio between the worst possible Nash equilibrium and the value of the optimum performance. Remarkable results have been given in [3, 10] establishing asymptotically tight bounds on the coordination ratio and also in [1] where a model similar to ours, so called the subset model, is considered. Anyway since, as remarked also in [16], it is not known any algorithm to find a Nash equilibrium given a network of links and agents, in [11] it has been noticed that if we are given a set of n *supports* (a vector of m binary variables indicating whether an agent routes its traffic with non zero probability or not on each of the m links) then at most one Nash equilibrium exists and this can be computed in $POLY(mn)$ time. In [13] necessary and sufficient conditions on the existence of *fully-mixed* Nash equilibria have been given, that is the case in which each agent has non zero probability to route its traffic on every link (to note that the conditions can be computed in $O(mn)$). Many other interesting results have been published: some of them (see [8, 9, 17]) concern cases of pure or fully-mixed strategies: either an agent has probability 1 to route its traffic on a particular link and probability 0 on all other links or it has non zero probability, different from 1, to use every link (in this case all supports are identical); the paper [4] concerns with the time the system takes to converge to a pure Nash Equilibrium, while [5] deals with the problem of converting any given non-equilibrium routing into a Nash equilibrium without increasing some measures of costs (see also [7] where similar results have been given for the model of unrelated links where each user is only allowed to ship its traffic on a subset of allowed links).

In this paper we aim at establishing conditions on the existence of Nash equilibria in presence of *mixed* supports. Actually our results fits in a setting of *constrained* routing: given a set of n supports, one for each agent, we investigate the case when an agent is *forbidden* to route its traffic on some links. This means that in the case of Nash equilibrium the game does not take into account the

---

[1] Note that, even in the case when all the links have equal capacity, the problem to minimize the maximum delay of the agents is an NP-complete problem (reduction from the *Partition* problem).

agent's cost of routing on the forbidden link. To make more clear the description and motivation of our results, let us consider an *equivalent* view of the network: each agent is connected to m sources and each source has a private link with a single destination. With this setting in mind, we address situations that may arise from a total congestion or a fault of some link between agents and sources, thus forbidding some of them to route their traffic through these links. The question we wish to answer here is: "under which conditions are there Nash equilibria when some agents are constrained not to route their traffic on some links?" We point out that given a network and a set of constraints, at most one Nash Equilibrium which fulfills this set of constraints may exist. We give necessary and sufficient conditions on the existence of equilibrium, relating the traffic weights and we also show which are the strategies of the agents, yielding thus the conditions effective and computable in $POLY(mn)$. More precisely, given a set of constraints describing on which links the agents may or may not route their traffic, we provide conditions for the existence of a Nash equilibrium providing also the unique strategy for each agent associated to the constraints. Moreover since the number of equilibria is, in general, exponential in m and n (the number of possible supports), our results are given for classes of constraints parameterized in the number of agents that will never route traffic on a link and in the number of links which will never carry traffic of some agents.

We also study a dynamic behaviour of the network: given n strategies, one for each agent, yielding a Nash equilibrium fulfilling a given set of constraints, we establish under which conditions the strategies are still a Nash equilibrium when the set of constraints changes. In particular we investigate the case when some of the constraints are removed and the agents are free to route their traffic (for a more extensive discussion of our results we refer the reader to Section 1.1).

Organization of the Paper. The paper is mostly self-contained and is organized as follows: in the next subsection we give some preliminary definitions and list our results. In Section 2, we first introduce some notation used throughout the paper and then give two preliminary results that will be exploited in Section 3 to show the necessary and sufficient conditions for the existence of Nash equilibria when the agents route their traffic subject to the constraints and when these are changed. Finally we give some short conclusions in Section 4.

1.1   The Model and Our Results

In this subsection we give some preliminary definitions, mostly following the notation of [3] and [11]. For all natural numbers x, y, $[x, y]$ denotes the set $\{x, x+1, \ldots, y\}$, when $x = 1$ we simply use $[y]$. Consider a network N constituted by m parallel links, each having capacity $s_1, \ldots, s_m$ respectively, on which n agents wish to route their unsplittable traffic, of length $w_1, \ldots, w_n$, respectively. If $s_1 = \ldots = s_n$, the system is called **uniform**. In what follows, m and n will always be the number of links and the number of agents in the network, respectively. A **strategy** P_i for an agent i, is a probability distribution P_i^j, $j \in [m]$, on the set of the m links. In particular a **pure** strategy is a strategy such that $P_i^j \in \{0, 1\}$ (that is agent i sends its traffic on a unique link with probability 1) and a **fully mixed** strategy P_i is a strategy such that $P_i^j \in (0, 1)$, for all $j \in [m]$. A **mixed set of strategies** $P = \{P_1, \ldots, P_n\}$ is a set of n strategies (one for each agent). Let us remark that the term *mixed* is used to underline that a set can contain strategies that are pure, fully mixed and strategies that are none of these two kinds, as well. In fact, as said above, almost all known results are for either pure or fully mixed strategies, while the present paper deals with solutions of mixed sets of strategies. A **support** S_i associated to a strategy P_i is a set of indicator (binary) variables S_i^j with $j \in [m]$ such that $S_i^j = 0$ if and only if $P_i^j = 0$. A **set of supports** $S = \{S_1, \ldots, S_n\}$ is a set of n supports (one for each agent). The **expected cost** of agent i on link j is $c_i^j = (\sum_{k \neq i} P_k^j w_k + w_i)/s_j$.

Our scenario, with the constraints on the routing of agents, is the following: a **constraint** F_i associated to agent i is a set of indicator (binary) variables F_i^j with $j \in [m]$ such that $F_i^j = 0$ if and only if agent i cannot (is forbidden to) route its traffic on link j. A **set of constraints** (one for each agent) $F = \{F_1, \ldots, F_n\}$ is a set of n constraints. A strategy P_i is *compatible* with a constraint F if whenever $F_i^j = 0$ then $P_i^j = 0$, for $j \in [m]$. Similarly a support S_i is *compatible* with a constraint F if whenever $F_i^j = 0$ then $S_i^j = 0$, for $j \in [m]$. In this setting, we can say that a set of strategies P is a Nash equilibrium if

$$P_i^j > 0 \text{ implies that } c_i^j \leq c_i^l \ i \in [n] \text{ and } j, l \in [m] \text{ s.t. } S_i^j = 1 \text{ and } F_i^l = 1. \quad (1)$$

Informally, a mixed set of strategies $P = \{P_1, \ldots, P_n\}$ on a network N, is a **Nash equilibrium** if there is no incentive for any agent to modify unilaterally its strategy. From this it follows that in a Nash equilibrium the costs c_i^j of an agent $i \in [n]$ are all equal on all links $j \in [m]$ such that $P_i^j > 0$. We can now define our problem.

Definition 1. *Given a network N, a set of constraints F and a set of supports S compatible with F, **Route**(N, F, S) **problem** is the problem of finding a Nash equilibrium P such that S is associated to P.*

In the rest of the paper, except where otherwise stated, we assume that a *Route(N, F, S)* problem is for uniform networks, that is a network with $s_1 = \ldots = s_m$. Hence a solution to a *Route(N, F, S)* problem is given by solving the following inequality system (if solution exist):

$$\begin{cases} c_i^j \leq c_i^l & \forall i \in [n], \ j, l \in [m] \ \text{s.t.} \ S_i^j = 1 \text{ and } F_i^l = 1 \\ P_i^j = 0 & \forall i \in [n], \ j \in [m] \quad \text{s.t.} \ S_i^j = 0 \\ P_i^j > 0 & \forall i \in [n], \ j \in [m] \quad \text{s.t.} \ S_i^j = 1 \\ \sum_{j=1}^{m} P_i^j = 1 \ \forall \ i \in [n] \end{cases} \quad (2)$$

Equivalently and more practically a solution to the *Route(N, F, S)* problem can be computed by first solving the following system of $m \ n$ equations in $m \ n$ unknowns:

$$\begin{cases} c_i^j = c_i^l & \forall\, i \in [n],\ l, j \in [m] \quad \text{s.t. } S_i^j = 1 \text{ and } S_i^l = 1 \\ P_i^j = 0 & \forall\, i \in [n],\ j \in [m] \quad\ \ \text{s.t. } S_i^j = 0 \\ \sum_{j=1}^m P_i^j = 1 \ \forall\, i \in [n] \end{cases} \tag{3}$$

and then by simply verifying whether the given solution is a probability distribution inducing a Nash equilibrium satisfying condition (1). From this it follows that $Route(N, F, S)$ problem has at most one solution.

We now define the case in which for a given link there exists only one agent with non zero probability on this link. Consider a $Route(N, F, S)$ problem, non necessarily uniform, and a mixed set of strategies. A link j is an **s-link** if and only if there exists $i \in [n]$ such that $P_i^j > 0$ and $P_k^j = 0$ for all $k \neq i$ (in [13] this is called a *solo link*). The agent i is called **s-agent** for the s-link j.

Consider a $Route(N, F, S)$ problem (not necessarily uniform) and a strategy P_i for an agent $i \in [n]$. In order to treat mixed sets of strategies we introduce two measures: a set of supports S is called a β-**set of supports** if and only if all the links on which the agents do not route their traffic is exactly β. A (α, β)-**set of supports** is a β-set of supports in which there are α supports that induce 1-strategies and $(n - \alpha)$ supports that induce fully-mixed strategies.

Given two sets of constraints, F and H, we introduce the following partial order relation on them: $F \preceq H$ if $F_i^j \leq H_i^j$, for all $i \in [n], j \in [m]$.

Our Results. In Section 2 we consider a $Route(N, F, S)$ problem, not necessarily uniform, with an s-link and give a necessary condition for the existence of a Nash equilibrium relating the capacity of the links to the traffic weights of the agents. From this result we obtain, as a corollary, a result presented in [6]. Then we present a lemma showing that, given a Nash equilibrium associated to a set of supports, if some agents are forbidden to route their traffic on a pair of links, then the remaining agents have the same probabilities to route their traffic on these two links. In Section 3 we give effective necessary and sufficient conditions for the existence of Nash equilibria in the case of (α, β)-set of supports, first in the case of $\beta = 1$ and $1 \leq \alpha \leq (n - 2)$ and then for $\alpha = \beta$, for a given set of constraints F. In these cases we show also that if the network constraints change and we are given a new set of constraints H, such that $F \preceq H$, then the conditions above still hold to get the same equilibrium.

2 Notation and Preliminary Results

Given a $Route(N, F, S)$ problem (not necessarily uniform), we will adopt the following notation:

- α is the number of agents with neither pure nor fully-mixed strategies;
- $L_i = \{j \in [m] \mid S_i^j = 0\}$, i.e. a link j is in L_i if agent i does not route its traffic on it. $L = \cup_{i \in [n]} L_i$ and $\beta = |L|$.
- for any set of agents G, w_{min}^G and w_{max}^G are, respectively, the minimum and the maximum weight of jobs of agents in G. Moreover, T_G is the sum of the jobs of the agents in the set G.

- for any set of agents G, G_l^h, where $l, h \in \{0, \cdots, |G|\}$, denotes the set obtained from G by deleting the agents with l lowest jobs and h highest jobs.
- $A^j = \{i \in [n] \mid S_i^j = 0\}$, contains the set of agents that do not route their traffic on link j. $[A^j] = \{l \in [m] \mid S_i^j = 0$ if and only if $S_i^l = 0$, for all $i \in [n]\}$, i.e. $[A^j]$ is an equivalence class containing links on which the same set of agents do not route their traffic. $A = \cup_{j \in [m]} A^j$ and $\alpha = |A|$.

Now we give two results which will be used in the next section. The first is an (effective) necessary condition for the existence of a solution to the $Route(N, F, S)$ problem such that S induces at least one s-link. The second result is a technical lemma which establishes when it is possible to reduce the dimension of system (3). Actually the so obtained system of equations, will be the starting point to describe (and compute) the solutions to the route problems presented in the rest of the paper.

Proposition 1. *Consider a* Route(N, F, S) *problem (not necessarily uniform) and a mixed set of strategies P associated to S. If there exists an **s-agent** i_l for an **s-link** l and a link $j \neq l$ such that $S_{i_l}^j = 1$ and $\frac{s_j}{s_l} \neq \frac{\sum_{t \neq i_l} P_t^j w_t}{w_{i_l}} + 1$, then the set of strategies P is not a Nash equilibrium.*

Proof. From definition of expected cost of an agent, it results that $c_{i_l}^l = w_{i_l}/s_l$. Therefore in a Nash equilibrium it must hold $c_i^j = c_{i_l}^l$ for all $j \in [m]$ such that $P_i^j > 0$. That is $(\sum_{t \neq i_l} P_t^j w_t + w_{i_l})/s_j = w_{i_l}/s_l$ for all $j \in [m]$ such that $P_{i_l}^j > 0$, thus yielding $(\sum_{t \neq i_l} P_t^j w_t)/w_{i_l} + 1 = s_j/s_l$ for all j such that $P_{i_l}^j > 0$ which completes the proof. $\qquad\square$

This proposition in fact can be specialized to the case of uniform links, getting so the following corollary (which resembles Proposition 1 of [6]).

Corollary 1. *Consider a uniform* Route(N, F, S) *problem. If there exists an s-agent i_l for an **s-link** l such that a link $j \neq l$ and an agent $k \neq i_l$ exist with $S_{i_l}^j = S_k^j = 1$, then there is no solution for the* Route(N, F, S) *problem.*

Given a set of supports $S = \{S_1, \ldots, S_n\}$, let us abuse in the notation (for clarity of exposition) and treat S and each S_i as ordered vectors, thus S can be seen as a boolean matrix with columns $S^j, j \in [m]$. The following lemma establishes that given a $Route(N, F, S)$ problem if $S^j = S^l$, $j, l \in [m]$ and a set of probability distributions (P_i^k) exists inducing a Nash equilibrium, then $P_i^j = P_i^l$ for all $i \in [n]$. In words, when agents $i_1, \ldots, i_k$ are forbidden to route their traffic on a pair of links j and l then, in a situation of Nash equilibrium, the remaining agents have all equal probabilities to route their traffic on links j and l.

Lemma 1. *Consider a* Route(N, F, S) *problem and let $P_i = \{P_i^1, \ldots, P_i^m\}$ be a Nash equilibrium associated to S, for all $i \in [n]$. Then, given two links $l \neq j \in [A^j] = [A^l]$ it holds that $P_i^j = P_i^l$ for all $i \in [n]$.*

Proof. The proof is by contradiction. Let $l, j \in [A^j]$ and assume, by way of contradiction, that agents i with $P_i^j \neq P_i^l$ exist (without loss of generality let $i \in [t]$ for some $t > 0$). Thus $P_i^l \neq P_i^j$ for $1 \leq i \leq t$ and $P_i^l = P_i^j$ for $(t + 1) \leq i \leq n$. Moreover, let $1, \ldots, z$ be the agents with non zero probabilities on links l and j. The proof for $t = 0$ and for $t = z$ is trivial. Thus consider the case $1 \leq t < z$ and let $P_i^l < P_i^j$, for $i \in [t]$. Because P_i^j is a Nash equilibrium, then $c_i^l = c_i^j$ must hold. Anyway from the assumptions $\sum_{k=t+1}^n P_k^l w_k = \sum_{k=t+1}^n P_k^j w_k$, thus $\sum_{k=1, k \neq i}^t P_k^l w_k = \sum_{k=1, k \neq i}^t P_k^j w_k$ and since $P_i^l < P_i^j$ then $\sum_{k=1}^t P_k^l w_k < \sum_{k=1}^t P_k^j w_k$, yielding that $c_r^l < c_r^j$ for all $t < r \leq z$, which is the desired contradiction. $\qquad \square$

From this lemma, we can show an equivalent reduced system of linear equations. Without loss of generality, we assume that if there are v different classes of equivalence, then $[A^1] \neq [A^2] \neq \ldots \neq [A^v]$. In other words, links $1, \ldots, v$ are all in different classes of equivalence.

$$\begin{cases} c_i^j - c_i^l = 0 & \forall j \in [v], \left(i \in \bigcap_{t=j+1}^v A^t, i \notin A^j \right), l \in [v] - L_i - \{j\} \\ P_i^j = P_i^l & \forall l \in [v], j \in [A^l] - \{l\}, i \notin A^l \\ \sum_{j \notin L_i} P_i^j = 1 \ \forall i \in [n] \end{cases} \qquad (4)$$

It easy to see that the solutions of this system have to be necessarily greater than zero in a Nash equilibrium.

3   Nash Equilibria

Consider a uniform *Route(N, F, S)* problem. In this section we will establish effective necessary and sufficient conditions on the existence of a solution to the problem. As said above a necessary condition is that the system (4) has all solutions in $(0, 1]$. Moreover, we consider the case in which some or all of the constraints are removed (a *Route(N, H, S)* problem with $F \preceq H$) and show sufficient conditions for the solution of the *Route(N, F, S)* problem are a solution for the *Route(N, H, S)* too. In particular, we study (α, β)-sets of supports in some particular cases. For simplicity, we assume in the rest of paper, that the α agents with zero probabilities are the agents $1, \ldots, \alpha$, that is $A = \{1, \ldots, \alpha\}$ and the links with zero probabilities are the links $1, \ldots \beta$, that is $L = \{1, \ldots, \beta\}$. In the following proofs, we make use of solutions of the system (4) which have been calculated using the software $Matlab^{\copyright}$ to found the base cases and then by guessing the solution for the generalized problem.

3.1 $(\alpha, 1)$-Sets of Supports, $1 \leq \alpha \leq n - 2$

All the α agents are forbidden to route the traffic on the first link.

Theorem 1. *Consider a* Route(N, F, S) *problem such that S is an $(\alpha, 1)$-set of supports with $1 \leq \alpha \leq (n - 2)$ and $F = S$. A solution to* Route(N, F, S) *exists if and only if*

$$T_{[\alpha]} < (m-1)[n-(\alpha+1)]w_{min}^{[\alpha+1,n]}. \tag{5}$$

In this case the Nash probabilities are

$$P_i^l = \begin{cases} 0 & l = 1 \ and \ i \in [\alpha] \\ \frac{1}{m-1} & l \in [2,m] \ and \ i \in [\alpha] \\ \frac{[n-(\alpha+1)]w_i + T_{[\alpha]}}{m[n-(\alpha+1)]w_i} & l = 1 \ and \ i \in [\alpha+1,n] \\ \frac{(m-1)[n-(\alpha+1)]w_i - T_{[\alpha]}}{m(m-1)[n-(\alpha+1)]w_i} & l \in [2,m] \ and \ i \in [\alpha+1,n] \end{cases} \tag{6}$$

Proof. In this case the system (4) has the values P_i^l given by (6) as solutions. As $m > 1$, $P_i^j \in (0,1)$ for all $i \in [\alpha+1,n]$ and obviously $P_i^l \in (0,1)$ for $i \in [\alpha]$. Moreover, as $P_i^l = \frac{1-P_i^1}{m-1}$ for $i \in [\alpha+1,n]$ and $l \in [2,m]$, then $P_i^l \in (0,1)$ if and only if $P_i^1 \in (0,1)$ for $i \in [\alpha+1,n]$ and $l \in [2,m]$. Therefore, from equations (6), the P_i^l are a set of strategies if

$$-[n-(\alpha+1)]w_i < T_{[\alpha]} < (m-1)[n-(\alpha+1)]w_i \quad \forall \ i \in [n] - [\alpha].$$

These $(n-\alpha)$ inequalities can be equivalently verified by just verifying the inequality $T_{[\alpha]} < (m-1)[n-(\alpha+1)]w_{min}^{[\alpha+1,n]}$ which gives the necessary condition.

The sufficient condition is trivial. Indeed, it is easy to see that if the relation (5) holds, then $P_i^1 \in (0,1)$ for all $i \in [\alpha+1,n]$ and therefore the P_i^l are a set of strategies. From this, as the P_i^l are solutions of the system (4), they are a Nash equilibrium. □

Removing some constraints we can get the following theorem.

Theorem 2. *Consider a* Route(N,F,S) *problem such that S is an $(\alpha,1)$-set of supports with $1 \le \alpha \le (n-2)$ and $F = S$, and a* Route(N,H,S) *such that $F \preceq H$. If the solution to the* Route(N,F,S) *exists, then also the solution to the* Route(N,H,S) *exists. The Nash probabilities for the* Route(N,H,S) *are those given in (6).*

Proof. If a solution to the *Route(N,F,S)* exists, then the relation (5) holds and obviously $P = \{P_i^l\}$ is a set of strategies then it remains only to prove that it is a Nash equilibrium. Since for the agents $i \in [n]$ such that $H_i^1 = 0$ or $i \in [\alpha+1,n]$, $P_i^j > 0$ for all $j \in [m]$ such that $H_i^j = 1$, let us concentrate on the agents i such that $F_i^1 = 0$ and $H_i^1 = 1$. For these agents it results:

$$c_i^1 = \frac{1}{m}\sum_{k>\alpha} w_k + \frac{(n-\alpha+mn-\alpha m-m)w_i}{m[n-(\alpha+1)]} + \frac{(n-\alpha)\sum_{k\le\alpha,k\ne i} w_k}{m[n-(\alpha+1)]} \ \text{and}$$

$$c_i^l = \frac{1}{m}\sum_{k>\alpha} w_k + \frac{\{m[n-(\alpha+1)]-n+\alpha\}\sum_{k\le\alpha,k\ne i} w_k}{m(m-1)[n-(\alpha+1)]} + \frac{\{m(m-1)[n-(\alpha+1)]-n+\alpha\}w_i}{m(m-1)[n-(\alpha+1)]}$$

for $l \in [m]/[1]$, from which $c_i^l - c_i^1 = -\frac{\sum_{k\le\alpha,k\ne i} w_k + (n-\alpha)w_i}{(m-1)[n-(\alpha+1)]} < 0$ for all $l > 1$, therefore the set of strategies P yield a Nash equilibrium for the *Route(N,H,S)*. □

3.2 (β, β)-Sets of Supports, $\beta = m > 2$

Agent i is forbidden to route its traffic on link i, $i \in [\alpha]$ and $\alpha = \beta = m$.

Theorem 3. *Consider a* Route(N, F, S) *problem such that S is a (β, β)-set of supports, $n > m = \beta = \alpha > 2$ and $F = S$. A solution to* Route(N, F, S) *exists if and only if:*

$$
\begin{aligned}
T_{[m]_1^1} &> (m-2)w_{max}^{[m]} - (m-2)[(m-1)n - (2m-1)]w_{min}^{[m]} \\
T_{[m]_2^0} &< (m-2)w_{min}^{[m]} + [(m-1)n - (2m-1)]w_{min}^{[m]_1^0} \\
T_{[m]_0^1} &> (m-1)w_{max}^{[m]} - (m-1)[(m-1)n - (2m-1)]w_{min}^{[m+1,n]} \\
T_{[m]_1^0} &< (m-1)w_{min}^{[m]} + [(m-1)n - (2m-1)]w_{min}^{[m+1,n]}
\end{aligned} \tag{7}
$$

In this case the Nash probabilities are:

$$
P_i^l = \begin{cases}
0 & i \in [m] \text{ and } l = i \\
\frac{[(m-1)n - (2m-1)]w_i + (m-2)w_l - T_{[m]-\{i,l\}}}{(m-1)[(m-1)n - (2m-1)]w_i} & i \in [m] \text{ and } l \in [m]/\{i\} \\
\frac{[(m-1)n - (2m-1)]w_i + (m-1)w_l - T_{[m]-\{l\}}}{m[(m-1)n - (2m-1)]w_i} & l \in [m] \text{ and } i \in [m+1, n]
\end{cases} \tag{8}
$$

Proof. In this case the system (4) has the values P_i^l given by (8) as solutions. A necessary condition for the existence of a Nash equilibrium is that system (4) has solutions in $(0, 1]$. In particular, as $m > 1$, $P_i^j \in (0, 1)$ for all $i \in [\alpha + 1, n]$, i.e.

$$
\begin{cases}
0 < P_i^l < 1 \; \forall \; i \in [m], \; l \in [m]/\{i\} \\
0 < P_i^l < 1 \; \forall \; i \in [m+1, n], \; l \in [m]
\end{cases}
$$

must hold, that is: for $i \in [m]$ and $l \in [m]/\{i\}$

$$
\begin{cases}
\sum_{k \leq m, k \neq i, l} w_k > (m-2)w_l - (m-2)[(m-1)n - (2m-1)]w_i \\
\sum_{k \leq m, k \neq i, l} w_k < (m-2)w_l + [(m-1)n - (2m-1)]w_i
\end{cases}
$$

and for $i \in [m+1, n]$ and $l \in [m]$

$$
\begin{cases}
\sum_{k \leq m, k \neq l} w_k > (m-1)w_l - (m-1)[(m-1)n - (2m-1)]w_i \\
\sum_{k \leq m, k \neq l} w_k < (m-1)w_l + [(m-1)n - (2m-1)]w_i
\end{cases}
$$

Now it is easy to see that to verify the first $m(m-1)$ inequalities is equivalent to verify just the following inequality

$$
T_{[m]_1^1} > (m-2)w_{max}^{[m]} - (m-2)[(m-1)n - (2m-1)]w_{min}^{[m]}
$$

since $[(m-1)n - (2m-1)] \geq 1$.

Analogously, the others inequalities can be verified by testing the following three ones:

$$
\begin{aligned}
T_{[m]_2^0} &< (m-2)w_{min}^{[m]} + [(m-1)n - (2m-1)]w_{min}^{[m]_1^0} \\
T_{[m]_0^1} &> (m-1)w_{max}^{[m]} - (m-1)[(m-1)n - (2m-1)]w_{min}^{[m+1,n]} \\
T_{[m]_1^0} &< (m-1)w_{min}^{[m]} + [(m-1)n - (2m-1)]w_{min}^{[m+1,n]}
\end{aligned}
$$

This completes the proof of the necessary condition.

The sufficient condition is trivial. In fact, it is easy to see that if relations (7) are verified, then $0 < P_i^l < 1$ for all $i \in [n]$ and $l \in [m]/\{i\}$ and, therefore, the P_i^l are a set of strategies. From this, as the P_i^l are solutions of the system (4), they are a Nash equilibrium. □

Removing some constraints we can get the following theorem.

Theorem 4. *Consider a* Route(N, F, S) *problem such that S is an (β, β)-set of supports with $n > m = \beta = \alpha > 2$ and $F = S$, and a* Route(N, H, S) *such that $F \preceq H$. If the solution to the* Route(N, F, S) *exists and*

$$T_{[m]_1^0} < [(m-1)n - m]w_{min}^{[m]} \tag{9}$$

then also the solution to the Route(N, H, S) *exists. The Nash probabilities for the* Route(N, H, S) *are those given in (8).*

Proof. If a solution to the *Route(N, F, S)* exists, then the relation (7) hold and, obviously, $P = \{P_i^l\}$ is a set of strategies. From this, it remains only to prove that this set of strategies gives a Nash equilibrium. Since for all the agents $i \in [n]$ such that $H_i^i = 0$ or $i \in [\beta + 1, n]$, $P_i^j > 0$ for all $j \in [m]$ such that $H_i^i = 1$ let us concentrate on the agents i such that $F_i^i = 0$ and $H_i^i = 1$. For these agents it results:

$$c_i^i = \sum_{k>m} \frac{[(m-1)n - (2m-1)]w_k + (m-1)w_i - \sum_{s \le m, s \ne i} w_s}{m[(m-1)n - (2m-1)]} + w_i +$$

$$+ \sum_{k \le m, k \ne i} \frac{[(m-1)n - (2m-1)]w_k + (m-2)w_i - \sum_{s \le m, s \ne k,i} w_s}{(m-1)[(m-1)n - (2m-1)]}$$

$$c_i^l = \sum_{k>m} \frac{[(m-1)n - (2m-1)]w_k + (m-1)w_l - \sum_{s \le m, s \ne l} w_s}{m[(m-1)n - (2m-1)]} + w_i +$$

$$+ \sum_{k \le m, k \ne i,l} \frac{[(m-1)n - (2m-1)]w_k + (m-2)w_l - \sum_{s \le m, s \ne k,l} w_s}{(m-1)[(m-1)n - (2m-1)]}$$

from which $c_i^l - c_i^i = -\frac{[(m-1)n - m]w_i - \sum_{k \le m, k \ne i} w_k}{(m-1)[(m-1)n - (2m-1)]} < 0$ as the (9) holds, therefore the set of strategies P give a Nash equilibrium for the *Route(N, H, S)* , which completes the proof of theorem. □

3.3 (β, β)-Sets of Supports, $\beta < m$

Each agent i is forbidden to route the traffic on link i, $i \in [\alpha]$ and $\alpha = \beta < m$.

Theorem 5. *Consider a* Route(N, F, S) *problem such that S is a (β, β)-set of supports with $m > \beta$ and $n > m > 2$ and $F = S$. A solution to* Route(N, F, S) *exists if and only if, said*

$$A = n - 2 \quad B = (m-1)n - (2m - \beta) \quad C = (m-2)n - (2m - (\beta + 2))$$
$$D = n - 1 \quad E = (m-1)n - (2m - 1) \quad F = (m-1)n - (2m - (\beta + 1))$$

the following inequalities hold $\forall\ k = 1,\dots,\beta$

$$0 < \frac{(AB-1)w_k+Cw_l-D*T_{[\beta]-\{k,l\}}}{EFw_k} < 1 \qquad \forall\ l \in [\beta]/\{k\}$$
$$0 < \frac{EFw_i+[(m-1)B+(\beta-1)]w_l-(m-1)D*T_{[\beta]-\{l\}}}{mEFw_i} < 1\ \forall\ i \in [\beta+1,n]\ and\ l \in [\beta]} \tag{10}$$

and the Nash probabilities are:

$$P_i^l = \begin{cases} 0 & \forall\ i \in [\beta],\ l = i \\ \frac{(AB-1)w_i+Cw_l-D*T_{[\beta]/\{i,l\}}}{EFw_i} & \forall\ i \in [\beta],\ l \in [\beta]/\{i\} \\ \frac{ABw_i-A*T_{[\beta]/\{i\}}}{EFw_i} & \forall\ i \in [\beta],\ l \in [\beta+1,m] \\ \frac{EFw_i+[(m-1)B+(\beta-1)]w_l-(m-1)D*T_{[\beta]/\{l\}}}{mEFw_i} & \forall\ i \in [\beta+1,n],\ l \in [\beta] \\ \frac{Fw_i-T_{[\beta]}}{mFw_i} & \forall\ i \in [\beta+1,n],\ l \in [\beta+1,m] \end{cases} \tag{11}$$

Proof. As in the previous theorems, the system (4) has the values of (11) as solutions. Now, for the necessary condition

$$\begin{cases} 0 < P_i^l < 1 \ \forall\ i \in [\beta],\ l \in [\beta]/\{i\} \\ 0 < P_i^l < 1 \ \forall\ i \in [\beta+1,n],\ l \in [\beta] \end{cases}$$

must hold, that is

$$\begin{cases} 0 < \frac{(AB-1)w_i+Cw_l-D*T_{[\beta]/\{i,l\}}}{EFw_i} < 1 & \forall\ i \in [\beta],\ l \in [\beta]/\{i\} \\ 0 < \frac{EFw_i+[(m-1)B+(\beta-1)]w_l-(m-1)D*T_{[\beta]/\{l\}}}{mEFw_i} < 1\ \forall\ i \in [\beta+1,n],\ l \in [\beta] \end{cases}$$

The sufficient condition is trivial. In fact, it is easy to see that if relations (10) are verified, then $0 < P_i^l < 1$ for all $i \in [n]$ and $l \in [m]/\{i\}$ and, therefore, the P_i^l are a set of strategies. From this, as the P_i^l are solutions of the system (4), they are a Nash equilibrium. $\qquad\square$

Removing some constraints we can get the following theorem.

Theorem 6. *Consider a* Route(N, F, S) *problem such that S is an* (β, β)-*set of supports with* $m > \beta$ *and* $n > m > 2$ *and* $F = S$, *and a* Route(N, H, S) *such that* $F \preceq H$. *If the solution to the* Route(N, F, S) *exists and*

$$w_{max}^{[\beta]} < \frac{m\beta-6mn-3+4m-mn\beta-2n^2+5n-\beta+2mn^2-n\beta}{mn\beta-4m^2+2m-3n+3\beta-\beta^2+mn^2-3mn+2m\beta+n\beta-m^2n^2+4m^2n}T_{[\beta]_0^1} \tag{12}$$

then also the solution to the Route(N, H, S) *exists. The Nash probabilities for the* Route(N, H, S) *are those given in (11).*

Proof. If a solution to the *Route(N, F, S)* exists, then the relation (10) hold and, obviously, $P = \{P_i^l\}$ is a set of strategies. From this, it remains only to prove that this set of strategies gives a Nash equilibrium. Since for all the agents $i \in [n]$ such that $H_i^i = 0$ or $i \in [\beta+1,n]$, $P_i^j > 0$ for all $j \in [m]$ such that $H_i^i = 1$ let us concentrate on the agents i such that $F_i^i = 0$ and $H_i^i = 1$. For these agents it results $c_i^l - c_i^i = (mn\beta - 4m^2 + 2m - 3n + 3\beta - \beta^2 + mn^2 - 3mn + 2m\beta + n\beta - m^2n^24m^2n)w_i+(-m\beta+6mn+3-4m+mn\beta+2n^2-5n+\beta-2mn^2-n\beta)T_{[\beta]/\{i\}} < 0$ as the (12) holds, therefore the set of strategies P gives a Nash equilibrium for the *Route(N, H, S)* , which completes the proof of theorem. $\qquad\square$

4   Conclusions

The framework studied in this paper can also be used to handle centralized networks in which a coordinator wishes to force the system to a particular equilibrium point. For this reason, he fix a particular set of supports (compatible with a set of constraints) to decide which links can be used by each agent. Once the system has reached the equilibrium, he could then also remove the constraints, thus freeing the links and leaving the system still in equilibrium.

Acknowledgments

We would like to thank Enzo Auletta and Pino Persiano for many useful comments and discussions on a preliminary version of the paper.

References

1. B. Awerbuch, Y. Azar, Y. Richter, D. Tsur, Trade-offs in worst-case equilibria, In proceeding of WAOA '03.
2. A. Czumaj, P. Krysta, B. Vocking, Selfish Traffic Allocation for Server Farms, In proceeding of 34th Annual ACM STOC '02, 287-296.
3. A. Czumaj, B. Vocking, Tight Bounds for Worst-Case Equilibria, In proceedings of the 13th annual ACM-SIAM SODA '02, 413-420.
4. E. Evel-Dar, A. Kesselman, Y. Mansur, Convergence time to Nash Equilibria, In proceeding of ICALP '03.
5. R. Feldmann, M. Gairing, T. Luecking, B. Monien, M. Rode, Nashification and the coordination ratio for a selfish routing game, In proceeding of ICALP'03, 514–526.
6. D. Fotakis, S. Kontogiannis, E. Koutsoupias, M. Mavronicolas, P. Spirakis, The Structure and Complexity of Nash Equilibria for a Selfish Routing Game, in proceedings of the 29th ICALP '02, 123-134, L.N.C.S. Vol. 2380.
7. M. Gairing, T. Luecking, M. Mavronicolas, B. Monien, Computing Nash Equilibria for Scheduling on Restricted Parallel Links, In proceeding of STOC '04.
8. Y. Korilis, A. Lazar, On the existence of equilibria in noncooperative optimal flow control, Journal of the ACM, Vol. 42, No. 3, 584-613, 1995.
9. Y. Korilis, A. Lazar, A. Orda, Architecting noncooperative network. IEEE Journal on selected Areas in Communications, Vol. 13 (7), 1241-1251, 1995.
10. E. Koutsoupias, M. Mavronicolas, P. Spirakis, Approximate Equilibria and Ball Fusion. In proceeding of TOCS '03, 683-693.
11. E. Koutsoupias, C.H. Papadimitriou, Worst-Case Equilibria, proceedings of the 16th Annual STACS '99, 404-413, L.N.C.S. Vol. 1563.
12. R.J. Lipton, E. Markakis, Nash Equilibria via Polynomial Equations, In proceeding of Latin '04.
13. M. Mavronicolas, P. Spirakis, The Price of Selfish Routing, proceedings of the 33th Annual STOC '01, 510-519.
14. J. Nash, Non-Cooperative Games, The Annals of Mathematics, Second Series Vol. 54 (2), 286-295, 1951.
15. M. J. Osborne, A. Rubinstein, A course in game theory, MIT Press, 1994.
16. C.H. Papadimitriou, Algorithms, games and the Internet. In proceedings 33th Annual ACM STOC '01, 749-753.
17. T. Roughgarden, É. Tardos, How bad is selfish routing?, proceedings of the 41st Annual IEEE Symposium on FOCS '00, 428-437.

Mobile Agents Rendezvous When Tokens Fail

Paola Flocchini[1], Evangelos Kranakis[2], Danny Krizanc[3], Flaminia L. Luccio[4],
Nicola Santoro[2], and Cindy Sawchuk[2]

[1] SITE, University of Ottawa, Ottawa, ON, Canada
flocchin@site.uottawa.ca
[2] School of Computer Science, Carleton University, Ottawa, ON, Canada
{kranakis,santoro,sawchuk}@scs.carleton.ca
[3] Department of Mathematics and Computer Science, Wesleyan University,
Middletown, Connecticut, USA
dkrizanc@wesleyan.edu
[4] Dipartimento di Scienze Matematiche, Università di Trieste, Trieste, Italy
luccio@dsm.univ.trieste.it

Abstract. The mobile agent rendezvous problem consists of $k \geq 2$ mobile agents trying to rendezvous or meet in a minimum amount of time on an n node ring network. Tokens and markers have been used successfully to achieve rendezvous when the problem is symmetric, e.g., the network is an anonymous ring and the mobile agents are identical and run the same deterministic algorithm. In this paper, we explore how token failure affects the time required for mobile agent rendezvous under symmetric conditions with different types of knowledge. Our results suggest that knowledge of n is better than knowledge of k in terms of achieving rendezvous as quickly as possible in the faulty token setting.

1 Introduction

1.1 The Problem

The general *rendezvous search* problem includes many situations that arise in the real world, e.g., searching for or regrouping animals, people, equipment, and vehicles. It has been extensively studied (see [1] for a detailed review), with operations research specialists producing the bulk of this research. In these investigations, the search domain is usually assumed to be *continuous*, and *symmetry* occurs largely because locations in the search domain are indistinguishable. The searchers break the symmetry by running the *same randomized algorithm* or by running *different deterministic algorithms*. In the latter case, for example, one searcher remains stationary while another searches the domain. Even when the searchers in a general rendezvous problem marked their starting points, as prescribed by Baston and Gal [3], the case where searchers ran the *same deterministic algorithm* was largely unexplored.

Computer scientists have been interested in the rendezvous search problem in the context of mobile agents and networks. After being dispersed in a network, mobile agents may need to gather or rendezvous in order to share information,

R. Králović and O. Sýkora (Eds.): SIROCCO 2004, LNCS 3104, pp. 161–172, 2004.
© Springer-Verlag Berlin Heidelberg 2004

receive new instructions, or download new programs that extend the mobile agents' capabilities. The *mobile agent rendezvous* problem differs from the general rendezvous problem in significant ways. Network topologies are *discrete* and include highly symmetric structures like rings, tori, and hypercubes. In addition, memory and time requirements are important in computer science problems but in the general rendezvous problem research the *memory* requirements are usually ignored.

In the mobile agent rendezvous problem, *symmetry* can occur at several levels (topological structure, nodes, agents, algorithm), each playing a key role in difficulty of the problem and of its resolution. The standard symmetric situation is indeed a *ring* network of n identical nodes in which k identical agents running the same deterministic protocol must rendezvous.

If no other computational power is given to the agents, rendezvous becomes impossible to achieve by deterministic means (eg. see [9]), and thus the problem is in general deterministically *unsolvable*.

The research has thus focused on what additional powers the agents need to perform the task. The aim is to identify the "weakest" possible condition. Some investigators have considered empowering the agents with unique *identifiers*; this is for example the case of of Yu and Yung [9], and of Dessmark, Fraigniaud, and Pelc [4] who also assume unbounded memory; note that having different identities allows each agent to execute different algorithms. Other researchers have empowered the agents with the ability to leave notes in each node they travel, i.e., *read/write* capacity; this is the model of Barrière *et al* [2], and of Dobrev *et al* [5].

In the third approach, originally developed by Baston and Gal [3] for the general rendezvous problem, the agents are given something much less powerful than distinct identities or than the ability to write in every node. Specifically, each agent has available a marker, called *token*, placed in its home base, visible to any agent passing by that node; the tokens are however all *identical*. Since the tokens are stationary, the original intertoken distances are preserved while the agents move and, with enough memory, could be used to break the symmetry of the setting. This is the approach taken by Kranakis *et al* [6] and Flocchini *et al* [7], and is the one we follow in this paper.

In the case of $k = 2$ mobile agents, Kranakis *et al* [6] proved that solving the rendezvous problem in an n node anonymous ring requires $\theta(\log \log n)$ memory; several algorithms were presented to demonstrate that when more memory is available, rendezvous time can be reduced.

Flocchini *et al* [7] studied the use of tokens to break symmetry when $k \geq 2$ and proved that when the agents run the same deterministic algorithm, rendezvous is impossible if the sequence of the original intertoken distances is periodic. They also presented several algorithms, with various memory and time requirements, that solve the mobile agent rendezvous problem for two or more mobile agents. Finally, they proved that a solution to the leader election in the $k \geq 2$ mobile agent case implies a solution to the mobile agent rendezvous problem, but the reverse is not true.

The tokens used in the research of Kranakis *et al* [6] and Flocchini *et al* [7] were not subject to failure. Once placed on a node, a token was always visible to any mobile agent on the same node.

In this paper, we are concerned with *fault tolerance*; we investigate how the mobile agent rendezvous problem changes when *tokens can fail*. Such failures may be the result of hostile nodes or malicious mobile agents and may occur either as soon as the token is placed by the mobile agent or at any time later.

1.2   Our Contribution

In this paper, we study the rendezvous of identical agents in an anonymous ring when tokens may fail. We are concerned with the failure of a token which will make it no longer visible by the agents, and investigate how time required to rendezvous changes under symmetric conditions with different knowledge available to the agents, in particular, knowledge of the number of agents and the size of the ring.

We consider two types of failures. We consider first the case when tokens can fail and thus disappear *upon their release*. We then consider the more complex situations arising when the tokens can fail *anytime*.

Some facts can be determined quite easily. If both the number k of agents and the size n of the ring are unknown, the problem is unsolvable [7]. If all tokens fail, clearly no rendezvous is possible. In this paper we assume that either n or k is known to all agents, that the number f of failed token is $f < k$. Furthermore, we assume that $\gcd(k', n) = 1$ for all $k' \leq k$. Under these conditions we show that, rendezvous is possible regardless of the number of failures and of when they occur. It follows from [7] that if this condition does not hold, there will always be initial conditions under which rendezvous is not possible.

The solutions and their time complexity are clearly different depending on the types of faults, as well as on whether the agents know the size n of the ring or their number k. The time required by our solution protocols is summarized in Table 1. When tokens are reliable and never fail, the existing algorithm [7] ensures rendezvous in $O(n)$ time when either k or n is known. We note that all of these algorithms require $O(k \log n)$ memory per mobile agent.

Table 1. The Cost of Token Failure

TokensFail	Knowledge	Time	Algorithm
never	n or k	$O(n)$	[7]
upon release	n	$O(n)$	$Meet1$
	k	$O(k\,n)$	$Meet2$
anytime	n	$O(k\,n)$	$Meet3$
	k	$O(k^2\,n)$	$Meet4$

If the tokens fail upon release, knowledge of n allows the agents to overcome all the failures; in fact, rendezvous takes place within $O(n)$ time units.

If the agents know k, we propose an algorithm that uses $O(kn)$ time units to rendezvous. Interestingly, this bound holds also if the ring is *asynchronous*.

If tokens can fail anytime, but the mobile agents know n, we show that the time required for rendezvous only increases by a factor of k over that required when no tokens fail. On the other hand, if the agents know only k, we describe an algorithm that uses $O(k^2n)$ time units to rendezvous. Both of these algorithms work in the synchronous setting.

For both types of failures, our results suggest that knowledge of n yields a time saving over knowledge of just k. We note that in the fault-free environment this does not hold, in that it is always possible to compute n from k, or vice versa, in $O(n)$ time.

2   Definitions and Basic Properties

We consider a collection of $k \geq 2$ identical mobile *agents* scattered on a synchronous ring of n identical nodes. The ring is *unoriented*, i.e., there is no globally consistent indication of "left", and the agents have no globally consistent notion of "clockwise". Each mobile agent owns a single, identical, stationary *token* that is comprised of one bit. A given node requires only enough memory to host a token and, at most, k mobile agents.

Each agent is initially in a different node (its *home base*) whose location is unknown to the others. The agents follow the same deterministic algorithm and begin execution at the same time. The task for the agents is to *rendezvous*, i.e., to meet in the same node in finite time; the goal is to do so in as little time as possible. The problem is to design an algorithm, the same for all agents, that will allow them to perform the task. An agent releases its token in the first step of any rendezvous algorithm; when an agent transits on a node with a token, it can see it. Moreover, agents can see each other when they are in the same node.

We call *intertoken distances* the sequence of distances between the fault-free tokens, once they have been released. Since the tokens are stationary, the original intertoken distances are maintained unless a token fails.

When a token fails it is no longer visible to any agent passing by the node. We will consider two types of failures: when tokens can fail and thus disappear *upon their release*, and when the tokens can fail *anytime*. Let f be the number of tokens that fail.

Some limiting facts can be determined quite easily.

Fact 1 *The problem is unsolvable:*
1. If both the number k of agents and the size n of the ring are unknown.
2. If all tokens fail; i.e., $f = k$.

Proof. Part (1) is due to [7]. Part (2) is well-known.   □

We assume that the agents have knowledge of either k or n and that the number f of failed tokens is less than k. We also assume that that $\gcd(k', n) = 1$ for all $k' \leq k$. It follows from [7] that if this condition does not hold, there will exist initial configurations under which rendezvous will not be possible.

3   Rendezvous When Tokens Fail Upon Release

First, we assume a token can fail only upon release, i.e., in the first step of a given algorithm. The agent that released the token is unaware that it failed. If neither n, the size of the ring, nor k, the number of agents, are known to the agent the problem is unsolvable [7]; thus, in the following we will consider two cases: when only n is known and when only k is known.

3.1 n Known, k Unknown

The algorithm is rather simple. Each agent releases its token, and moves along the ring computing the distances between the successive tokens it encounters, and returns to its homebase. Let $s_1, \ldots, s_h$ be the sequence it computed, where s_1 and s_h are the distances between the homebase and the first encountered token, and between the last token and the homebase, respectively. If its own token did not fail, $h = k - f$ and this sequence is exactly the (circular) sequence $S = d_1, \ldots, d_h$ of the intertoken distances of the tokens that did not fail. If however the agent finds that its token has failed, then $h = k - f + 1$ and the agent must adjust the sequence to determine the actual (circular) sequence of intertoken distances: $S = d_1, \ldots, d_{h-1} = s_1 + s_h, s_2, \ldots, s_{h-1}$.

Since the failures happen before the agents start to move around, the circular sequence computed in this way is the same for all agents within a rotation (due to the fact that the ring is not oriented).

The agents will then agree on a meeting node by exploiting the asymmetry of the sequence, which cannot be periodic since $\gcd(k', n) \neq 1$ for all $k' \leq k$.

Consider two sequences of intertoken distances, $A = \{i_1, \ldots, i_n\}$ and $B = \{j_1, \ldots, j_n\}$, in a given ring with n nodes. The two sequences are lexicographically equal if $i_m = j_m$ for m, where $1 \leq m \leq n$. Sequence A is lexicographically greater than B if $i_m > j_m$ for some m where $1 \leq m \leq n$ and for all $q < m$, $i_q = j_q$. Let R_A denote the rotations of the sequence A. The lexicographically maximum rotation of A is the sequence L_A in R_A such that no sequence in R_A is lexicographically larger than L_A.

Each mobile agent calculates a sequence of intertoken distances S. Let S^R denote the reverse of S. Each mobile agent will compute the lexicographically maximum sequence in both S and S^R. If both strings start at the same node, that will become the meeting point. Otherwise, the meeting point will be the midpoint in the odd path between the starting nodes of the two strings.

The routine to identify the meeting point, which will be used in all our algorithms, is described in Table 2, where $lexi(someSequence)$ denotes the lexicographically maximum rotation of $someSequence$. The overall algorithm is described in Table 3. The proof of the following theorem is straightforward.

Theorem 1. *When the agents have $O(k \log n)$ memory, n is known, $\gcd(k', n)$ $= 1$ holds for all $k' \leq k$, $f \leq (k - 1)$ tokens fail, and tokens can fail only upon release, then the mobile agent rendezvous problem can be solved in less than $2n$ time units.*

Table 2. Calculation of the meeting point

Meetingpoint(string)
1. Compute $forward = lexi(string)$ and $reverse = lexi(string^R)$.
2. Let x and y denote the nodes at the beginning of $forward$ and $reverse$ respectively.
3. If $x = y$, then x is the rendezvous point.
4. If $x \neq y$, then the rendezvous point is the midpoint of the odd path
   between x and y (such a path must exist because n is odd).

Table 3. Failure at the beginning, n known

Let $m = k - f$ be the number of remaining tokens.
Algorithm MEET1
1. Release the token at the starting node.
2. Choose a direction and walk once around the ring.
3. Compute the sequence of intertoken distances $s_1, \ldots, s_h$.
   If a token is present at this node: $\mathcal{S} = d_1, \ldots, d_m = s_1, \ldots, s_h$.
   If no token is present at this node: $\mathcal{S} = d_1, \ldots, d_m = s_1 + s_h, s_2, \ldots, s_{h-1}$.
4. Move to $Meetingpoint(\mathcal{S})$.

Notice that, for this algorithm to work, the system does not need to be synchronous; in fact, each agent could independently make its calculations and move to the rendezvous point even if the ring is *asynchronous*.

3.2 k Known, n Unknown

More interesting is the case when n is unknown. As in the previous case, the agents will walk around the ring long enough to be able to calculate a sequence of intertoken distances; this sequence must be such that it covers the entire ring and contains an asymmetry that can be exploited by the agents to agree on a meeting point.

Unlike the previous case, however, since n is not known, the circular sequences computed by the agents using solely knowledge of k, are not necessarily identical. In fact, each agent, to construct its sequence, will fully traverse the ring the same (unknown) number of times but, depending on the starting point and on the number of failures, it will also traverse a portion of the ring which is in general different for different agents.

Although the sequences are not the same, we will show that the agents can still identify an unique point where to meet by calculating $3k$ intertoken distances. The Algorithm Meet2 is described in Table 4.

Let $\mathcal{S}^R$ denote the reverse of $\mathcal{S}$. Since the tokens fail only upon release, $\mathcal{S}^R$ can be partitioned as follows:

$$\mathcal{S}^R = Q^q + d_\gamma, \ldots, d_1 \tag{1}$$

where Q^q is the concatenation of q copies of a unique aperiodic subsequence Q, $+$ is the concatenation operator, and $d_\gamma, \ldots, d_1$ is a subsequence such that $\gamma < |Q|$. Upon identifying the subsequence Q, the agents can identify a unique node upon which to rendezvous.

Table 4. Failure at the beginning, k known

Algorithm MEET2
1. Release the token at the starting node.
2. Choose a direction and start walking.
3. Compute the sequence of $3k$ intertoken distances i.e., $S = d_1, d_2, \ldots, d_{3k}$.
4. Let S^R be the reverse of S.
5. Find the shortest aperiodic subsequence Q that starts with the first element of S^R and is repeated such that $S^R = Q^q + d_\gamma, \ldots, d_1$ where $\gamma < |Q|$.
6. Move to $Meetingpoint(Q)$

Theorem 2. *When the agents have $O(k \log n)$ memory, k is known, $\gcd(k', n) = 1$ holds for all $k' \leq k$, $f \leq (k-1)$ tokens fail, and tokens can fail only upon release, then the mobile agent rendezvous problem can be solved in $O(kn)$ time.*

Proof. Let $m = k - f$ be the number of tokens that do not fail. Let $A = \delta_1, \ldots, \delta_m$ be the sequence of the m intertoken distances that exist after the f tokens have failed; clearly $\sum_{i=1}^{m} \delta_i = n$. Let $S(a)$ be the sequence of $3k$ intertoken distances calculated by a given agent a in step 3 of Algorithm MEET2. Let $S^R(a)$ be the reverse of $S(a)$ and A^R be the reverse of A. For all agents, the $3k$ intertoken distances are of the form

$$S^R(a) = (A^R)^\rho + d_\gamma, \ldots, d_1 = (\delta_m, \ldots, \delta_1)^\rho + d_\gamma, \ldots, d_1. \tag{2}$$

where $(A^R)^\rho$ is the concatenation of ρ copies of the aperiodic subsequence A^R, $+$ is the concatenation operator, and $d_\gamma, \ldots, d_1$ is a subsequence such that $\gamma < m$. Thus, there exists at least one aperiodic subsequence, namely A^R, that satisfies equation 1. Note that A^R is aperiodic as it has k or fewer intertoken distances, and by assumption $\gcd(k', n) = 1$ for all $k' \leq k$.

If A^R is the shortest subsequence that satisfies equation 1, the agents discover A^R in step 5 of Algorithm MEET2. Otherwise, the agents discover a shorter aperiodic subsequence, Q, that satisfies equation 1.

The subsequence discovered in step 5 of Algorithm MEET2 is unique. If the shortest subsequence has z elements, these elements are the first z elements of S^R. Any other subsequence of the same length that satisfies equation 1 is also comprised of the first z elements of S^R and thus the subsequence discovered in step 5 is unique. This implies that all the agents identify the same rendezvous node in the remaining steps of Algorithm MEET2 and rendezvous occurs.

Calculating S, the sequence of $3k$ intertoken distances requires $O(k \log n)$ memory and requires $O(kn)$ time. Identifying the appropriate subsequence in step 5, determining the rendezvous node, and walking to the rendezvous node takes $O(kn)$ time, so the overall time requirement is $O(kn)$. This completes the proof of Theorem 2.   □

When the tokens only fail upon release, Algorithm MEET2 also solves the mobile agent rendezvous problem when the ring is *asynchronous*. Once an agent has calculated S^R, it can identify the smallest aperiodic sequence that satisfies equation 1. Since this sequence is unique, the agent can then identify the unique rendezvous node, walk there, and wait until all k agent arrive. The algorithm does not depend on timing but rather on the agents' ability to count and then identify the rendezvous node.

4 Rendezvous When Tokens Can Fail at Any Time

We now consider the situation when token failures can occur at any time. In this case, for the algorithms to work the system must be synchronous.

The idea is still to make use of the intertoken distances to agree on a meeting point. In this case, however, the problem is made much more complicated by the fact that these distances vary unpredictably during the algorithm. To cope with that, our algorithm works in rounds, in each round, the agents compute some intertoken distances and try to meet in a node. In a round, however, the rendezvous may fail because the agents might have computed different intertoken distances. In such a case, only groups of agents (the ones that have computed the same intertoken distances) might meet (in a "false" rendezvous point).

To make the algorithms work, the rounds must be synchronized so that the agents start them within some bounded time interval (if not simultaneously). In other words, an agent arriving at its meeting point will have to know how long to wait before declaring that point a false rendezvous point and correctly start the next round, or before realizing that such a point is a true meeting point.

4.1 n Known k Unknown

First of all notice that since k is unknown, a group of agents finding themselves on the same node cannot determine whether rendezvous has been accomplished simply by counting how many they are. A different strategy will have to be used.

A round is composed of three distinct steps. In *step*1 an agent travels around the ring to compute the intertoken distances; it then identifies the lexicographical maximum string. In *step*2 the agent moves to the computed meeting point (let us assume that it takes t time units to go there), and waits $n - t$ units to synchronize with the others for the next step. Notice that, in general, the agent cannot understand at this point if rendezvous is accomplished just by counting the other agents on the meeting node. As we will prove, if this is the third time that the same string has been calculated, the agent can be sure that this is a true meeting point. In this case, in fact, the agent knows that everybody else has seen this string at least in the previous and in the current rounds and is now in its own meeting point. *Step*3 is used for notifying all agents that indeed the meeting point is the correct one. The agents who know, go around the ring. The agents who do not know, wait for n units: if nothing happens, they go to the next round; if a notifying agent arrives, they terminate the algorithm.

The Algorithm is described in Table 5.

Table 5. Failure at any time, n known

Algorithm MEET3
1. Release token.
2. Set $S_1 = S_2 = S_3 = \emptyset$
3. Set $r = 0$, choose a direction and begin walking.
4. Travel for n time units computing the intertoken distances $S = (s_1, \ldots, s_h)$.
       If a token is present at the last node: $S = d_1, \ldots, d_m = s_1, \ldots, s_h$.
       If no token is present at the last node: $S = d_1, \ldots, d_m = s_1 + s_h, \ldots, s_{h-1}$.
5. Set $t = 0$.
6. Walk to $Meetingpoint(S)$ and increment t for each node traveled.
7. Wait $n - t$ clock ticks.
8. Set $S_1 = S_2;\ S_2 = S_3;\ S_3 = S$.
9. If $S_1 = S_2 = S_3$
10.   become(notifying), go around the ring, and then terminate
11. Else wait n units
12.   If, while waiting, a notifying agent arrived, terminate.
        /* Rendezvous has occurred. */
13.   Else, after waiting, set $r = r + 1$ and repeat from step 3.

Lemma 1. *When an agent sees the same string for the third time, all the other agents have seen it at least twice (in the previous and in the current round).*

Theorem 3. *When the agents have $O(k \log n)$ memory, know n, $\gcd(k', n) = 1$ holds for all $k' \leq k$, $f \leq (k - 1)$ tokens fail, and tokens can fail upon release or later, then the mobile agent rendezvous problem can be solved in $O(kn)$ time.*

Proof. By lemma 1 we know that when an agent sees the same string for the third time the meeting point was the true rendezvous point. All the agents met at that point and after n time units everybody will terminate. In each round, every agent has spent n time units to compute the intertoken distances, n additional time units to move to the meeting point and wait for the restart, n time units for the termination check. Thus, all agents are perfectly synchronized and start each round simultaneously. After, at most $k + 2$ rounds rendezvous is accomplished and the total time is then $O(kn)$. □

4.2 k Known n Unknown

Lack of knowledge of n makes the situation more complicated; the main problem is to achieve synchronization.

In the following algorithm, if more than one but fewer than k agents meet on a given node, they *merge* and act as one agent for the remainder of the algorithm. Merged agents will be perceived by other agents with their multiplicity; thus, if an agent meets a group of merged agents, it will "see" how many they are. In this algorithm, before starting the next round, every group that has met in a false meeting point, merge and behave like one agent.

Since the agents, not knowing n, cannot travel around the ring, they will travel trying to guess the ring size at each round. Initially they compute k intertoken distances and they guess n to be the sum of those distances (they use this guessed value for synchronization purposes); at each subsequent round they compute one less intertoken distance and they change their guess of n.

The algorithm is designed in such a way that 1) it is guaranteed that agents are always in at most one round apart, 2) after at most $k-1$ rounds rendezvous is accomplished. The Algorithm is described in Table 6.

Table 6. Failure at any time, k known

Algorithm MEET4
1. Release token.
2. Set $r = 0$, where r denotes a round of the algorithm.
3. Choose a direction and begin walking.
4. Upon meeting another agent, merge with it.
5. Calculate the first $k - r$ intertoken distances, i.e., $S = (d_1, \ldots, d_{k-r})$.
6. Estimate n as $\hat{n} = \sum_{i=1}^{k-r} d_i$.
7. If S is periodic, wait $2\hat{n}$ steps, set $r = r + 1$, and repeat from step 3.
8. Calculate S_{LMR}, the lexicographically maximum rotation of S.
9. Set $t = 0$.
10. Walk to the node that starts S_{LMR} and increment t for each node traveled.
11. Wait $2\hat{n}$ - t clock ticks.
12. If there are k agents or their merged equivalent on the current node, stop.
/* Rendezvous has occurred. */
13. Else if there are $1 < v < k$ agents on the current node, then merge.
14. Set $r = r + 1$ and repeat from step 3.

The following three lemmata are used in the proof of Theorem 4. Lemma 2 proves that the agents are always less than a round apart and thus an agent need only wait $\frac{3\hat{n}}{2}$ clock ticks for any agents that saw the same view.

Lemma 2. *Let a be the first agent to finish estimating n in round r, where $0 \leq r \leq k-1$, and let τ denote the time at which agent a finishes the estimation. All other agents will either finish round $r-1$ or merge with a on or before time τ.*

Lemma 3. *Mobile agents that see the same sequence S, up to a rotation, of intertoken distances are in the same round. If S is aperiodic, then these agents will rendezvous at the end the given round. However, if S is periodic, then the agents cannot rendezvous during this round and must proceed to the next round.*

Lemma 4. *If at most f tokens have failed, then the agents will not execute round $r = f + 1$ in Algorithm MEET4.*

Theorem 4. *When the agents have $O(k \log n)$ memory, know k $\gcd(k', n) = 1$ holds for all $k' \leq k$, $f \leq (k - 1)$ tokens fail, and tokens can fail upon release or later, then the mobile agent rendezvous problem can be solved in $O(k^2 n)$ time.*

Proof. Let $f \leq k - 1$ be the number of tokens that actually fail. Lemma 4 implies that no agent will execute more than $f + 1$ rounds of Algorithm MEET4. Suppose that rendezvous has not occurred by the end of round $r = f$. Let a^* denote the first agent that begins round $r = f + 1$ and let t_0 denote the time that a^* starts round $r = f + 1$. Since f tokens have failed, a^*'s estimate for n in round $r = f + 1$ will be correct, i.e., $\hat{n} = n$. The remaining agents see the same aperiodic sequence, up to a rotation, of intertoken distances as a^* and thus Lemma 3 implies that rendezvous occurs at the end of round f.

The number of tokens that fail, f, is at most $k - 1$ so, as mentioned above, at most k rounds of Algorithm MEET4 are executed. A round takes at most $k(n - 1)$ time, i.e., the product of the number of intertoken distances measured and the maximum intertoken distance possible. As a result, the time required is $O(k^2 n)$. Because at most k intertoken distances are calculated and the maximum intertoken distance is $n - 1$, the memory required is $O(k \log n)$. This completes the proof of Theorem 4. □

5   Conclusion

The effect of token failure on the time required to rendezvous suggests that it would be interesting to explore other sources of failure in the mobile agent rendezvous problem. For example, what are the implications for rendezvous when mobile agents fail? Mobile agent failure could be partial, such as not merging when appropriate, or absolute, such as not operating at all. It would also be interesting to determine the impact of network problems, such as heavy traffic, on mobile agent rendezvous.

Acknowledgements

This research is supported in part by NSERC (Natural Sciences and Engineering Research Council of Canada) and MITACS (Mathematics of Information Technology and Complex Systems) grants, and by an OGS (Ontario Graduate Scholarship).

References

1. S. Alpern and S. Gal. The Theory of Search Games and Rendezvous, Kluwer Academic Publishers, Norwell, Massachusetts, 2003.
2. L. Barrière, P. Flocchini, P. Fraigniaud, and N. Santoro. Election and rendezvous in fully anonymous systems with sense of direction. In 10th Symposium on Structural Information and Communication Complexity (SIROCCO '03), 2003.

3. V. Baston and S. Gal. Rendezvous search when marks are left at the starting points. Naval Research Logistics, 38, pp. 469-494, 1991.
4. A. Dessmark, P. Fraigniaud, and A. Pelc. Deterministic rendezvous in graphs. In European Symposium on Algorithms (ESA '03), pp. 184-195, 2003.
5. S. Dobrev, P. Flocchini, G. Prencipe, and N. Santoro. Multiple agents rendezvous in a ring in spite of a black hole. In Symposium on Principles of Distributed Systems (OPODIS '03), LNCS, 2003.
6. E. Kranakis, D. Krizanc, N. Santoro, and C. Sawchuk. Mobile agent rendezvous problem in the ring. In International Conference on Distributed Computing Systems (ICDCS '03), pp. 592-599, 2003.
7. P. Flocchini, E. Kranakis, D. Krizanc, N. Santoro, and C. Sawchuk. Multiple mobile agent rendezvous in the ring. In Latin American Conference of Theoretical Informatics (LATIN '04), 2004.
8. C. Sawchuk. Mobile Agent Rendezvous in the Ring. Ph.D Thesis, Carleton University, January 2004.
9. X. Yu and M. Yung. Agent rendezvous: A dynamic symmetry-breaking problem. In International Colloquium on Automata, Languages, and Programming (ICALP '96), LNCS 1099, pp. 610-621, 1996.

Time Efficient Gossiping in Known Radio Networks

Leszek Gąsieniec[1], Igor Potapov[1], and Qin Xin[1]

Department of Computer Science, University of Liverpool, Liverpool L69 7ZF, UK
{leszek,igor,qinxin}@csc.liv.ac.uk

Abstract. We study here the gossiping problem (all-to-all communication) in known radio networks, i.e., when all nodes are aware of the network topology. We start our presentation with a deterministic algorithm for the gossiping problem that works in at most n units of time in any radio network of size n. This is an optimal algorithm in the sense that there exist radio network topologies, such as: a line, a star and a complete graph in which the radio gossiping cannot be completed in less then n units of time. Furthermore, we show that there isn't any radio network topology in which the gossiping task can be solved in time $< \lfloor \log(n-1) \rfloor + 2$. We show also that this lower bound can be matched from above for a fraction of all possible integer values of n; and for all other values of n we propose a solution admitting gossiping in time $\lceil \log(n-1) \rceil + 2$. Finally we study asymptotically optimal $O(D)$-time gossiping (where D is a diameter of the network) in graphs with max-degree $\Delta = O(\frac{D^{1-1/(i+1)}}{\log^i n})$, for any integer constant $i \geq 0$ and D large enough.

1 Introduction

The two classical problems of information dissemination in computer networks are the *broadcasting* problem and the *gossiping* problem. In the broadcasting problem, we want to distribute a particular message from a distinguished *source* node to all other nodes in the network. In the gossiping problem, each node v in the network initially holds a message m_v, and we wish to distribute all messages m_v to all nodes in the network. In both problems, we very often like to minimise the time needed to complete the tasks.

A radio network can be modelled as an undirected graph $G = (V, E)$, where V represents the set of nodes of the network and a set E contains (unordered) pairs of vertices, s.t., a pair $(v, w) \in E$, for some $v, w \in V$ iff nodes v and w can communicate directly. We say that all neighbours of a vertex $v \in V$ form the *range* of v. One of the radio network properties is that at any time step each processor is either in the transmitting mode or in the receiving mode. We also assume that all processors work synchronously, and if a processor u transmits a message m at time step t, the message reaches all nodes within its range at the same time step. However any of its neighbours v will receive message m successfully iff u is the only node that transmits to v (has v in its range) at time step t. Otherwise a *collision* occurs and the message m does not reach the node v.

In this paper we focus on gossiping algorithms that use the entire information about the network topology. Such topology-wise communication algorithms are useful in radio networks that have relatively stable topology/infrastructure. And as long as no

R. Králović and O. Sýkora (Eds.): SIROCCO 2004, LNCS 3104, pp. 173–184, 2004.
© Springer-Verlag Berlin Heidelberg 2004

changes occur in the network topology during the actual execution of the algorithm, the task of gossiping is completed successfully. Another interesting aspect of deterministic communication in known radio networks is its close relation to randomised communication in unknown radio networks.

1.1   Previous Work

Most of the work devoted to radio networks is focused on the broadcasting problem. In the model with known radio network topology Gaber and Mansour [8] showed that the broadcasting task can be completed in time $O(D + \log^5 n)$, where D is the diameter of the network. In other work of Diks *et al.* [7] we find efficient radio broadcasting algorithms for (various) particular types of network topologies. However in the general case, it is known that the computation of an optimal (radio) broadcast schedule for an arbitrary network is $NP - hard$, even if the underlying graph of connections is embedded in the plane [2, 12]. The gossiping problem was not studied in the context of known radio networks until very recent work of Gąsieniec and Potapov [10]. In that paper one can find a study on the gossiping problem in known radio networks, where each node transmission is limited to unit messages. In this model they proposed several optimal and almost optimal $O(n)$-time gossiping algorithms in various standard network topologies, including: lines, rings, stars and free trees. They also proved that there exists a radio network topology in which the gossiping (with unit messages) requires $\Omega(n \log n)$ time. A similar work, in model with messages of logarithmic size, can be found in [1], where Bar-Yehuda *et al.* study randomised multiple point-to-point radio communication, and in [5], where Clementi *et al.* consider simultaneous execution of multiple radio broadcast procedures. So far, the gossiping problem was mostly studied in the context of *ad-hoc* radio networks, where the topology of connections is unknown in advance. In this model, Chrobak *et al.* [4] proposed fully distributed deterministic algorithm that completes the gossiping task in time $O(n^{3/2} \log^3 n)$. For small values of diameter D, the gossiping time was later improved by Gąsieniec and Lingas [9] to $O(nD^{1/2} \log^3 n)$. Another interesting $O(n^{3/2})$-time algorithm, a tuned version of the gossiping algorithm from [4] can be found in [14]. A study on deterministic gossiping in unknown radio networks with messages of limited size can be found in [3]. The gossiping problem in *ad-hoc* radio networks attracted also studies on efficient randomised algorithms. In [4], Chrobak et al. proposed a $O(n \log^4 n)$ time gossiping procedure. This time was later reduced in [11] to $O(n \log^3 n)$, and very recently in [6] to $O(n \log^2 n)$.

1.2   Our Results

We start here with a proof that in any known radio network of size n, the gossiping task can be completed in at most n units of time. Please note here that we are interested in exact complexity (as oppose to the asymptotic complexity) since the design of an $O(n)$-time gossiping procedure is rather trivial. Our new algorithm is optimal in the sense that there exist radio network topologies including: lines, stars, but also complete graphs, in which the gossiping task cannot be completed in time $< n$. Further, we show that there isn't any radio network topology in which the gossiping task can be solved in time $<$

$\lfloor \log(n-1) \rfloor + 2$. We also show that this bound can be matched from above for a fraction of all possible integer values of n; and for all other values of n we propose a solution admitting gossiping in time $\lceil \log(n-1) \rceil + 2$. The second part of the paper is devoted to efficient gossiping in arbitrary graphs (as oppose to the worst case) graphs. This work is done in relation to [8], where the authors proposed an asymptotically optimal $O(D + \log^5 n)$-time broadcasting procedure in known radio networks. For obvious reasons, the diameter D is a natural lower bound for both the broadcasting and gossiping problems. In this paper we study a non-trivial class of graphs in which the gossiping can be done in time $O(D)$, i.e., in optimal asymptotic time. We first show that the gossiping can be performed in time $(2D - 1)\Delta + 1$ in graphs with the maximal degree Δ. This result admits $O(D)$-time gossiping in all graphs with $D = O(1)$. Later we show how to perform gossiping in time $O(D)$ in any graph with $\Delta = O(\frac{D^{1-1/(i+1)}}{\log^i n})$, for any integer constant $i \geq 0$ and D large enough.

2   The Gossiping Algorithm in Time $\leq n$

Let the graph $G = (V, E)$ be the underlying graph of connections in a given radio network, where $n = |V|$ stands for the size (number of nodes) of the network. We assume that initially each node $v \in V$ holds a unique message m_v. The gossiping task (all-to-all communication) is performed in discrete *units of time*, also called *time steps*. At any unit of time, a node can be in one of the two transmission modes: either in the *receiving mode* or in the *transmitting mode*. A gossiping algorithm is understood to be a well defined transmission/reception procedure, for each node of the network. We assume here that during a single time step a transmitting node is allowed to send a combined message which includes all messages that it received so far. The running time of a gossiping algorithm is the smallest number of time steps in which the all-to-all communication is completed.

We say that a radio network has a radius k if there exists at least one node in V which is at distance $\leq k$ from all other nodes in the network, and there isn't any node in V at distance $< k$ from all other nodes in the network. We call this special node a *central node* and we denote it by c. Let m_v be a gossip message originated in a node $v \in V$. At any time step of the gossiping process M_v denotes the set of messages acquired by v until now. E.g., initially $M_v = \{m_v\}$. We also distinguish a set T of nodes that have not transmitted (at all) since the beginning of the gossiping process. Initially $T = V$. In a radio network with a radius k and a central node c, we can partition the set of nodes T into disjoint subsets $N_0, N_1, .., N_k$, such that the set $N_i = \{v : dist_G(v, c) = i\}$, for $i = 0, 1, 2, ..., k$, where $dist_G(u, v)$ stands for a length of a shortest path between nodes u and v in G. In other words, N_0 contains only the central node c, N_1 contains all neighbours of the central node, etc. A reasoning presented here is based on the notion of a *minimal covering set*. The minimal (in terms of inclusion) covering set C_i is a subset of N_i, s.t., every node in N_{i+1} is connected to some node in C_i, and a removal of any node from C_i destroy this property.

Observation 1 *Each node $v \in C_i$ is connected to some node $u \in N_{i+1}$, such that u is not connected to any other node in $C_i - \{v\}$.*

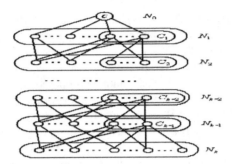

Fig. 1. Radio network with central node c and radius k

2.1   The Outline of the Algorithm

The general idea of the gossiping algorithm is as follows. Initially we point out that in any radio network with radius 1 and size n the gossiping task can be completed in time n. In radio networks with larger radii we show that there always exist four distinct nodes v, v', w and w', s.t., v can transmit (its current content) M_v to v' and w can transmit M_w to w' in the same time step, and that the removal of both v and w does not disconnect the remaining part of the network. We call this approach a *2-vertex reduction*. The gossiping algorithm uses the new approach for the purpose of moving all n messages to some connected subnetwork (of the original network) with radius 1 and size $n - 2k$, in exactly k units of time. This is followed by the gossiping in the subnetwork in at most $n - 2k$ units of time. Finally, we use the last k units of time to distribute the messages to all other nodes in the network by reversing the transmission direction in the initial k units of time. This means that the gossiping in any radio network of size n can be completed in time $\leq n$.

2.2   Gossiping in Radio Networks with Radius 1

We show here that in any radio network topology with radius 1 the gossiping algorithm can be completed in time n, where n is the size of the network. W.l.o.g., we assume that $n \geq 2$. According to the definition of a radio network with radius 1, we know that there exists a central node c which is at distance ≤ 1 from any other node in the network. In this case, we can transmit all messages m_v, s.t., $v \in V - \{c\}$ to the central node c, one by one, in time $n - 1$. When this stage is completed, $M_c = \{m_v : v \in V\}$. And we need one more time step to disseminate the content of M_c to all other nodes in the radio network.

2.3   Gossiping in Radio Networks with Higher Radii

We show here an efficient reduction of a network with an arbitrary radius k to its connected subnetwork with radius 1. This reduction is based on the concept of 2-vertex reduction, introduced at the beginning of this section.

Reduction 2→1. We prove here that any radio network with radius 2 can be efficiently reduced (by a sequence of 2-vertex reductions) to its subnetwork with radius 1. Given a radio network with radius 2. Initially, we choose a central node c and we construct sets N_0, N_1, N_2 and a covering set C_1. As long as $|C_1| \geq 2$, we can transmit successfully the contents of at least two nodes $v, w \in N_2$ to some nodes $v', w' \in C_1$ at each time step, thanks to the covering property of C_1. Please note that after each transmission nodes v, w are "removed" temporarily from the network and the content of the covering set C_1 is recomputed. When eventually $|C_1| < 2$ two cases apply. When $|C_1| = 0$ the radius of the subnetwork becomes 1, and the reduction is completed. If $|C_1| = 1$, we call the node remaining in C_1 an *essential node* e_1. Let Y_1 be the set of nodes in N_1 that are not neighbours of e_1. If both $|Y_1|, |N_2| > 0$, we can always chose $v \in Y_1$ and $w \in N_2$, where $v' = c$ and $w' = e_1$. If eventually $|N_2| = 0$ also $|C_1| = 0$, the reduction is completed. Alternatively when $|Y_1| = 0$, the essential node e_1 becomes a new central node in a subnetwork with radius 1.

Reduction k+1→k. We prove here that any radio network with radius $k + 1$ can be efficiently reduced (by a sequence of 2-vertex reductions) to its subnetwork with radius k, for any $k \geq 2$.

In a radio network with radius $k+1$, we initially choose a central node c and we construct the layer sets $N_0, N_1, N_2, \ldots, N_k, N_{k+1}$ and the covering sets $C_1, C_2, C_3, \ldots, C_{k-1}, C_k$. Let $\tilde{C}_i$ be a set $N_i - C_i$, for all $i = 1, 2, 3, \ldots, k - 1, k$. Note, that as long as $|C_k| \geq 2$, we can transmit successfully the contents of at least two nodes $v, w \in N_{k+1}$ to some nodes $v', w' \in C_k$ during each time step, thanks to the covering property of C_k. When eventually $|C_k| < 2$ two cases apply. When $|C_k| = 0$, the radius of the subnetwork becomes k, and the reduction is completed. If $|C_k| = 1$, we call the node remaining in C_k an *essential node* e_k. By Y_k we denote the set of nodes in N_k that are not neighbours of e_k. If now $|\tilde{C}_1| + |\tilde{C}_2| + |\tilde{C}_3| +, \ldots, |\tilde{C}_{k-2}| + |\tilde{C}_{k-1}| + |Y_k| > 0$ and $|N_{k+1}| > 0$ we match any transmission from a node $v \in N_{k+1}$ to a node $v' \in C_k$ with a transmission from a node $w \in \tilde{C}_1 \cup \tilde{C}_2 \cup \tilde{C}_3 \cup, \ldots, \tilde{C}_{k-2} \cup \tilde{C}_{k-1} \cup Y_k$ to its neighbour w' in the layer closer to the central node. This reduction process will terminate when eventually either: the set N_{k+1} is empty, which means that the radius of the subnetwork has been reduced to k; or at some point $|\tilde{C}_1| + |\tilde{C}_2| + |\tilde{C}_3| +, \ldots, |\tilde{C}_{k-2}| + |\tilde{C}_{k-1}| + |Y_k| = 0$. But then we know that the essential node e_k is a neighbour of all nodes in the set $N_k - \{e_k\}$. In this case, the essential node e_k becomes a new central node since its distance from any other node in the network is $\leq k$. The following theorem follows:

Theorem 1. *The gossiping task can be solved in any radio network of size n in at most n units of time.*

3   Optimal Topology for Gossiping in Radio Networks

In this section we present a radio network topology in which radio gossiping can be performed in time $\lceil \log(n - 1) \rceil + 2$. We later present a simple argument that radio gossiping cannot be completed in time $< \lfloor \log(n - 1) \rfloor + 2$. We conclude this section with presentation of more complex topology that allows to perform radio gossiping in time $\lfloor \log(n - 1) \rfloor + 2$, for a fraction of all possible integer values of n.

3.1 Upper Bound

A topology of "gossiping-friendly" radio network supports the following transmission strategy. Initially collect (gather) all gossip messages in one distinguished node c and then distribute the messages to all other nodes as quickly as its is possible (i.e., in a single time unit). In order to achieve this goal we select a node (center) $c \in V$ and we connect it with any other node in V. The remaining $n - 1$ vertices in $V - \{c\}$ are organised in an *optimal broadcasting tree* (OBT) used in the matching model, see, e.g., [13], with a root r.

We point out here, that an OBT of size m rooted in node r can also serve the purpose of gathering messages (in radio network model) in root r in the optimal time $\lceil \log m \rceil$. This is done by reversing all transmissions in time. I.e., if during broadcasting (in matching model) node v transmits to node w in step $i = 1, .., \lceil \log m \rceil$, in gathering algorithm (in radio network) node w transmits all messages it collected so far in step $\lceil \log m \rceil - i + 1$. Using this observation we show that we can collect all messages from $V - \{c\}$ in root r in time $\lceil \log(n - 1) \rceil - i + 1$. These messages are later passed onto the central node c, who has all messages now. In the last step node c transmits a combined message (containing all messages) to all other nodes in one time step (since c is connected to every other node in the network). And the gossiping process is completed in time $\lceil \log(n - 1) \rceil + 2$.

Lemma 1. *There exists a radio network topology in which the gossiping task can be completed in time* $\lceil \log(n - 1) \rceil + 2$, *for any integer* n.

3.2 Lower Bound

On the other hand, note that during each consecutive round knowledge (a number of possessed messages) in each node can at most double. This means that after step i knowledge of any node is limited to 2^i original messages. Thus, after initial $\lfloor \log(n-1) \rfloor$ steps of any gossiping algorithm in any radio network topology, non of the nodes is completely informed, since $2^{\lfloor \log(n-1) \rfloor} < n$. Note also that during the last round of the gossiping process the only nodes that are permitted to transmit, are those who already possess all messages, since a transmitting node cannot receive messages at the same time. The following lemma follows.

Lemma 2. *The completion of the gossiping task in any radio network requires at least* $\lfloor \log(n - 1) \rfloor + 2$ *steps.*

3.3 Tightening the Gap

We have just presented both the upper and the lower bounds for the most suitable topology for radio gossiping. Please note that the upper bound and the lower bound meet each other when $n = 2^k + 1$, for any integer k. For all other values of n the gap between the bounds is 1. This poses an interesting question, i.e., which of the two: $\lfloor \log(n - 1) \rfloor + 2$ or $\lceil \log(n - 1) \rceil + 2$ is the correct exact bound? In this section we show that the latter one is not. We propose more sophisticated radio network topology in which, for large enough n, s.t., $n \leq 2^k + 2^{k-3} - O(2^{\frac{k}{2}})$, the gossiping can be done in time $k + 2 = \lfloor \log(n-1) \rfloor + 2$.

Consider a network H which is composed of three components, see Figure 2:

- a tree T_1 with a root r_1 (and its two exact copies r_2 and r_3 including adjacent edges),
- a tree T_2 with a root h_3, and
- a group of *special* nodes: three *roots* r_1, r_2, r_3, three *central nodes* c_1, c_2, c_3, and three *helpers* h_1, h_2, h_3.

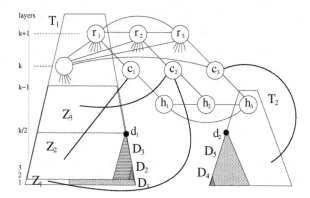

Fig. 2. The network topology for the optimal gossiping in time $\lfloor \log(n-1) \rfloor + 2$

The structure of trees T_1 and T_2 is very much based on the structure of the optimal broadcasting (in our case gathering) tree OBT(d) which is of size 2^d and it contains $d + 1$ (time) levels. The nodes in T_1 and T_2 that transmit in time step 1 during the gathering process are understood to be at layer 1, those that transmit in time step 2 at layer 2, etc, see Figure 2. Thus, layers in the trees T_1 and T_2 are enumerated in reverse order, comparing to the numbering of layers in standard OBTs. The tree T_1 is obtained from OBT(k) via deletion of three sets of nodes D_1, D_2, D_3 (to be defined later) and all edges connected to them. The tree T_2 is obtained from OBT($k - 3$) via deletion of two sets of nodes D_4, D_5 and all edges connected to these nodes. The content of each D_i, for $i = 1, .., 5$ is defined as follows:

- D_1 is a set of leaves in T_1 that are children of the nodes at layers $(\frac{k}{2} + 1), .., k - 1$,
- D_2 is a set of nodes, from layers $1, 2, 3$ and 4 in T_1 that belong to subtrees rooted in children and grandchildren of the root r_1,
- D_3 is a set of nodes of a subtree of T_1 rooted in node d_1, where d_1 is a child of the root r_1 at layer $k/2$,
- D_4 contains 3 nodes in T_2: two children x_1 and x_2 of the root h_3 at layer 1 and 2 respectively, and the child of x_2 at layer 1, and
- D_5 is a set of nodes in T_2 that form a subtree rooted in node d_2, where d_2 is a child of the root h_3 at layer $k/2$.

Another important component of the network is a set of special nodes. This set includes, the root r_1 of T_1 and its entirely equivalent copies r_2 and r_3, which are connected to the same nodes (as r_1 is) in the tree T_1. On the top of this, the roots r_1, r_2 and r_3 are mutually connected. The roots will be used to sent messages collected from the tree T_1 to the central nodes c_1, c_2 and c_3 in one time step (i.e., step $k + 1$).

The set of special nodes includes also three central nodes c_1, c_2 and c_3. The central nodes share direct connections to all other nodes in the network H. In fact, the direct connection from the central nodes form a partition of other nodes. It means that after we gather all messages in each of the nodes c_1, c_2 and c_3, we are able to distribute the messages to all other nodes (i.e., complete the gossiping process) in a single time step. We show later how to inform all central nodes in at most $k + 1$ rounds. In particular, the center c_1 is connected to all nodes in T_1 at layers $2, .., \frac{k}{2}$, to the root r_1, and the helper h_1. The center c_2 is connected to all nodes in T_1 at layer 1 and layers $(\frac{k}{2} + 1), .., (k - 1)$, the root r_2, and the helper h_2. The center c_3 is connected to all nodes in T_2 (including the helper h_3), the child of the root in T_1 at layer k, and the root r_3.

The last group of special nodes contains three helpers h_1, h_2 and h_3, where h_3 is the root of T_2. They are mutually connected and their purpose is to exchange original messages from the central nodes and to acquire messages gathered in the tree T_2.

Once the construction of the network H is completed we show that gossiping in H can be performed in $k + 2 = \lfloor log(n - 1) \rfloor + 2$ rounds. The expression $a \rightarrow b, c, d, ...$ is used to denote that a node a sends its all current knowledge to nodes $b, c, d, ...$ and $W_1 \| W_2$ means that transmissions W_1 and W_2 are performed simultaneously.

Steps	Transmissions
(1)	$c_2 \rightarrow h_2 \| c_3 \rightarrow h_3 \|$ (all nodes at layer 1 in T_1 and T_2 transmit)
(2)	$r_1 \rightarrow r_2, r_3 \| h_2 \rightarrow h_3, h_1 \|$
	(all nodes at layer 2 in T_1 transmit) $\|$ (all nodes at layer 1 in T_2 transmit)
(3)	$r_2 \rightarrow r_1, r_3 \|$ (all nodes at layer 3 in T_1 transmit) $\|$
	(all nodes at layer 2 in T_2 transmit)
(4)	$r_3 \rightarrow r_1, r_2 \|$ (all nodes at layer 4 in T_1 transmit) $\|$
	(all nodes at layer 3 in T_2 transmit)
...	
$(\frac{k}{2})$	$c_1 \rightarrow h_1 \|$ (all nodes at layer $\frac{k}{2}$ in T_1 transmit) $\|$
	(all nodes at layer $\frac{k}{2} - 1$ in T_2 transmit)
$(\frac{k}{2} + 1)$	$h_1 \rightarrow h_2, h_3 \|$ (all nodes at layer $\frac{k}{2} + 1$ in T_1 transmit) $\|$
	(all nodes at layer $\frac{k}{2}$ in T_2 transmit)
...	
(k-2)	(all nodes at layer $k - 2$ in T_1 transmit) $\|$
	(all nodes at layer $k - 3$ in T_2 transmit)
(k-1)	$h_3 \rightarrow c_3 \| h_3 \rightarrow h_2 \| h_3 \rightarrow h_1 \|$ (all nodes at layer $k - 1$ in T_1 transmit)
(k)	$h_1 \rightarrow c_1 \| h_2 \rightarrow c_2 \|$ (a node at layer k in T_1 transmits to r_1, r_2 and r_3)
(k+1)	$r_1 \rightarrow c_1 \| r_2 \rightarrow c_2 \| r_3 \rightarrow c_3$
(k+2)	c_1, c_2, c_3 transmit to all their neighbours.

During the gossiping process (in the network H) we first collect all messages in central nodes c_1, c_2 and c_3 in time $k + 1 = \lfloor log(n - 1) \rfloor + 1$. The main idea behind the removal of sets D_1 through D_5 is to avoid collisions when the special nodes act, i.e., when they transmit and listen. The loss of nodes caused by removal of the sets D_1, D_2 and D_3 from T_1 is compensated by the nodes available in the tree T_2. In fact, the size of H formed of trees T_1 and T_2 and a few more special nodes is $2^k + 2^{k-3} - O(2^{k/2})$. This is due to the fact that the cardinality of each D_i, for $i = 1, ..., 5$ is $O(2^{k/2})$. The following lemma holds:

Lemma 3. *There exists a radio network topology in which the gossiping task can be completed in time* $\lfloor \log(n-1) \rfloor + 2$, *for any integer* $n = 2^k + 2^{k-3} - O(2^{k/2})$, *and* k *large enough.*

In particular, we conclude that we know how to build the optimal (in terms of gossiping) radio network topology for a fraction of all integer values of n.

4   Gossiping in Time $O(D)$

We discuss here a class of graphs admitting radio gossiping in time $O(D)$.

4.1   Gossiping in Time $(2D - 1)\Delta + 1$

The general idea of the algorithm is as follows. Initially, we chose a central node c and we partition all nodes into disjoint subsets, *layers* l_i, where $0 \leq i \leq D$. This is followed by gathering stage when all (other $n - 1$) messages are moved to the central node c, layer by layer. Finally, a combined message (including all original messages) is distributed from c to all other nodes, also layer by layer. In what follows we show that all messages that reside at layer l_k can be moved to a neighbouring layer l_{k-1} (or l_{k+1}) in at most Δ units of time.

Lemma 4. *All messages available at layer* l_k *can be moved to a neighbouring layer* l_{k-1} *in at most* Δ *units of time, where* $1 \leq k \leq D$.

Proof. We use here notation introduced in Section 2. Let $N_k^0 = l_k$, and C_{k-1}^0 (subset of l_{k-1}) be the minimal covering set for N_k^0. I.e., every node in N_k^0 is connected to some node in C_{k-1}^0, and removal of any node from C_{k-1}^0 destruct this property. Note, that every node $v \in C_{k-1}^0$ is connected to some node $u \in N_k^0$, s.t., u is not connected to any other node in $C_{k-1}^0 - \{v\}$; otherwise we could remove v from C_{k-1}^0. Thus, during a single time step, every node $v \in C_{k-1}^0$ receives a message m_u transmitted from its unique node $u \in N_k^0$. Later, node u is removed from N_k^0, which means that a (virtual) degree of each node in C_{k-1}^0 is decreased by one. After removal of all us involved in the transmissions we end up with a new set N_k^1, and its new covering set $C_{k-1}^1 \subset C_{k-1}^0$. We repeat the whole process at most Δ times, since the degree of nodes in the covering set is decreased by 1 during each round of transmissions.

This means that the gossiping task in any radio network with diameter D and maximum degree Δ can be completed in time $\leq (2D - 1)\Delta + 1$, where $D\Delta$ comes from the gathering stage and $1 + (D - 1)\Delta$ from the broadcasting stage.

Theorem 2. *In any graph* G, *with a diameter* D *and a constant maximum degree, the gossiping task can be completed in time* $O(D)$.

4.2   Gossiping in Graphs with Larger Max-Degree

An algorithm presented in this section is based on a concept of efficient broadcasting $O(D + \log^5 n)$-time procedure presented in [8]. We use here very similar partition of a (network) graph into clusters and super-levels. The main difference lies in greater complexity of the gossiping problem.

Cluster Graph and Tree of Clusters.   Assume we have a graph $G = (V, E)$, where $|V| = n$ and a distinguished node $s \in V$. Assume also that the diameter of G is at most D. A layer l_i in G is formed by nodes that are at (same) distance i from s, for $i = 1, .., D$. All layers in G are grouped in x super-levels, s.t., the jth super-level is formed of layers $l_{\frac{D(j-1)}{x}+1}, .., l_{\frac{Dj}{x}}$, for $j = 1, .., x$. Each super-level is covered by the set of clusters, s.t., (1) each cluster has diameter $O(\frac{D \log n}{x})$, (2) the union of the clusters covers the super-level, and (3) the clusters graph can be coloured with $O(\log n)$ colours, where the clusters graph is obtained by treating each cluster as a node, and introducing an edge between two nodes if in the original graph there is some edge that connects nodes from the corresponding clusters or if the clusters share a common node. Note, that the number of clusters does not exceed n; otherwise we would be able to remove at least one (redundant) of them. It also follows from construction presented in [8] that each cluster at super-level i has a direct connection (an edge in the cluster graph) with some clusters at super-levels $i - 1$ and $i + 1$. This property allows to define a *tree of clusters*, which is a BFS tree rooted in a cluster that contain a distinguished node s. The broadcasting procedure proposed in [8] uses two types of information transfer in clusters, from the top layer through the bottom layer of a super-level. I.e., within each cluster we have either *slow* or *fast* transfers. The slow transfer is implemented by non-optimal broadcasting procedure, while the fast transfer is performed along a single path of length $\frac{D}{x}$. It is known, that transfers in the tree of clusters can be organised, s.t., on a path from any leaf to the root of the cluster tree there is at most $O(\log n)$ clusters operating slow transfers. In our gossiping algorithms the slow transfers are implemented by *limited gossiping* (defined below), and fast transfers (as in broadcasting) are performed along simple paths.

In our algorithm we use 3 types of communication procedures:

1. LIMITED GOSSIPING – where each node distributes its (currently possessed) message to all nodes within some radius r. Note that if $r = D$, the limited gossiping coincides with the gossiping problem. Note also that slow transfers are based on limited gossiping;
2. BETWEEN SUPER-LEVELS – where information residing at the top layer of a lower (further from the root cluster) super-level to the bottom layer of an upper super-level. This communication procedure is used when at least one cluster of the neighbouring super-levels is involved in slow transfers;
3. FAST TRANSFER – where information is moved across one cluster by fast pipelining along a single (simple) path.

The gossiping algorithm is implemented in 3 stages.

1. Initially messages in each cluster are collected in a distinguished node (possibly belonging to a fast route) in the top layer of each cluster. This is done by LIMITED GOSSIPING, where $r = \frac{D \log n}{x}$, i.e., maximal diameter of each cluster. Since the cluster graph can be coloured with $O(\log n)$ colours, all limited gossipings performed simultaneously in each cluster (at all super-levels of the cluster tree) can be preformed simultaneously with at most $\log n$-time slowdown. I.e., if $T_\Delta^n(D)$ stands for the time complexity of limited gossiping in a graph with n nodes, max-degree Δ, and diameter D, the contribution of the first stage to the time complexity of our gossiping algorithm is $O(T_\Delta^n(\frac{D \log n}{x}) \log n)$.
2. Messages from each cluster are delivered to the root cluster and in particular to the distinguished node s. During this stage the execution of three types of communication procedures is performed in separate (interleaved) time steps. I.e., LIMITED GOSSIPING in time steps $i = 0 \pmod 3$, BETWEEN SUPER-LEVEL in time steps $i = 1 \pmod 3$, and FAST TRANSFER in time steps $i = 2 \pmod 3$. Also, the execution of single rounds of communication procedures (within each type) is synchronised across all super-levels. E.g., the procedure LIMITED GOSSIPING starts and ends exactly at the same time in each cluster and each super-level. Moreover, when new messages arrive at the bottom of a super-level (e.g., delivered by the BETWEEN SUPER-LEVEL communication procedure) they are buffered at the bottom layer and allowed to traverse towards the upper layers only when the new round of LIMITED GOSSIPING or FAST TRANSFER is about to begin.
   A contribution to the time complexity of our gossiping algorithm of each the communication procedures is as follows:
   (a) Since executions of LIMITED GOSSIPINGs are synchronised across all super-levels, simultaneous execution of a single round of the LIMITED GOSSIPING procedure is done in time $T_\Delta^n(\frac{D \log n}{x})$. And since each message, traversing towards the root cluster, experiences at most $O(\log n)$ slow transfers (based on LIMITED GOSSIPING) the total contribution to the slowdown of each message is bounded by $O(T_\Delta^n(\frac{D \log n}{x}) \log n)$.
   (b) A single execution of one round of BETWEEN SUPER-LEVELS procedure can be implemented in time Δ. This is a consequence of Lemma 4. Since each message has to pass at most x borders between super-levels the contribution of this type of communication to the total time complexity is bounded by $x\Delta$.
   (c) A simultaneous execution of FAST TRANSFER in potentially many clusters on the same super-level results in a need of pipelined transmission of messages according to the color of a cluster. I.e., during one round of a fast transfer, messages that traverse along a path in clusters coloured with number 1 start their journey immediately, in all clusters coloured with number 2 messages are released three time steps later (in order to avoid collisions between layers), in clusters coloured with number 3 – six time steps later, etc. Thus finally in clusters coloured with the largest number – $O(\log n)$ time steps later. After message is released at the bottom layer it reaches the upper one in exactly $\frac{D}{x}$ time steps (a property of a fast transfer). Thus the contribution of this type of communication (across all super-levels) to the total time complexity is bounded by $O(x(\frac{D}{x} + \log n))$.

3. Eventually, after all messages are successfully gathered in the distinguished node s the combined message (containing all original messages) is broadcasted to all other nodes in the graph. This can be done by reversing the gathering process presented above, where the time complexity remains the same.

Theorem 3. *The time complexity of our gossiping procedure can be expressed by the recursive equation:* $T_\Delta^n(D) = O(T_\Delta^n(\frac{D \log n}{x}) \log n + x(\Delta + \log n) + D)$, *where x is the number of super-levels in the Cluster Graph.*

After iteration of the recursive equation from Theorem 3 $i - 1$ times, we obtain:
$T_\Delta^n(D) = O(T_\Delta^n(\frac{D \log^i n}{x^i}) \log^i n + x(\Delta + \log n)(\log^{i-1} + ... + \log n + 1) + D(\frac{\log^{2(i-1)} n}{x^{i-1}} + ... + \frac{\log^2 n}{x} + 1))$. After further substitution of the recursive component by the complexity $(2D - 1)\Delta + 1$ (see the gossiping algorithm presented in Section 4.1) and taking $x = D^{\frac{1}{i+1}} \log n$, we get:

Corollary 1. $T_\Delta^n(D) = O(D)$, *for all graphs with* $\Delta = O(\frac{D^{1-1/i+1}}{\log^i n})$ *and* $D = \Omega(\log^{i+1} n)$, *for all constant integers $i \geq 0$.*

References

1. R. Bar-Yehuda, A. Israeli, A. Itai, Multiple Communication in Multihop Radio Networks, SIAM Journal on Computing, 22(4), 1993, pp 875-887.
2. I. Chlamtac and S. Kutten, On broadcasting in radio networks-problem analysis and protocol design, IEEE Transactions on Communications 33, 1985, pp 1240-1246.
3. M. Christersson, L. Gąsieniec and A. Lingas, Gossiping with bounded size messages in ad-hoc radio networks, 29th Int. Colloq. on Automata, Lang. and Prog., ICALP'02, pp 377-389.
4. M. Chrobak, L. Gąsieniec and W. Rytter, Fast Broadcasting and Gossiping in Radio Networks, Journal of Algorithms 43(2), 2002, pp 177-189.
5. A.E.F. Clementi, A. Monti, R. Silvestri, Distributed multi-broadcast in unknown radio networks, 20th ACM Symp. on Principles of Distributed Computing, PODC'01, pp 255-264.
6. A. Czumaj and W. Rytter, Broadcasting Algorithms in Radio Networks with Unknown Topology, 44th Symposium on Foundations of Computer Science, FOCS'03, pp 492-501.
7. K. Diks, E. Kranakis, A. Pelc, The impact of knowledge on broadcasting time in radio networks, 7th European Symposium on Algorithms, ESA'99, pp 41-52.
8. I. Gaber and Y. Mansour, Broadcast in radio networks, 6th Annual ACM-SIAM Symposium on Discrete Algorithms, SODA'95, pp 577-585.
9. L. Gąsieniec and A. Lingas, On adaptive deterministic gossiping in ad hoc radio networks, Information Processing Letters 2(83), 2002, pp 89-94.
10. L. Gąsieniec and I. Potapov, Gossiping with unit messages in known radio networks, 2nd IFIP International Conference on Theoretical Computer Science, TCS'02, pp 193-205.
11. D. Liu and M. Prabhakaran, On Randomized Broadcasting and Gossiping in Radio Networks, 8th Int. Conf. on Computing and Combinatorics, COCOON'02, pp 340-349.
12. A. Sen and M.L. Huson, A new model for scheduling packet radio networks, 15th Annual Joint Conference of the IEEE Computer and Communication Societies, 1996, pp 1116-1124.
13. P.J. Slater, E.J.Cockayne and S.T. Hedetniemi, Information Dissemination in Trees, SIAM Journal on Computing, 10, 1981, pp 892-701.
14. Y. Xu, An $O(n^{1.5})$ deterministic gossiping algorithm for radio networks, Algorithmica, 36(1), 2003, pp 93-96.

Long-Lived Rambo:
Trading Knowledge for Communication*

Chryssis Georgiou[1], Peter M. Musial[2], and Alexander A. Shvartsman[2,3]

[1] Department of Computer Science, University of Cyprus, Nicosia, Cyprus
[2] Department of Computer Science & Engineering, University of Connecticut, Storrs, CT, USA
[3] CSAIL, Massachusetts Institute of Technology, Cambridge, MA, USA

Abstract. Shareable data services providing consistency guarantees, such as atomicity (linearizability), make building distributed systems easier. However, combining linearizability with efficiency in practical algorithms is difficult. A reconfigurable linearizable data service, called RAMBO, was developed by Lynch and Shvartsman. This service guarantees consistency under dynamic conditions involving asynchrony, message loss, node crashes, and new node arrivals. The specification of the original algorithm is given at an abstract level aimed at concise presentation and formal reasoning about correctness. The algorithm propagates information by means of gossip messages. If the service is in use for a long time, the size and the number of gossip messages may grow without bound. This paper presents a consistent data service for *long-lived* objects that improves on RAMBO in two ways: it includes an incremental communication protocol and a leave service. The new protocol takes advantage of the local knowledge, and carefully manages the size of messages by removing redundant information, while the leave service allows the nodes to leave the system gracefully. The new algorithm is formally proved correct by forward simulation using levels of abstraction. An experimental implementation of the system was developed for networks-of-workstations. The paper also includes analytical and preliminary empirical results that illustrate the advantages of the new algorithm.

1 Introduction

This paper presents a practical algorithm implementing long-lived, survivable, atomic read/write objects in dynamic networks, where participants may join, leave, or fail during the course of computation. The only way to ensure survivability of data is through redundancy: the data is replicated and maintained at several network locations. Replication introduces the challenges of maintaining *consistency* among the replicas, and managing *dynamic participation* as the collections of network locations storing the replicas change due to arrivals, departures, and failures of nodes.

A new approach to implementing atomic read/write objects for dynamic networks was developed by Lynch and Shvartsman [10] and extended by Gilbert *et al.* [6]. This memory service, called RAMBO (Reconfigurable Atomic Memory for Basic Objects) maintains atomic (linearizable) readable/writable data in highly dynamic environments. In order to achieve availability in the presence of failures, the objects are replicated at

---

* This work is supported in part by the NSF Grants 9984778, 9988304, 0121277, and 0311368.

R. Královič and O. Sýkora (Eds.): SIROCCO 2004, LNCS 3104, pp. 185–196, 2004.

several locations. In order to maintain consistency in the presence of small and transient changes, the algorithm uses *configurations* of locations, each of which consists of a set of *members* plus sets of *read-* and *write-quorums*. In order to accommodate larger and more permanent changes, the algorithm supports *reconfiguration*, by which the set of members and the sets of quorums are modified. Obsolete configurations can be removed from the system without interfering with the ongoing read and write operations. The algorithm tolerates arbitrary patterns of asynchrony, node failures, and message loss. Atomicity is guaranteed in any execution of the algorithm [10, 6].

The original RAMBO algorithm is formulated at an abstract level aimed at concise specification and formal reasoning about the algorithm's correctness. Consequently the algorithm incorporates a simple communication protocol that maintain very little protocol state. The algorithm propagates information among the participants by means of gossip messages that contain information corresponding to the sender's state. The number and the size of gossip message may in fact grow without bound. This renders the algorithm impractical for use in *long-lived* applications.

The gossip messages in RAMBO include the set of participants, and the size of these messages increases over time for two reasons. First, RAMBO allows new participants to join the computation, but it does not allow the participants to leave gracefully. In order to leave the participants must pretend to crash. Given that in asynchronous systems failure detection is difficult, it may be impossible to distinguish departed nodes from the nodes that crash. Second, RAMBO gossips information among the participants without regard for what may already be known at the destination. Thus a participant will repeatedly gossip substantial amount of information to others even if it did not learn anything new since the last time it gossiped. While such redundant gossip helps tolerating message loss, it substantially increases the communication burden. Given that the ultimate goal for this algorithm is to be used in long-lived applications, and in dynamic networks with unknown and possibly infinite universe of nodes, the algorithm must be carefully refined to substantially improve its communication efficiency.

Contributions. The paper presents a new algorithm for reconfigurable atomic memory for dynamic networks. The algorithm, called LL-RAMBO, makes implementing atomic survivable objects practical in long-lived systems by managing the knowledge accumulated by the participants and the size of the gossip messages. Each participating node maintains a more complicated protocol state and, with the help of additional local processing, this investment is traded for substantial reductions in the size and the number of gossip messages. Based on [6, 10], we use Input/Output Automata [11] to specify the algorithm, then prove it correct in two stages by forward simulation, using levels of abstraction. We include analytical and preliminary empirical results illustrating the advantages of the new algorithm. In more detail, our contributions are as follows.

(1) We develop L-RAMBO that implements an atomic memory service and includes a *leave* service (Sect. 3). We prove correctness (safety) of L-RAMBO by forward simulation of RAMBO, hence we show that every trace of L-RAMBO is a trace of RAMBO.

(2) We develop LL-RAMBO by refining L-RAMBO to implement *incremental gossip* (Sect. 4). We prove that LL-RAMBO implements the atomic service by forward simulation of L-RAMBO. This shows that every trace of LL-RAMBO is a trace of L-RAMBO, and thus a trace of RAMBO. The proof involves subtle arguments relating the knowledge extracted from the local state to the information that is *not* included in gossip messages. We present the proof in two steps for two reasons: (i) the presentation matches the intuition that the leave service and the incremental gossip are independent, and (ii) the resulting proof is simpler than a direct simulation of RAMBO by LL-RAMBO.

(3) We show (Sect. 5) that LL-RAMBO consumes smaller communication resources than RAMBO, while preserving the same read and write operation latency, which under certain steady-state assumptions is at most $8d$ time, where d is the maximum message delay unknown to the algorithm. Under these assumptions, in runs with periodic gossip, LL-RAMBO achieves substantial reductions in communication.

(4) We implemented all algorithms on a network-of-workstations. Preliminary empirical results complement the analytical comparison of the two algorithms (Sect. 5).

Background. Several approaches can be used to implement consistent data in (static) distributed systems. Many algorithms used collections of intersecting sets of object replicas to solve consistency problems, e.g., [2, 14, 15]. Extension with reconfigurable quorums have been explored [4], but this system has limited ability to support long-lived data when the longevity of processors is limited. Virtual synchrony [3], and group communication services (GCS) in general [1], can be used to implement consistent objects, e.g., by using a global totally ordered broadcast. The universe of nodes in a GCS can evolve, however forming a new view is indicated after a single failure and can take a substantial time, while reads and writes are delayed during view formation.

The work on reconfigurable atomic memory [4, 10, 6] results in algorithms that are more dynamic because they place fewer restrictions on the choice of new configurations and allow for the universe of processors to evolve arbitrarily. However these approaches are based on abstract communication protocols that are not suited for long-lived systems. Here we provide a long-lived solution by introducing graceful processor departures and incremental gossip. The idea of incrementally propagating information among participating nodes has been previously used in a variety of different settings, e.g., [7, 12]. Incremental gossip is also called anti-entropy [5, 13] or reconciliation [8]; these concepts are used in database replication algorithms, however due to the nature of the application they assume stronger assumptions, e.g., ordering of messages.

Document structure. In Section 2 we review RAMBO. In Section 3 we specify and prove correct the graceful leave service. Section 4 presents the ultimate system, with leave and incremental gossip, and proves it correct. Section 5 gives the analysis and experimental results. (Complete proofs and analysis are found in MIT/LCS/TR-943.)

2 Reconfigurable Atomic Memory for Basic Objects (RAMBO)

We now describe the RAMBO algorithm as presented in [10], including the rapid configuration upgrade as given in [6]. The algorithm is given for a single object (atomicity is preserved under composition, and multiple objects can be composed to yield a complete shared memory). For the detailed Input/Output Automata code see [10, 6]. In order to achieve fault tolerance and availability, RAMBO replicates objects at several network locations. In order to maintain memory consistency in the presence of small and transient changes, the algorithm uses *configurations*, each of which consists of a set of *members* plus sets of *read-quorums* and *write-quorums*. The quorum intersection property requires that every read-quorum intersect every write-quorum. In order to accommodate larger and more permanent changes, the algorithm supports *reconfiguration*, by which the set of members and the sets of quorums are modified. Any quorum configuration may be installed, and atomicity is preserved in all executions.

The algorithm consists of three kinds of automata: (i) *Joiner* automata, handling join requests, (ii) *Recon* automata, handling reconfiguration requests and generating a totally ordered sequence of configurations, and (iii) *Reader-Writer* automata, handling read and write requests, manage configuration upgrades, and implement

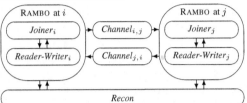

Fig. 1. RAMBO architecture depicting automata at nodes i and j, the channels, and the *Recon* service.

gossip messaging. The overall systems is the composition of these automata with the automata modelling point-to-point communication channels, see Fig. 1. The *Joiner* automaton simply sends a join message when node i joins, and sends a join-ack message whenever a join message is received. The *Recon* automaton establishes a total ordering of configurations (for details see [10]).

The external signature of the service is in Fig. 2. A client at node i uses join$_i$ action to join the system. After receiving join-ack$_i$, the client can issue read$_i$ and write$_i$ requests, which result in read-ack$_i$ and write-ack$_i$ responses. The client can issue a recon$_i$ request a reconfiguration. The fail$_i$ action models a crash at node i.

Domains: I, a set of processes; V, a set of legal values; and C, a set of configurations, each consisting of members, read-and write-quorums

Input: join(rambo, $J)_i$, $J \subseteq I - \{i\}, i \in I$,
    such that if $i = i_0$ then $J = \emptyset$
    read$_i$, $i \in I$
    write$(v)_i$, $v \in V, i \in I$
    recon$(c, c')_i$, $c, c' \in C, i \in$ members$(c), i \in I$
    fail$_i$, $i \in I$

Output: join-ack(rambo)$_i, i \in I$
    read-ack$(v)_i, v \in V, i \in I$
    write-ack$_i, i \in I$
    recon-ack$(b)_i, b \in \{ok, nok\}, i \in I$
    report$(c)_i, c \in C, i \in I$

Fig. 2. RAMBO: External signature.

Every node of the system maintains a *tag* and a *value* for the data object. Every time a new value is written, it is assigned a unique tag, with ties broken by process-ids. These tags are used to determine an ordering of the write operations, and therefore determine the value that a read operation should return. Read and write operations has two phases, *query* and *propagation*, each accessing certain quorums of replicas. Assume the operation is initiated at node i. First, in the query phase, node i contacts read quorums to determine the most recent known tag and value. Then, in the propagation phase, node i contacts write quorums. If the operation is a read operation, the second phase propagates the largest discovered tag and its associated value. If the operation is a write operation, node i chooses a new tag, strictly larger than every tag discovered in the query phase. Node i then propagates the new tag and the new value to a write quorum. Note that every operation accesses both read and write quorums.

Configurations go through three stages: proposal, installation, and upgrade. First, a configuration is *proposed* by a recon event. Next, if the proposal is successful, the

Recon service achieves consensus on the new configuration, and notifies participants with decide events. When every non-failed member of the prior configuration has been notified, the configuration is *installed*. The configuration is *upgraded* when every configuration with a smaller index has been removed. Upgrades are performed by the configuration upgrade operations. Each upgrade operation requires two phases, a query phase and a propagate phase. The first phase contacts a read-quorum and a write-quorum from the old configurations, and the second phase contacts a write-quorum from the new configuration. All three operations, *read, write,* and *configuration upgrade*, are implemented using gossip messages.

The *cmap* is a mapping from integer indices to configurations $\cup\{\perp, \pm\}$, initially mapping every integer to $\perp$. It tracks which configurations are active, which have not yet been created, indicated by $\perp$, and which have already been removed, indicated by $\pm$. The total ordering on configurations determined by *Recon* ensures that all nodes agree on which configuration is stored in each position in *cmap*. We define $c(k)$ to be the configuration associated with index k.

The record *op* is used to store information about the current phase of an ongoing read or write operation, while *upg* is used for information about an ongoing configuration upgrade operation. A node can process read and write operations concurrently with configuration upgrade operations. The *op.cmap* subfield records the configuration map associated with the operation. For read or write operations this consists of the node's *cmap* when a phase begins, augmented by any new configurations discovered during the phase. A phase completes when the initiator has exchanged information with quorums from every valid configuration in *op.cmap*. The *pnum* subfield records the phase number when the phase begins, allowing the initiator to determine which responses correspond to the phase. The *acc* subfield records which nodes from which quorums have responded during the current phase.

Finally, the nodes communicate via asynchronous unreliable point-to-point channels. We denote by $Channel_{i,j}$ the channel from node i to node j.

3 RAMBO with Graceful Leave

Here we augment RAMBO with a *leave service* allowing the participants to depart gracefully. We prove that the new algorithm, called L-RAMBO, implements atomic memory.

Nodes participating in RAMBO communicate by means of gossip messages containing the latest object value and bookkeeping information that includes the set of known participants. RAMBO allows participants to fail or leave without warning. Since in asynchronous systems it is difficult or impossible to distinguished slow or departed nodes from crashed nodes, RAMBO implements gossip to all known participants, regardless of their status. In highly dynamic systems this leads to (a) the size of gossip messages growing without bounds, and (b) the number of messages sent in each round of gossip increasing as new participants join the computation.

L-RAMBO allows graceful node departures by letting a node that wishes to leave the system to send notification messages to an arbitrary subset of known participants. When another node receives such notification, it marks the sender as departed, and stops gossiping to that node. The remaining nodes propagate the information about

Signature:
As in RAMBO, plus new actions:
Input : leave$_i$, recv(leave)$_{j,i}$
Output : send(leave)$_{i,j}$

State:
As in RAMBO, plus new states:
leave-world, a finite subset of I, initially $\emptyset$
departed, a finite subset of I, initially $\emptyset$

$ig \in IGMap$, initially $\forall k \in I$,
$ig(k).wk = \emptyset, ig(k).w\text{-}ua = \emptyset,$
$ig(k).dk = \emptyset, ig(k).d\text{-}ua = \emptyset,$
$ig(k).p\text{-}ack = 0$

Transitions at i:

Input recv($\langle W, D, v, t, cm, pns, pnr \rangle)_{j,i}$
Effect:
  if $\neg failed \wedge status \neq$ idle then
    $status \leftarrow$ active
    $world \leftarrow world \cup W$
    $departed \leftarrow departed \cup D$

> $[h]hr\text{-}W(j,i,pnr) \leftarrow W$
> $[h]hr\text{-}D(j,i,pnr) \leftarrow D$
> $ig(j).wk \leftarrow ig(j).wk \cup W$
> $ig(j).w\text{-}ua \leftarrow ig(j).w\text{-}ua - W$
> $ig(j).dk \leftarrow ig(j).dk \cup D$
> $ig(j).d\text{-}ua \leftarrow ig(j).d\text{-}ua - D$
> if $pnr > ig(j).p\text{-}ack$ then
> $ig(j).wk \leftarrow ig(j).wk \cup ig(j).w\text{-}ua$
> $ig(j).w\text{-}ua \leftarrow world - ig(j).wk$
> $ig(j).dk \leftarrow ig(j).dk \cup ig(j).d\text{-}ua$
> $ig(j).d\text{-}ua \leftarrow departed - ig(j).dk$
> $ig(j).p\text{-}ack \leftarrow pnum1$

  if $t > tag$ then $(value, tag) \leftarrow (v, t)$
  $cmap \leftarrow update(cmap, cm)$
  $pnum2(j) \leftarrow \max(pnum2(j), pns)$
  if $op.phase \in \{$query, prop$\} \wedge pnr \geq op.pnum$ then
    $op.cmap \leftarrow extend(op.cmap, truncate(cm))$
    if $op.cmap \in Truncated$ then
      $op.acc \leftarrow op.acc \cup \{j\}$
    else
      $op.acc \leftarrow \emptyset$
      $op.cmap \leftarrow truncate(cmap)$
  if $upg.phase \in \{$query, prop$\} \wedge pnr \geq upg.pnum$ then
    $upg.acc \leftarrow upg.acc \cup \{j\}$

Input recv(leave)$_{j,i}$
Effect:
  if $\neg failed \wedge status =$ active then
    $departed \leftarrow departed \cup \{j\}$

Output send($\langle W, D, v, t, cm, pns, pnr \rangle)_{i,j}$
Precondition:
  $\neg failed$
  $status =$ active
  $j \in (world - departed)$
  $\langle W, D, v, t, cm, pns, pnr \rangle =$
    $\langle world \boxed{-ig(j).wk}, departed \boxed{-ig(j).dk},$
    $value, tag, cmap, pnum1, \boxed{pnum2(j)} \rangle$
Effect:
  $pnum1 \leftarrow pnum1 + 1$
  $[h]h\text{-}msg \leftarrow h\text{-}msg \cup$
    $\langle\langle W, D, v, t, cm, pns, pnr \rangle, i, j \rangle$

> $[h]hs\text{-}world(i,j,pns) \leftarrow world$
> $[h]hs\text{-}departed(i,j,pns) \leftarrow departed$
> $[h]hs\text{-}wk(i,j,pns) \leftarrow ig(j).wk$
> $[h]hs\text{-}dk(i,j,pns) \leftarrow ig(j).dk$
> $[h]hs\text{-}wua(i,j,pns) \leftarrow ig(j).w\text{-}ua$
> $[h]hs\text{-}dua(i,j,pns) \leftarrow ig(j).d\text{-}ua$
> $[h]hs\text{-}pack(i,j,pns) \leftarrow ig(j).p\text{-}ack$

input leave$_i$
Effect:
  if $\neg failed$ then
    $failed \leftarrow$ true
    $departed \leftarrow departed \cup \{i\}$
    $leave\text{-}world \leftarrow world - departed$

output send(leave)$_{i,j}$
Precondition:
  $j \in leave\text{-}world$
Effect:
  $leave\text{-}world \leftarrow leave\text{-}world - \{j\}$

Fig. 3. Modification of *Reader-Writer*$_i$ for L-RAMBO, and for LL-RAMBO (the $\boxed{\text{boxed}}$ code).

the departed nodes to other participants, eventually eliminating gossip to nodes that departed gracefully.

Specification of L-RAMBO. We interpret the fail$_i$ event as synonymous with the leave$_i$ event – both are inputs from the environment and both result in node i stopping to participate in all operations. The difference between fail$_i$ and leave$_i$ is strictly internal: leave$_i$ allows a node to leave gracefully. The well-formedness conditions of RAMBO and the specifications of *Joiner*$_i$ and *Recon* remain unchanged. The introduction of the leave service affects only the specification of the *Reader-Writer*$_i$ automata. These changes for L-RAMBO are given in Fig. 3, except for the $\boxed{\text{boxed}}$ segments of code that should be disregarded until the ultimate long-lived algorithm LL-RAMBO is presented in Sect. 4 (we combine the two specifications in the interest of space).

The signature of *Reader-Writer*$_i$ automaton is extended with actions recv(leave)$_{j,i}$ and send(leave)$_{i,j}$ used to communicate the graceful departure status. The state of *Reader-Writer*$_i$ is extended with new state variables: *departed*$_i$, the set of nodes that left the system, as known at node i, $leave\text{-}world_i$, the set of nodes that node i can inform of its own departure, once it decides to leave and sets $leave\text{-}world_i$ to $world - departed$.

The key algorithmic changes involve the actions $recv(m)_{j,i}$ and $send(m)_{i,j}$. The original RAMBO algorithm gossips message m includes: W the $world$ of the sender, v the object and its tag t, cm the $cmap$, pns the phase number of the sender, and pnr the phase number of the receiver that is known to the sender. The gossip message m in L-RAMBO also includes D, a new parameter, equal to the $departed$ set of the sender.

We now detail the leave protocol. Assume that nodes i and j participate in the service, and node i wishes to depart following the $leave_i$ event, whose effects set the state variable $failed_i$ to $true$ in $Joiner_i$, $Recon_i$, and $Reader\text{-}Writer_i$. The $leave_i$ action at $Reader\text{-}Writer_i$ (see Fig. 3) also initializes the set $leave\text{-}world_i$ to the identifiers found in $world_i$, less those found in $departed_i$. Now $Reader\text{-}Writer_i$ is allowed to send one leave notification to any node in $leave\text{-}world_i$. This is done by the $send(leave)_{i,j}$ action that arbitrarily chooses the destination j from $leave\text{-}world_i$. Note that node i may nondeterministically choose the original $fail_i$ action, in which case no notification messages are sent (this is the "non-graceful" departure).

When $Reader\text{-}Writer_i$ receives a leave notification from node j, it adds j to its $departed_i$ set. Node i sends gossip messages to all nodes in the set $world_i - departed_i$, which including information about j's departure. When $Reader\text{-}Writer_i$ receives a gossip message that includes the set D, it updates its $departed_i$ set accordingly.

Atomicity of L-RAMBO *service.* The L-RAMBO system is the composition of all $Reader\text{-}Writer_i$ and $Joiner_i$ automata, the $Recon$ service, and $Channel_{i,j}$ automata for all $i, j \in I$. We show atomicity of L-RAMBO by forward simulation that proves that any trace of L-RAMBO is also a trace of RAMBO, and thus L-RAMBO implements atomic objects. The proof uses history variables, annotated with the symbol $[h]$ in Fig. 3.

For each i we define $h\text{-}msg_i$ to be the history variable that keeps track of all messages sent by $Reader\text{-}Writer_i$ automata. Initially, $h\text{-}msg_i = \emptyset$ for all $i \in I$. Whenever a message m is sent by i to some node $j \in I$ via $Channel_{i,j}$, we let $h\text{-}msg_i \leftarrow h\text{-}msg_i \cup \{\langle m, i, j \rangle\}$. We define $h\text{-}MSG$ to be $\bigcup_{i \in I} h\text{-}msg_i$. (The remaining history variables are used in reasoning about LL-RAMBO, see Sect. 4).

The following lemma states that only good messages are sent.

Lemma 1. *In any execution of* L-RAMBO, *if* m *is a message received by node* i *in a* $recv(m)_{i,j}$ *event, then* $\langle m, j, i \rangle \in h\text{-}MSG$, *and* $m = \langle W, D, v, t, cm, pns, pnr \rangle$ *or* $m = $ leave *or* $m = $ join.

Next we show that L-RAMBO implements RAMBO, assuming the environment behavior as (informally) described in Sect. 2. Showing well-formedness is straightforward by inspecting the code. The proof of atomicity is based on a forward simulation relation [9] from L-RAMBO to RAMBO.

Theorem 1. L-RAMBO *implements atomic read/write objects.*

4 RAMBO **with Graceful Leave and Incremental Gossip**

Now we present, and prove correct, our ultimate algorithm, called LL-RAMBO (Long-Lived RAMBO). The algorithm is obtained by incorporating incremental gossip in L-RAMBO, so that the size of gossip messages is controlled by eliminating redundant in-

formation. In L-RAMBO (resp. RAMBO) the gossip messages contain sets corresponding to the sender's *world* and *departed* (resp. *world*) state variables at the time of the
sending (Fig. 3). As new nodes join the system and as participants leave the system,
the cardinality of these sets grows without bound, rendering RAMBO and L-RAMBO
impractical for implementing long-lived objects. The LL-RAMBO algorithm addresses
this issue. The challenge here is to ensure that only the certifiably redundant information
is eliminated from the messages, while tolerating message loss and reordering.

Specification of LL-RAMBO. We specify the algorithm by modifying the code of L-
RAMBO. In Fig. 3 the $\boxed{\text{boxed}}$ segments of code specify these modifications. The new
gossip protocol allows node i to gossip the information in the sets $world_i$ and $departed_i$
incrementally to each node $j \in world_i - departed_i$. Following j's acknowledgment,
node i never again includes this information in the gossip messages sent to j, but will
include new information that i has learned since the last acknowledgment by j.

To describe the incremental gossip in more detail we consider an exchange of a
gossip messages between nodes i and j, where i is the sender and j is the receiver. The
sets *world* and *departed* are managed independently and similarly, and we illustrate
incremental gossip using just the set *world*. First we define new data types. Let an
incremental gossip identifier be the tuple $\langle wk, dk, w\text{-}ua, d\text{-}ua, p\text{-}ack \rangle$, where wk, dk,
$w\text{-}ua$, and $d\text{-}ua$ are finite subsets of I, and $p\text{-}ack$ is a natural number. Let IG denote
the set of all *incremental gossip identifiers*. Finally, let *IGMap* be the set of *incremental
gossip maps*, defined as the set of mappings $I \rightarrow IG$. We extend the state of the *Reader-
Writer$_i$* automaton with $ig_i \in IGMap$. Node i uses $ig(j)_i$ tuple to keep track of the
knowledge it has about the information already in possession of, and currently being
propagated to, node j (see Fig.3). Specifically, for each $j \in world_i$, $ig(j)_i.wk$ is the set
of node identifiers that i is assured is a subset of $world_j$, $ig(j)_i.w\text{-}ua$ is the set of node
identifiers, a subset of $world_i$, that j needs to acknowledge. The components $ig(j)_i.dk$
and $ig(j)_i.d\text{-}ua$ are defined similarly for the *departed* set. Lastly, $ig(j)_i.p\text{-}ack$ is the
phase number of i when the last acknowledgment from j was received. Initially each of
these sets is empty, and $p\text{-}ack$ is zero for each $ig(j)_i$ with $j \in I$.

Node j *acknowledges* a set of identifiers by including this set in the gossip message,
or by sending a phase number of i such that node i can deduce that node j received
this set of identifiers in some previous message from i to j. Messages that include i's
phase number that is larger than $ig(j)_i.p\text{-}ack$ are referred to as *fresh* or *acknowledgment*
messages, otherwise they are referred to as *late* messages. (This is discussed later.)

The lines annotated with $[h]$ in Fig. 3 deal with history variables that are used only
in the proof of correctness.

In RAMBO, once node i learns about node j, it can gossip to j at any time. We now
examine the send($\langle W, D, v, t, cm, pns, pnr \rangle)_{i,j}$ action. The world component, W, is
set to the difference of $world_i$ and the information that i knows that j has, $ig(j)_i.wk$, at
the time of the send. Remaining components of the gossip message are the same as in
L-RAMBO. The effect of the send action causes phase number of the sender to increase;
this ensures that each message sent is labeled with a unique phase number of the sender.

Now we examine recv($\langle W, D, v, t, cm, pns, pnr \rangle)_{i,j}$ action at j (note that we
switch i and j relative to the code in Fig. 3 to continue referring to the interaction of the
sender i and receiver j). The component W contains a subset of node identifiers from
j's *world*. Hence W is always used to update $world_j$, $ig(i)_j.wk$, and $ig(i)_j.w\text{-}ua$. The

update of $world_j$ is identical to that in L-RAMBO. By definition $ig(i)_j.wk$ is the set of node identifiers that j is assured that i has, hence we update it with information in W. Similarly, by definition $ig(i)_j.w\text{-}ua$ is the set of node identifiers that j is waiting for i to acknowledge. It is possible that i has learned some or all of this information from other nodes and it is now a part of W, hence we remove any identifiers in W that are also in $ig(i)_j.w\text{-}ua$ from $ig(i)_j.w\text{-}ua$; these identifiers do not need further acknowledgment.

What happens next in the effect of recv depends on the value of pnr (the phase number that i believes j to be in). First, if $pnr \leq ig(i)_j.p\text{-}ack$, this means that this message is a late message since there must have been a prior message from i to j that included this or higher pnr. Hence, no updates take place. Second, if $pnr > ig(i)_j.p\text{-}ack$, this message is considered to be an acknowledgment message. By definition $ig(i)_j.p\text{-}ack$ contains the phase number of j when last acknowledgment from i was received. Following last acknowledgment, phase number of j was incremented, $ig(i)_j.p\text{-}ack$ was assigned the new value of phase number of j, and lastly new set of identifiers to be propagated was recorded. Since node i replied to j with phase number larger than $ig(i)_j.p\text{-}ack$ it means that j and i exchanged messages where i learned about the new phase number of j, by the same token i also learned the information included in these messages. (We show formally that $ig(i)_j.w\text{-}ua$ is always a subset of each message component W that is sent to i by j.) Hence, it is safe for j to assume that i at least received the information in $ig(i)_j.w\text{-}ua$ and to add it to $ig(i)_j.wk$.

Since the choice of i and j is arbitrary, gossip from j to i is defined identically.

Atomicity of LL-RAMBO. We show that any trace of LL-RAMBO is a trace of L-RAMBO, and thus a trace of RAMBO. We start by defining the remaining history variables used in the proofs. These variables are annotated in Fig. 3 with a $[h]$ symbol.

- For every tuple $\langle m, i, j \rangle \in h\text{-}msg_i$, where $m = \langle W, D, v, t, cm, pns, pnr \rangle$ and $pns = p$, the history variable $hs\text{-}W(i, j, p)$ is a mapping from $I \times I \times \mathbb{N}$ to $2^I \cup \{\bot\}$. This variable records the *world* component of the message, W, when i sends message m to j, and i's phase number is p. Similarly, we define a derived history variable $hs\text{-}D(i, j, p)$, a mapping from $I \times I \times \mathbb{N}$ to $2^I \cup \{\bot\}$. This history variable records the *departed* component of the message, D, when i sends message m to j, and i's phase number is p.

Now we list history variables used to record information for each $send(\langle W, D, v, t, cm, pns, pnr \rangle)_{i,j}$ event.

- Each of the following variables is a mappings from $I \times I \times \mathbb{N}$ to $2^I \cup \{\bot\}$. $hs\text{-}world(i, j, pns)$ records the value of $world_i$, $hs\text{-}departed(i, j, pns)$ records the value of $departed_i$, $hs\text{-}wk(i, j, pns)$ records the value of $ig(j)_i.wk$, $hs\text{-}dk(i, j, pns)$ records the value of $ig(j)_i.dk$, $hs\text{-}wua(i, j, pns)$ records the value of $ig(j)_i.w\text{-}ua$, and $hs\text{-}dua(i, j, pns)$ records the value of $ig(j)_i.d\text{-}ua$.
- $hs\text{-}pack(i, j, pns)$ is a mapping from $I \times I \times \mathbb{N}$ to $\mathbb{N}$. It records the value of $ig(j)_i.p\text{-}ack$.

The last history variables record information in messages at each $recv(\langle W, D, v, t, cm, pns, pnr \rangle)_{j,i}$ event.

- Each of the following is a mapping from $I \times I \times \mathbb{N}$ to $2^I \cup \{\bot\}$. $hr\text{-}W(j, i, pns)$ records the component W (*world*) and $hr\text{-}D(j, i, pns)$ records the component D (*departed*).

We begin by showing properties of messages delivered by *Reader-Writer* processes.

Lemma 2. *Consider a step* $\langle s, e, s' \rangle$ *of an execution* α *of* LL-RAMBO, *where* $e = $ recv($\langle\langle W, D, v, t, cm, p_j, p_i \rangle\rangle_{j,i}$ *for* $i, j \in I$, *and* $p_i > s.ig(j)_i.p\text{-}ack$. *Then,*
(a) $s.ig(j)_i.p\text{-}ack = s.hs\text{-}pack(i, j, p_i)$, (b) $s.ig(j)_i.w\text{-}ua \subseteq s.hs\text{-}wua(i, j, p_i)$,
(c) $s.ig(j)_i.d\text{-}ua \subseteq s.hs\text{-}dua(i, j, p_i)$.

Invariant 1 is used in proving the correctness of LL-RAMBO. In Invariant 1, parts (a) to (e) and Lemma 1 are used to show the key parts (f) and (g).

Invariant 1 *For all states* s *of any execution* α *of* LL-RAMBO:
(a) $\langle\langle W, D, v, t, cm, pns, pnr\rangle, i, j\rangle \in s.h\text{-}MSG \Rightarrow W \subseteq s.world_i \wedge D \subseteq s.departed_i$,
(b) $\forall i, j \in I : s.ig(j)_i.w\text{-}ua \subseteq s.world_i - s.ig(j)_i.wk$,
(c) $\forall i, j \in I : s.ig(j)_i.d\text{-}ua \subseteq s.world_i - s.ig(j)_i.dk$,
(d) $\langle\langle W, D, v, t, cm, p, pnr\rangle, i, j\rangle \in s.h\text{-}MSG \Rightarrow s.hs\text{-}wua(i, j, p) \subseteq W$,
(e) $\langle\langle W, D, v, t, cm, p, pnr\rangle, i, j\rangle \in s.h\text{-}MSG \Rightarrow s.hs\text{-}dua(i, j, p) \subseteq D$,
(f) $\forall i, j \in I : s.ig(j)_i.wk \subseteq s.world_j$, *and*
(g) $\forall i, j \in I : s.ig(j)_i.dk \subseteq s.departed_j$.

Parts (f) and (g) of Invariant 1 show that no node overestimates the knowledge of another node about its *world* and *departed* sets. Finally we show the atomicity of objects implemented by LL-RAMBO by proving that it simulates L-RAMBO, i.e., by showing that every trace of LL-RAMBO is a trace of L-RAMBO (hence of RAMBO).

Theorem 2. LL-RAMBO *implements atomic read/write objects.*

Proof. (Sketch) We define a relation R to map (a) a state t of LL-RAMBO to a state s of L-RAMBO so that every "common" state variable has the same value (e.g., for $i \in I$, $t.world_i = s.world_i$, $t.pnum1_i = s.pnum1_i$, etc.) and (b) a message $m = \langle W, D, v, t, cm, pns, pnr\rangle$ in the *Channel* automaton of LL-RAMBO to a message $m' = \langle W, D, v, t, cm, pns, pnr\rangle$ in the *Channel* automaton of L-RAMBO so that: $m.v = m'.v$, $m.t = m'.t$, $m.cm = m'.cm$, $m.pns = m'.pns$, $m.pnr = m'.pnr$, $m.W = hs\text{-}world(i, j, pns) - hs\text{-}wk(i, j, pns)$ and $m'.W = hs\text{-}world(i, j, pns)$, and $m.D = hs\text{-}departed(i, j, pns) - hs\text{-}dk(i, j, pns)$ and $m'.D = hs\text{-}departed(i, j, pns)$. Using the specifications of the two algorithms and Invariant 1, we show that R is a simulation mapping from LL-RAMBO to L-RAMBO. Since L-RAMBO implements atomic objects per Theorem 1, so does LL-RAMBO.

5   LL-RAMBO **Implementation and Performance**

We developed proof-of-concept implementations of RAMBO and LL-RAMBO on a network-of-workstations. In this section we presents preliminary experimental results and overview conditional analysis of algorithms.

Experimental Results. We developed the system by manually translating the Input/Output Automata specification to Java code. To mitigate the introduction of errors during translation, the implementers followed a set of precise rules that guided the derivation of Java code. The platform consists of a Beowulf cluster with ten machines running Linux. The machines are various Pentium processors up to 900 MHz inter-connected via a 100 Mbps Ethernet switch. The implementation of the two algorithms share most of the code and all low-level routines, so that any difference in performance is traceable to the distinct *world* and *departed* set management and the gossiping disci-pline encapsulated in each algorithm.

We are interested in long-lived applications and we assume that the number of par-ticipants grows arbitrarily. Given the limited number of physical nodes, we use majority quorums of the these nodes, and we simulate a large number of other nodes that join the system by including such node identifiers in the *world* sets. Using non-existent nodes approximates the behavior of a long-lived system with a large set of participants. How-ever, when using all-to-all gossip that grows quadratically in the number of participants, it is expected that the differences in RAMBO and LL-RAMBO performance will become more substantial when using a larger number of physical nodes.

The experiment is designed as follows. There are ten nodes that do not leave the system. These nodes perform concurrent read and write operations using a single con-figuration (that does not change over time), consisting of majorities, i.e., six nodes. Figure 4 compares (a) the average latency of gossip messages and (b) the average la-tency of read and write operations in RAMBO and LL-RAMBO, as the cardinality of *world* sets grows from 10 to 7010.

LL-RAMBO exhibits substantially better gossip message latency than RAMBO (Fig. 4(a)). In fact the average gossip latency in LL-RAMBO does not vary noticeably. On the other hand, the gossip latency in RAMBO grows substantially as the cardinality of the *world* sets increases. This is expected due to the smaller incremental gos-sip messages of LL-RAMBO, while in RAMBO, the size of the gossip messages is always pro-portional to the cardinality of the *world* set. LL-RAMBO trades local resources (computation and memory) for smaller and fewer gossip messages. We observe that the read/write operation latency is slightly lower for RAMBO when the cardinal-ity of the *world* sets is small (Fig. 4(b)). As the size of the *world* sets grows, the operation la-tency in LL-RAMBO becomes substantially bet-ter than in RAMBO.

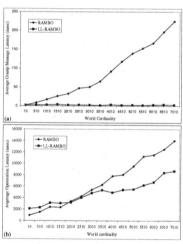

Fig. 4. Preliminary empirical results: (a) gossip message latency, (b) read and write latency.

Performance analysis. We briefly summarize our performance analysis of LL-RAMBO. Here we assume that the participating nodes perpetually gossip with a period d. We show that the latency of read and write operations matches that of RAMBO [10, 6]. Specifically, if d is the maximum message delay, then read and write operations take at most $8d$ time, when reconfigurations are not too frequent. We analyze the communication of LL-RAMBO in the following scenario. We consider the scenario where, once an object is created, several nodes join the system, such that, together with the creator, there are n nodes. Then l nodes leave the system, such that the number of remaining active nodes is $a = n - l$. We show that in this case, after r rounds of gossip, the savings in gossip messages for LL-RAMBO are between $\Omega(r \cdot n)$ and $O(r \cdot n^2)$.

6   Discussion and Future Work

We presented an algorithm for long-lived atomic data in dynamic networks. Prior solutions for dynamic networks [10, 6] did not allow the participants to leave gracefully and relied on gossip that involved sending messages whose size grew with time. The new algorithm, called LL-RAMBO improves on prior work by supporting graceful departures of participants and implementing incremental gossip. The algorithm substantially reduces the size and the number of gossip messages, leading to improved performance of the read and write operations. Our improvements are formally specified and proved.

Acknowledgements. The work of Chryssis Georgiou was performed at the University of Connecticut. The authors thank Nancy Lynch and Seth Gilbert for many discussions.

References

1. Special Issue on Group Communication Services, vol. 39(4) of Comm. of the ACM, 1996.
2. Attiya, H., Bar-Noy, A., Dolev, D.: Sharing Memory Robustly in Message Passing Systems. Journal of the ACM 42(1):124–142, 1996.
3. Birman, K., Joseph, T.: Exploiting virtual synchrony in distributed systems. In Proceedings of the 11th ACM Symposium on Operating Systems Principles, December 1987.
4. Englert, B., Shvartsman, A.A.: Graceful quorum reconfiguration in a robust emulation of shared memory. In Proc. of Inter. Conf. on Dist.d Computer Systems, pp. 454–463, 2000.
5. Golding, R.A.: Weak-consistency group communication and membership. PhD Thesis, University of California, 1992.
6. Gilbert, S.,Lynch, N., Shvartsman, A.A.: RAMBO II: Rapidly reconfigurable atomic memory for dynamic networks. In Proc. of Inter. Conf. on Dependable Systems and Networks, pp. 259–268, 2003.
7. Ghorbani, A., Bhavsar, V.: Training artificial neural networks using variable precision incremental communication. In Proc. of IEEE World Congress On Computational Intelligence, 3:1409–1414, 1994.
8. Guy, R.G., Heidemann, J.S., Mak, W., Page Jr., T.W., Popek, G.J., Rothmeier, D.: Implementation of the Ficus Replicated File System. In Proc. of Summer USENIX Conference, pp. 63-71, 1990.
9. Lynch, N.A.: Distributed Algorithms. Morgan Kaufmann Publishers, 1996.
10. Lynch, N., Shvartsman, A.A.: RAMBO: A reconfigurable atomic memory service for dynamic networks. In Proc. of 16th Inter. Symp. on Dist. Comp., pp. 173–190, 2002.
11. Lynch, N.A., Tuttle, M.: Hierarchical correctness proofs for distributed algorithms. LCS/TR-387, MIT, 1987.
12. Minsky, Y.: Spreading rumors cheaply, quickly, and reliably. Ph.D Thesis, Cornell University, 2002.
13. Rabinovich, M., Gehani, N., Kononov, A.: Efficient update propagation in epidemic replicated databases. In Proc. of 5th Int. Conf. on Extending Database Tech., pp. 207-222, 1996.
14. Upfal, E., Wigderson, A.: How to share memory in a distributed system. Journal of the ACM 34(1):116–127, 1987.
15. Vitanyi, P., Awerbuch, B.: Atomic shared register access by asynchronous hardware. In Proc. of 27th IEEE Symposium on Foundations of Computer Science, pp. 233–243, 1986.

Fault Tolerant Forwarding and Optical Indexes: A Design Theory Approach

Arvind Gupta[1,*], Ján Maňuch[1,**], and Ladislav Stacho[2,***]

[1] School of Computing, Simon Fraser University, Canada
{arvind,jmanuch}@sfu.ca
[2] Department of Mathematics, Simon Fraser University, Canada
lstacho@sfu.ca

Abstract. We study the problem of designing fault tolerant routings in complete and complete bipartite optical networks. We show that this problem has strong connections to various fundamental problems in design theory. Using design theory approach, we find optimal f-tolerant arc-forwarding indexes for complete networks of a prime power order, and all complete bipartite networks. Similarly, we find almost exact values for f-tolerant optical indexes for these networks. Our work motivates an interesting relaxation of an extensively studied problem on mutually orthogonal Latin squares.

1 Introduction

Routing communication demands is one of the fundamental problems in the area of networking. One of the most recognized recent applications of the problem is in the area of optical networking [1]. Most of the research concentrates on determining two basic invariants of a given optical network – the *arc-forwarding* and *optical indexes* [3]. In [8], fault-tolerant issues of optical networks were considered and the two invariants were generalized into so called *f-tolerant arc-forwarding* and *f-tolerant optical indexes*. The parameter f represents the number of faults that are tolerated in the optical network. To determine the arc-forwarding index of a network, one has to design a delicate path system which uses every link in the network evenly. Unfortunately, since most of the known constructions of such systems are ad hoc, possible connections of this problem with other areas of computer science and mathematics remain hidden.

The aim of this paper is twofold. First, we try to understand why path layout problems are hard even for very simple networks. In particular, we explore connections between the construction of path layouts in simple networks and some hard problems in design theory. This leads us to postulate an interesting conjectures on Latin squares. Second, using tools from design theory, we determine

---

* Research supported in part by NSERC (Natural Science and Engineering Research Council of Canada) grant.
** Research supported in part by PIMS (Pacific Institute for Mathematical Sciences.
*** Research supported in part by NSERC (Natural Science and Engineering Research Council of Canada) grant, and VEGA grant No. 2/3164/23 (Slovak grant agency).

f-tolerant arc-forwarding and (almost determine) f-tolerant optical indexes of complete and complete bipartite networks for many values of f.

We model an all-optical network as a *symmetric* directed graph G with vertex set $V(G)$ and arc set $A(G)$, i.e., if $(u, v) \in A(G)$ then $(v, u) \in A(G)$. Let $P(u, v)$ denote a directed path in G from u to v. The set

$$\mathcal{R}_f(G) = \{P_i(u, v) : (u, v) \in V(G), \ u \neq v, \ i = 0, \ldots, f\}$$

where for each ordered pair $(u, v) \in V(G) \times V(G)$ of distinct nodes, the paths $P_0(u, v), \ldots, P_f(u, v)$ are internally node disjoint is called f-*fault tolerant routing*. As noted in [2], when links and/or nodes may fail in a network, it is important to establish connections for a required communication demand, that guarantee fault-free transmission. Assuming that at most f links/nodes may fail in the network, the set $\mathcal{R}_f(G)$ will obviously provide such a routing. The routing $\mathcal{R}_f(G)$ can be viewed as an embedding of a complete f-multi-digraph on n nodes into G where n is the cardinality of $V(G)$. Most of the known work concentrated on 0-fault tolerant routings $\mathcal{R}_0(G)$.

In practical applications the number of different signals a link can carry, is limited. Therefore, the goal is to design a routing which minimizes the maximum load on arcs. Let $\pi(\mathcal{R}_f(G))$ denote the maximum load on arcs, that is, the maximum number of times an arc of G appears in directed paths of $\mathcal{R}_f(G)$. Then

$$\pi_f(G) = \min_{\mathcal{R}_f(G)} \pi(\mathcal{R}_f(G))$$

is called the f-*tolerant arc-forwarding index* of G. Note that the 0-tolerant arc-forwarding index is the well-known arc-forwarding index, cf. [9].

In an optical network, a request is serviced by sending a signal on a specific wavelength over the entire path assigned to the request by the routing. Therefore, whenever two paths of the routing share an arc, they must be assigned to different wavelengths. Let $w(\mathcal{R}_f(G))$ be the smallest number of wavelengths needed to assign to the directed paths of $\mathcal{R}_f(G)$ so that no two paths that share an arc receive the same wavelength. Then

$$w_f(G) = \min_{\mathcal{R}_f(G)} w(\mathcal{R}_f(G))$$

is called the f-*tolerant optical index* of G. Again, the 0-tolerant optical index is equivalent to the well-studied optical index.

Consider an f-fault tolerant routing $\mathcal{R}_f(G)$. We say that the routing $\mathcal{R}_f$ is *optimal* if its congestion achieves the f-tolerant arc-forwarding index $\pi_f(G)$, and it is *balanced* if the difference between congestions of any two arcs is at most one. For any $i = 0, \ldots, f$, the *level* i of the routing $\mathcal{R}_f$ is the set of paths $P_i(u, v) \in \mathcal{R}_f$, for all $u \neq v$. It follows that for any $f' < f$, the subrouting $\mathcal{R}_{f'}(G)$ consisting of levels $0, \ldots, f'$, is f'-fault tolerant. We say that an optimal balanced routing $\mathcal{R}_f(G)$ is *levelled* if every one of its subroutings is also optimal and balanced. A feature of a levelled f-tolerant routing is that it can easily be reconfigured when there is a change in the requirement on the number of faults the routing should tolerate.

In subsequent sections, we establish a close connection between levelled f-fault tolerant routings in the complete digraph $\boldsymbol{K}_n$ and the complete bipartite digraph $\boldsymbol{K}_{n,n}$ and the existence of $f + 1$ mutually orthogonal Latin squares. Since the later is a very hard problem, this indicates that determining $\boldsymbol{\pi}_f(G)$ for $f > 0$ is much harder than determining $\boldsymbol{\pi}_0(G)$ even for very simple graphs G. Using tools from design theory, we calculate $\boldsymbol{\pi}_f(\boldsymbol{K}_n)$ for every f up to the maximal possible value $n - 2$ when n is a power of prime, while for other n, the maximal value of f depends on some number-theoretic properties of n. Further, we calculate $\boldsymbol{\pi}_f(\boldsymbol{K}_{n,n})$ for every f up to its maximal value $n - 1$. We also determine almost optimal values for the f-tolerant optical indexes for these graphs. We believe that this design theory approach can also be used for other graphs G. Note that design theory approach has already been used in a related problem – minimizing the number of ADM switches in WDM ring networks, see for example [4].

2   Preliminaries

Let p be a path in a graph G. The *length* of p is the number of arcs of p. The *distance* of two nodes u and v of G is the length of a shortest path connecting u and v. We generalize distance as follows. The k-*distance* of u and v, denoted by $\mathrm{d}_k(u, v)$, is the minimal sum of lengths of k internally node disjoint paths connecting u and v.

We start with the following obvious lower bound for the f-fault tolerant forwarding index.

Proposition 1. *For any graph* $G = (V, E)$ *and* $f \leq |V| - 2$,

$$\boldsymbol{\pi}_f(G) \geq \left\lceil \frac{1}{|E|} \sum_{u,v \in V} \mathrm{d}_{f+1}(u, v) \right\rceil .$$

Let G be a symmetric digraph. To upper bound the optical index of the graph G, take an f-fault tolerant routing $\mathcal{R} = \mathcal{R}_f(G)$ of G. We can construct an undirected *path graph* $\mathcal{G}_{\mathcal{R}} = G(\mathcal{R}, E(\mathcal{R}))$ as follows: two paths $P, P' \in \mathcal{R}$ are connected by an edge if P and P' share an arc. Obviously, the minimum number of wavelength needed for $\mathcal{R}$ is the same as the chromatic number of the path graph $\mathcal{G}_{\mathcal{R}}$ which is at least $\boldsymbol{\pi}_f(G)$. Hence, we have the following lower bound for the optical index.

Proposition 2. *For any symmetric digraph* G *with connectivity* k, *and any* $0 \leq f \leq k$,

$$\boldsymbol{\pi}_f(G) \leq \boldsymbol{w}_f(G).$$

It is a well known open problem for 0-fault tolerant routings to show that $\boldsymbol{\pi}_0(G) = \boldsymbol{w}_0(G)$ for any symmetric digraph G. This is the case for many extensively studied interconnection networks and was recently also proved for symmetric trees, cf. [7]. We believe the same relation is true for f-tolerant routings and repeat here our conjecture from [8].

Conjecture 1. Let G be a symmetric digraph with connectivity k. For any $0 \leq f < k$,

$$\pi_f(G) = \boldsymbol{w}_f(G).$$

3   Complete Digraphs

In this section, we consider complete digraphs $\boldsymbol{K}_n$ and determine their f-tolerant arc-forwarding and (almost determine) their f-tolerant optical indexes. This is achieved by reformulating these problems as design theory problems. In particular, we show that these problems are very closely related to the existence of mutually orthogonal Latin squares, and propose some interesting conjectures about Latin squares.

Consider a complete digraph $\boldsymbol{K}_n$ with $V(\boldsymbol{K}_n) = V$ where V is a set of cardinality n and an f-fault tolerant routing $\mathcal{R}_f(\boldsymbol{K}_n)$. Necessarily, $f \leq n - 2$.

Take a pair of distinct nodes $u, v \in V$. Consider the $f + 1$ paths connecting u and v. At most one of these paths can have length 1. Therefore, $d_{f+1}(u, v) \geq 2f + 1$. Since the number of pairs (u, v) is the same as the number of arcs, by Proposition 1, $\pi_f(\boldsymbol{K}_n) \geq 2f + 1$.

In what follows, we would like to design a levelled f-fault tolerant routing attaining this lower bound. However, this is not always possible. For example, an exhaustive computer search has shown that there is no levelled f-fault tolerant routing for $n = 6$ and $f > 2$.

First, we will show that for every prime power n there is a levelled $(n-2)$-fault tolerant routing.

Let us start with three definitions on Latin squares.

Definition 1. *A* Latin square *of order n is an $n \times n$ array with entries from a set S of cardinality n such that every row (resp. column) contains each symbol of S exactly once.*

Formally, a Latin square *of order n is a pair (S, L) where L is a mapping $L : S \times S \to S$ such that for any $u, w \in S$, the equation*

$$L(u, v) = w \quad \text{(resp. } L(v, u) = w)$$

has a unique solution $v \in S$.

In other words the row mapping $R_u^L : S \to S$ and the column mapping $C_u^L : S \to S$ defined as

$$R_u^L(v) = L(u, v) \quad \text{and} \quad C_u^L(v) = L(v, u) \quad \text{for all } u, v \in S,$$

are permutations.

A Latin square is normally written as an $n \times n$ array for which the cell in row u and column v contains the symbol $L(u, v)$.

Definition 2. *We say that a* Latin square (S, L) *is* idempotent *if for every $u \in S$, $L(u, u) = u$.*

Definition 3. *We say that two Latin squares (S, L_1) and (S, L_2) are independent if for all $u \neq v$, $L_1(u, v) \neq L_2(u, v)$.*

Note that the property of independence of two idempotent Latin squares (S, L_1) and (S, L_2) is weaker than the property of *orthogonality* which has been studied extensively, cf. [6]. Two Latin squares (S, L_1) and (S, L_2) are orthogonal if

$$\{[L_1(u, v), L_2(u, v)]; \quad u, v \in S\} = S \times S.$$

Indeed, orthogonal idempotent Latin squares (S, L_1) and (S, L_2) are also independent, since if $L_1(u, v) = L_2(u, v) = w$ for some $u \neq v$ then

$$[L_1(u, v), L_2(u, v)] = [w, w] = [L_1(w, w), L_2(w, w)]$$

which contradicts orthogonality.

It has been shown in [12] that there are no 2 orthogonal Latin squares of order 6. Note that it is a fundamental problem of design theory to find the maximum number of mutually orthogonal Latin squares for each order n. On the other hand, it is not hard to find two independent idempotent Latin squares of order 6. Hence, it should be easier to find $n - 2$ mutually independent idempotent Latin squares than mutually orthogonal Latin squares.

If n is a prime power, an elementary construction in [11] shows that there exist $n - 2$ mutually orthogonal idempotent Latin squares of order n, and hence $n - 2$ mutually independent idempotent Latin squares.

Lemma 1 ([11]). *Let $n = p^r$ be a prime power. Then there exist $n - 2$ mutually orthogonal idempotent Latin squares.*

Let us recall the construction.

Construction 1. *Let $(\mathbb{F}, \cdot, +)$ be a finite field of order $n = p^r$. Then for every $i \in \mathbb{F} - \{0, 1\}$ and $u, v \in \mathbb{F}$ define*

$$L_i^*(u, v) = i \cdot u + (1 - i) \cdot v.$$

Theorem 1. *If n is a prime power we can construct a levelled $(n - 2)$-fault tolerant routing for K_n in polynomial time. In particular, $\pi_f(K_n) = 2f + 1$ for all $f \leq n - 2$.*

Proof. Assume that n and f are chosen so that there exists a levelled f-fault tolerant routing $\mathcal{R}_f(K_n)$ with congestion $2f + 1$. We show that such routing exists when n is a prime power and $f \leq n-2$. Obviously, the subrouting $\mathcal{R}_0(K_n)$ of $\mathcal{R}_f(K_n)$ contains only paths of length 1, and each consecutive level contains only paths of length 2. Therefore the levelled routing $\mathcal{R}_f(K_n)$ can be described using a system of functions $F_i : V \times V \to V$, $i = 1, \ldots, f$ defined as the union of the following sets:

$$\begin{aligned}
\mathcal{P}_0 &= \cup_{u \neq v}\{P_0(u, v) = u \to v\}, \\
\mathcal{P}_i &= \cup_{u \neq v}\{P_i(u, v) = u \to F_i(u, v) \to v\}, \text{ for every } i = 1, \ldots, f,
\end{aligned} \qquad \text{(P)}$$

where the system of functions $\{F_i\}_{i=1}^f$ satisfies conditions (1), (2) and (3) described bellow.

First, we have to guarantee that the P_i's are paths, i.e. that for all i and for all $u \neq v$,

$$F_i(u, v) \neq u, v. \tag{1}$$

Next, for every $i = 1, \ldots, f$ and every pair of distinct nodes $u, v \in V$, consider all paths of level i using the arc $u \to v$. Since, our system is levelled, there are exactly two such paths. We can divide them into two groups: $M_i^{(1)}(u, v) = \{P_i(u, z); \ F_i(u, z) = v\}$ containing the paths which use the arc $u \to v$ as the first arc of the path, and $M_i^{(2)}(u, v) = \{P_i(z, v); \ F_i(z, v) = u\}$ containing the paths which use the arc $u \to v$ as the second arc. The second condition is that

$$|M_i^{(1)}(u, v)| + |M_i^{(2)}(u, v)| = 2 \tag{2}$$

for every $u \neq v \in V$ and $i = 1, \ldots, f$.

Finally, we have to ensure that the paths from u to v on different levels are internally node disjoint: for every $u \neq v$ and $i \neq j$,

$$F_i(u, v) \neq F_j(u, v). \tag{3}$$

To find the required routing $\mathcal{R}_f(K_n)$, it is enough to find a system of functions $\{F_i\}_{i=1}^f$ satisfying conditions (1), (2) and (3). To make the system more symmetric, and hence easier to design, we will replace condition (2) by the stronger condition that

$$|M_i^{(1)}(u, v)| = |M_i^{(2)}(u, v)| = 1 \tag{2'}$$

for all u, v and i. Note that we no longer require $u \neq v$. Conditions (2') and (1) imply that

$$F_i(u, u) = u \tag{1'}$$

for all $u \in V$. On the other hand, conditions (2') and (1') imply (1). Hence, it is enough to find a system of functions $\{F_i\}_{i=1}^f$ satisfying conditions (1'), (2') and (3). Such a system of functions corresponds to a sequence of f mutually independent idempotent Latin squares. This follows from the following two claims.

Claim. If a system of functions $\{F_i\}_{i=1}^f$ satisfies condition (2') then (V, F_i) is a Latin square for each $i = 1, \ldots, f$.

Proof. Fix an i. For all u, v we have the following two implications:

$$|M_i^{(1)}(u, v)| = 1 \quad \Longrightarrow \quad \text{there exists exactly one } z \text{ such that } F_i(u, z) = v,$$
$$|M_i^{(2)}(u, v)| = 1 \quad \Longrightarrow \quad \text{there exists exactly one } z \text{ such that } F_i(z, v) = u,$$

which implies that (V, F_i) is a Latin square.

Claim. The system of functions $\{F_i\}_{i=1}^f$ which satisfies conditions (1'), (2') and (3) corresponds to the sequence of f mutually independent idempotent Latin squares.

Proof. Obviously, for a system of functions $\{F_i\}_{i=1}^f$ which satisfies (1') we have that (S, F_i) is an idempotent Latin square, for every $i = 1, \ldots, f$.

Obviously, for a system of functions $\{F_i\}_{i=1}^f$ which satisfies (3), we have that the sequence of Latin squares $(S, F_1), \ldots, (S, F_f)$ is mutually independent.

Now, the proof of the theorem follows from Claim 3 and Lemma 1 immediately.

If n is not a prime power then there may not exist $n-2$ mutually independent idempotent Latin squares of order n. We observed by an exhaustive computer search that:

- for small even n's which are not powers of 2, there do not exist $n-2$ such Latin squares;
- for small odd n's, there exist $n-2$ such Latin squares, and the number of such sequences grows very quickly.

This fact motivates us to pose the following conjecture.

Conjecture 2. For each odd $n \geq 3$, there exist $n-2$ mutually independent idempotent Latin squares of order n.

In the full version, we propose an approach for solving the conjecture and reformulate it as a very interesting problem about permutations. For general n we can use the direct product composition of Latin squares to construct $p^r - 1$ mutually orthogonal Latin squares, where p^r is the smallest maximal prime power dividing n, i.e, $p^r = \min\{p_1^{r_1}, p_2^{r_2}, \ldots, p_k^{r_k}\}$, where $n = p_1^{r_1} p_2^{r_2} \ldots p_k^{r_k}$ and $p_1, p_2, \ldots, p_k$ are distinct primes, cf. [10]. Using the same technique we can give a similar result directly formulated for mutually independent idempotent Latin squares. Before proving the result we need the following definition.

Definition 4. *The direct product $A \times B$ of two Latin squares of orders m and n respectively is the Latin square of order mn defined by*

$$(A \times B)([a_1, b_1], [a_2, b_2]) = [A(a_1, a_2), B(b_1, b_2)].$$

Lemma 2. *Let $A_1, \ldots, A_k$ be k mutually independent idempotent Latin squares of order m and let $B_1, \ldots, B_k$ be k mutually independent idempotent Latin square of order n. Then $A_1 \times B_1, \ldots, A_k \times B_k$ are k mutually orthogonal idempotent Latin squares of order mn.*

The proof will appear in the full version.

Now, we can combine Lemma 2 with Claim 3 to obtain the following.

Theorem 2. *Let $n = p_1^{r_1} \ldots p_k^{r_k}$ where $p_1, \ldots, p_k$ are distinct primes and let $p^r = \min\{p_1^{r_1}, \ldots, p_k^{r_k}\}$. We can construct a levelled $(p^r - 2)$-fault tolerant routing for K_n in polynomial time. In particular, $\pi_f(K_n) = 2f + 1$ for all $f \leq p^r - 2$.*

In most practical applications this is sufficient, since an f-tolerant system requires at least $2f+1$ wavelengths to send on one optical link but the bandwidth of an optical cable is limited.

3.1   An Upper Bound for the Optical Index

Consider a levelled f-fault tolerant routing $\mathcal{R}_f(\boldsymbol{K}_n) = \cup_{i=0}^{f} \mathcal{P}_i$ described in (P), where $F_1, \ldots, F_f$ are independent idempotent Latin squares. We will consider the subroutings $\mathcal{P}_0, \ldots, \mathcal{P}_f$ separately. For each we build the path graph and upper bound its chromatic number. Then, by assigning different sets of wavelength for the paths in different subroutings, we have the following upper bound for the wavelength number of $\mathcal{R}_f(\boldsymbol{K}_n)$:

$$w(\mathcal{R}_f(\boldsymbol{K}_n)) \le \sum_{i=0}^{f} \chi(\mathcal{G}_{\mathcal{P}_i}).$$

Lemma 3. *Consider a levelled f-fault tolerant routing $\cup_{i=0}^{f}\mathcal{P}_i$ described in* (P), *where $f \le n-2$ and $F_1, \ldots, F_f$ are independent idempotent Latin squares. Then $\chi(\mathcal{G}_{\mathcal{P}_0}) = 1$, and for every $i = 1, \ldots, f$, $\chi(\mathcal{G}_{\mathcal{P}_i}) \le 3$.*

Proof. The path graph $\mathcal{G}_{\mathcal{P}_0}$ contains no edges, hence its chromatic number is 1. Consider the path graph $\mathcal{G}_{\mathcal{P}_i}$ for some $i = 1, \ldots, f$. By (2), the degree of each node of $\mathcal{G}_{\mathcal{P}_i}$ is exactly 2, i.e., the path graph is a collection of cycles. Obviously, the path graph can be coloured using 3 colours.

Note that it can happen that the cycles of the path graph are of odd lengths, hence two colours are not enough to color the path graph $\mathcal{G}_{\mathcal{P}_i}$. We have the following upper bound for the optical index of $\boldsymbol{K}_n$.

Theorem 3. *Let $n = p_1^{r_1} \ldots p_k^{r_k}$ where $p_1, \ldots, p_k$ are distinct primes and let $p^r = \min\{p_1^{r_1}, \ldots, p_k^{r_k}\}$. For every $f = 0, \ldots, (p^r - 2)$, there exists a levelled f-fault tolerant routing $\mathcal{R}_f(\boldsymbol{K}_n)$ with wavelength number $w(\mathcal{R}_f(\boldsymbol{K}_n)) \le 3f+1$. Consequently,*
$$2f + 1 \le w_f(\boldsymbol{K}_n) \le 3f + 1.$$

4   Complete Bipartite Digraphs

Consider a complete bipartite digraph $\boldsymbol{K}_{n,n}$ with node partitioning $U = \{u_0, \ldots, u_{n-1}\}$ and $V = \{v_0, \ldots, v_{n-1}\}$. Let $\mathcal{R}_f(\boldsymbol{K}_{n,n})$ be an f-fault tolerant routing. Necessarily, $f \le n - 1$.

Take a pair of distinct nodes $x, y \in U \cup V$. If x and y belong to the same partition then every path from x to y has length at least 2, and thus we have $d_{f+1}(x, y) \ge 2(f + 1)$. If x and y belong to different partitions then there is only one path connecting x and y of length 1, and all other paths have length at least 3. Hence, in this case, $d_{f+1}(x, y) \ge 3f + 1$. By Proposition 1, we have

$$\pi_f(\boldsymbol{K}_{n,n}) \geq \left\lceil \frac{1}{2n^2} \left(2n(n-1) \cdot 2(f+1) + 2n^2(3f+1)\right) \right\rceil$$

$$= \begin{cases} 5f+3, & \text{if } f < \frac{n}{2} - 1, \\ 5f+2, & \text{if } \frac{n}{2} - 1 \leq f < n-1, \\ 5f+1, & \text{if } f = n-1. \end{cases}$$

We will show a construction of an $(n-1)$-fault tolerant levelled routing achieving this lower bound. We split the paths of the routing into the following sets of paths

$$\mathcal{A}_i = \bigcup_{x \neq y} \{u_x \to v_{A_i(x,y)} \to u_y, \quad v_x \to u_{A_i'(x,y)} \to v_y\},$$

$$\mathcal{B}_0 = \bigcup_{x,y} \{u_x \to v_y, \quad v_x \to u_x\},$$

$$\mathcal{B}_i = \bigcup_{x,y} \{u_x \to v_{B_i(x,y)} \to u_{C_i(x,y)} \to v_y, \quad v_x \to u_{B_i'(x,y)} \to v_{C_i'(x,y)} \to u_y\},$$

where $A_i, A_i', B_i, C_i, B_i', C_i'$ are mappings $\mathbb{Z}_n \times \mathbb{Z}_n \to \mathbb{Z}_n$. We choose A_i, A_i', B_i, C_i, B_i', C_i' so that both sets $\mathcal{A} = \cup_{i \in \mathbb{Z}_n} \mathcal{A}_i$ and $\mathcal{B} = \cup_{i \in \mathbb{Z}_n} \mathcal{B}_i$ will be levelled routings for $(\boldsymbol{K}_{n,n}, U \hat{\times} U \cup V \hat{\times} V)$ and $(\boldsymbol{K}_{n,n}, U \times V \cup V \times U)$, respectively, where $S \hat{\times} S = \{(s,t) : s,t \in S, \ s \neq t\}$. The symbol (G, W) denotes the communication demand in which only pairs of nodes in W needs to be joined by a directed path.

The following lemma describes the set of paths $\mathcal{A}$.

Lemma 4. *There exists a levelled $(n-1)$-fault tolerant routing $\mathcal{A} = \cup_{i \in \mathbb{Z}_n} \mathcal{A}_i$, where*

$$\mathcal{A}_i = \bigcup_{x \neq y} \{u_x \to v_{A_i(x,y)} \to u_y, \quad v_x \to u_{A_i'(x,y)} \to v_y\}, \quad i = 0, \ldots, n-1,$$

connecting nodes from the same partitions of $\boldsymbol{K}_{n,n}$ which can be constructed in quadratic time. In particular, mappings A_i and A_i' can be chosen in the following way, for all $i = 0, \ldots, n-1$,

$$A_i(x,y) = L(-x, -y) + \phi(i)$$
$$A_i'(x,y) = L'(-x, -y) + \phi(i) + 1, \quad \text{where}$$

$$\phi(i) = \begin{cases} 2i+1 & \text{if } n \text{ is even and } i \geq \frac{n}{2}, \\ 2i & \text{otherwise}, \end{cases}$$

and $(\mathbb{Z}_n, L)$ and $(\mathbb{Z}_n, L')$ are two idempotent Latin squares.

The proof will appear in the full version. A very simple construction of idempotent Latin squares of order $n \geq 3$ is presented in [10].

Next lemma describes the set of paths $\mathcal{B}$. We will need the following definition.

Definition 5. *A Latin square is a* column-normalized, *if its columns are sorted in increasing order by the values in the first row.*

Lemma 5. *We can construct a levelled $(n-1)$-fault tolerant routing $\mathcal{B} = \cup_{i \in \mathbb{Z}_n} \mathcal{B}_i$, where*

$$\mathcal{B}_0 = \bigcup_{x,y} \{u_x \to v_y, \quad v_x \to u_x\},$$

$$\mathcal{B}_i = \bigcup_{x,y} \{u_x \to v_{B_i(x,y)} \to u_{C_i(x,y)} \to v_y, \quad v_x \to u_{B'_i(x,y)} \to v_{C'_i(x,y)} \to u_y\}$$

connecting nodes from different partitions of $\boldsymbol{K}_{n,n}$ in quadratic time. In particular, mappings B_i, C_i, B'_i, C'_i can be chosen as follows:

$$B_i(x,y) = L_B(i,y), \quad C_i(x,y) = L_C(i,x),$$
$$B'_i(x,y) = L_{B'}(i,y), \quad C'_i(x,y) = L_{C'}(i,x),$$

where $(\mathbb{Z}_n, L_B), (\mathbb{Z}_n, L_C), (\mathbb{Z}_n, L_{B'}), (\mathbb{Z}_n, L_{C'})$ are any column-normalized Latin squares.

The proof will appear in the full version.

Lemmas 4 and 5 give the main result of this section.

Theorem 4. *We can construct a levelled $(n-1)$-fault tolerant routing for $\boldsymbol{K}_{n,n}$ in quadratic time. In particular, for all $f \leq n-1$,*

$$\pi_f(\boldsymbol{K}_{n,n}) = \begin{cases} 5f+3, & \text{if } f < \frac{n}{2}-1, \\ 5f+2, & \text{if } \frac{n}{2}-1 \leq f < n-1, \\ 5f+1, & \text{if } f = n-1. \end{cases}$$

4.1   An Upper Bound for the Optical Index

Consider a levelled f-fault tolerant routing $\mathcal{R}_f(\boldsymbol{K}_{n,n})$ as described in Lemmas 4 and 5. We will consider the subroutings $\mathcal{A}_0, \ldots, \mathcal{A}_f$ and $\mathcal{B}_0, \ldots, \mathcal{B}_f$ separately. For each we build the path graph and upper bound its chromatic number. Then, by assigning different sets of wavelength for the paths in different subroutings, we have the following upper bound for the wavelength number of $\mathcal{R}_f(\boldsymbol{K}_{n,n})$:

$$w(\boldsymbol{K}_{n,n}, \hat{D}, \mathcal{R}_f(\boldsymbol{K}_{n,n})) \leq \sum_{i=0}^{f} \chi(\mathcal{G}_{\mathcal{A}_i}) + \sum_{i=0}^{f} \chi(\mathcal{G}_{\mathcal{B}_i}).$$

For set of paths $\mathcal{A}_i$, we show that the corresponding path graph is bipartite. For set of paths $\mathcal{B}_i$ we first derive a simple upper bound on the maximum degree of the corresponding path graph. Next we show that by a special choice of Latin squares (used in construction of the system) this upper bound can be improved. Upper bounds on optical indexes will follow from Brook's theorem which gives an upper bound on chromatic number in terms of maximum degree of the graph.

Lemma 6. *Consider a levelled $(n-1)$-fault tolerant routing $\cup_{i \in \mathbb{Z}_n} \mathcal{A}_i$ described in Lemma 4. For every $i = 0, \ldots, n-1$, $\chi(\mathcal{G}_{\mathcal{A}_i}) \leq 2$.*

Proof. It is enough to prove that the path graph $\mathcal{G}_{\mathcal{A}_i}$ is a bipartite graph. We have two types of paths: UU-paths which start and end in U and VV-paths which start and end in V. It is easy to see that there are no edges between the paths of the same type. Indeed, assume for instance that for $x_1 \neq y_1$, $x_2 \neq y_2$, $(x_1, y_1) \neq (x_2, y_2)$, the paths $u_{x_1} \to v_{L(n-x_1, n-y_1)+\phi(i)} \to u_{y_1}$ and $u_{x_2} \to v_{L(n-x_2, n-y_2)+\phi(i)} \to u_{y_2}$ share the first arc. Then $x_1 = x_2 = n - x$ and $L(x, n - y_1) = L(x, n - y_2)$. Since L is a Latin square, $y_1 = y_2$, a contradiction.

Hence, for paths in $\cup_{i=0}^{f} \mathcal{A}_i$ we need $2(f+1)$ wavelengths. This is optimal for $f < n/2 - 1$, and differs from the optimal value by at most two for other values of f.

Lemma 7. *Consider a levelled $(n-1)$-fault tolerant routing $\cup_{i \in \mathbb{Z}_n} \mathcal{B}_i$ described in Lemma 5. Then $\chi(\mathcal{G}_{\mathcal{B}_0}) = 1$, and for every $i = 1, \ldots, n-1$, $\chi(\mathcal{G}_{\mathcal{B}_i}) \leq 6$.*

The proof will appear in the full version.

This does not give a very good upper bound for the optical index. The congestion induced by paths in $\mathcal{B}_i$ is 3, therefore the colouring we found might be twice the optimal. To improve the colouring we will further restrict the class of routings, by taking special mappings B_i, C_i, B_i', C_i'. We need the following definition.

Definition 6. *Let (S, L) be a Latin square. The row-inverse of (S, L) is a pair $(S, \bar{L})$, where $\bar{L} : S \times S \to S$ is a mapping such that $\bar{L}(u, v) = (R_u^L)^{-1}(v)$. Recall that the row mappings R_u^L are bijective.*

The next lemma checks that the row-inverse of a Latin square is also a Latin square.

Lemma 8. *Let (S, L) be a Latin square. The row-inverse $(S, \bar{L})$ is a Latin square. If $(\mathbb{Z}_n, L)$ is column-normalized then so is $(\mathbb{Z}_n, \bar{L})$.*

Proof. Obviously, the row mappings of $(S, \bar{L})$ are permutations. Hence, assume that for some $x, x', y \in S$, $x \neq x'$, we have $\bar{L}(x, y) = \bar{L}(x', y)$. Then $(R_x^L)^{-1}(y) = (R_{x'}^L)^{-1}(y) = z$. This yields

$$R_x^L(z) = R_x^L((R_x^L)^{-1}(y)) = y = R_{x'}^L(z).$$

i.e., $L(x, z) = L(x', z)$, a contradiction.

The second part of the lemma follows by a simple observation that a Latin square is column-normalized if and only if the row mapping R_0^L is the identity mapping.

Lemma 9. *Consider a levelled $(n-1)$-fault tolerant routing $\cup_{i \in \mathbb{Z}_n} \mathcal{B}_i$ described in Lemma 5 such that $L_{B'}$ is a row-inverse of L_C and $L_{C'}$ is a row-inverse of L_B (the column-normalized Latin squares L_B and L_C can be chosen arbitrary). Then for all $i = 1, \ldots, n-1$, $\chi(\mathcal{G}_{\mathcal{B}_i}) \leq 4$.*

The proof will appear in the full version.

As a conclusion we have the upper bound for the f-tolerant optical index of $\boldsymbol{K}_{n,n}$.

Theorem 5. *For every* $f = 0, \ldots, n - 1$, *there exists a levelled f-fault tolerant routing* $\mathcal{R}$ *for* $\boldsymbol{K}_{n,n}$ *with the wavelength number* $\boldsymbol{w}(\boldsymbol{K}_{n,n}, \hat{D}, \mathcal{R}) \leq 6f + 3$. *Consequently,*

$$5f + 1 \leq \boldsymbol{w}_f(\boldsymbol{K}_{n,n}) \leq 6f + 3.$$

References

1. A. Aggarwal, A. Bar-Noy, D. Coppersmith, R. Ramaswami, B. Schieber and M. Sudan, Efficient routing in optical networks, J. of the ACM **46** (1996) 973–1001.
2. B. Beauquier, All-to-all communication for some wavelength-routed all-optical networks, Networks **33** (1999) 179–187.
3. B. Beauquier, C. J. Bermond, L. Gargano, P. Hell, S. Perennes, and U. Vaccaro, Graph Problems arising from wavelength-routing in all-optical networks, in Proc. of the 2nd Workshop on Optics and Computer Science, part of IPPS'97, 1997.
4. J.-C. Bermond and D. Coudert, Traffic grooming in unidirectional WDM ring networks using design theory In ICC 2003, IEEE International Conference on Communications, 11-15 May 2003.
5. C.J. Bermond, L. Gargano, S. Perennes, A. Rescigno and U. Vaccaro, Efficient collective communications in optical networks, Theoretical Computer Science **233** (2000) 165–189.
6. C.J. Colbourn and J.H. Dinitz, Mutually orthogonal Latin squares: a brief survey of constructions, Journal of Statistical Planning and Inference **95** (2001) 9–48.
7. L. Gargano, P. Hell and S. Pérennes, Colouring paths in directed symmetric trees with applications to optical networks, J. Graph Theory **38** (2001) 183–186.
8. J. Maňuch and L. Stacho, On f-wise arc forwarding index and wavelength allocations in faulty all-optical hypercubes, Theor. Informatics and Appl. **37** (2003) 255–270.
9. Y. Manoussakis and Z. Tuza, The forwarding index of directed networks, Discrete Appl. Math. **68** (1996) 279–291.
10. C.C. Lindner, C.A. Rodger, Design theory, CRC Press Series on Discrete Mathematics and its Applications (1997).
11. N.S. Mendelsohn, On maximal sets of mutually orthogonal idempotent Latin squares, Canad. Math. Bull. **14** (1971) 449.
12. G. Tarry, Le problème de 36 officiers, Assoc. Franc. Av. Sci. **29** (1900) 170–203.

Tighter Bounds on Feedback Vertex Sets in Mesh-Based Networks

Flaminia L. Luccio[1] and Jop F. Sibeyn[2]

[1] Dipartimento di Scienze Matematiche, Università di Trieste, Italy
luccio@dsm.univ.trieste.it
[2] Institut für Informatik, Martin-Luther-Universität Halle, Germany
jopsi@informatik.uni-halle.de

Abstract. In this paper we consider the minimum feedback vertex set problem in graphs, i.e., the problem of finding a minimal cardinality subset of the vertices, whose removal makes a graph acyclic. The problem is $\mathcal{NP}$-hard for general topologies, but optimal and near-optimal solutions have been provided for particular networks. We improve the upper bounds of [11] both for the two-dimensional mesh of trees, and for the pyramid networks. We also present upper and lower bounds for other topologies: the higher-dimensional meshes of trees, and the trees of meshes networks. For the two-dimensional meshes of trees the results are optimal; for the higher-dimensional meshes of trees and the tree of meshes the results are asymptotically optimal. For the pyramid networks, the presented upper bound almost matches the lower bound of [11].

1 Introduction

The Problem. For a graph $G = (V, E)$, a subset $V' \subseteq V$ is called a *feedback vertex set*, if the subgraph of G induced by $V - V'$ has no cycles. If V' has minimum cardinality among all such subsets, then V' is called a *minimum vertex feedback set* and is denoted by $\overline{V}$. In this paper we study the *minimum feedback vertex set problem (MFP)* which is the problem of finding minimum feedback vertex sets.

Old and New Results. MFP is known to be $\mathcal{NP}$-hard for general networks [6], but optimal and near-optimal solutions have been provided for particular topologies [1, 3–5, 7, 10, 11].

In this paper we consider MFP for certain undirected graphs of bounded degree, namely two- and higher-dimensional meshes of trees networks, trees of meshes and pyramid networks [2, 8, 9, 13–16]. Upper and lower bounds for the two-dimensional mesh of trees and pyramid networks have been presented in [11]. For the two-dimensional mesh of trees we improve the upper bound of [11] obtaining a tight bound. Using a different construction, asymptotically optimal results are obtained for all higher dimensions. We then consider a topology for which MFP was not been studied before, the tree of meshes TM_n, and we show that at least $2/3 \cdot n^2 \cdot \log n - 5/9 \cdot n^2 + n - 4/9$ vertices must be removed and

R. Král* and O. Sýkora (Eds.): SIROCCO 2004, LNCS 3104, pp. 209–220, 2004.
© Springer-Verlag Berlin Heidelberg 2004

construct a set V'_n with $|V'_n| = 2/3 \cdot n^2 \cdot \log n - 25/72 \cdot n^2 + 2/9$. Finally, we consider the pyramid networks and we improve the upper bound of [11] using a construction that is quite involved. Nevertheless we do not get asymptotically tight results: there is a ratio $135/128 \simeq 1.055$ between upper and lower bounds.

When considering lower bounds, we will refer to a result of [1] applied to undirected graphs in which the maximum degree of the vertices is r. The number $|\overline{V}|$ of vertices to remove is: $|\overline{V}| \geq \lceil (|E| - |V| + 1)/(r - 1) \rceil$.   (1)

2   Meshes of Trees

2.1   Two-Dimensional Meshes of Trees

Definition 1. A $p \times q$ mesh is a graph $M_{p,q} = (V, E)$ with $v_{i,j} \in V$, $0 \leq i \leq p-1$, $0 \leq j \leq q - 1$, and E contains exactly the edges $(v_{i,j}, v_{i,j+1})$, $j \neq q - 1$, and $(v_{i,j}, v_{i+1,j})$, $i \neq p - 1$.

Lemma 1 ([10]). A minimum feedback vertex set of a $p \times q$ mesh $M_{p,q}$ has size at least $\lceil ((p - 1) \cdot (q - 1) + 1)/3 \rceil$.

Definition 2 ([8, 15, 16]). A height-k complete binary tree $T_k = (V, E)$ is a graph composed of $2^{k+1} - 1$ vertices labeled with all different binary strings of length $1 \leq i \leq k$. The root at level 0 is labeled with a distinct value ϵ, the 2^i vertices at level i, $1 \leq i \leq k$, are labeled with all possible distinct binary strings of length i in lexicographical order. An edge connects any vertex x lying at level i, $1 \leq i \leq k$ with label $(b_1, \ldots, b_{i-1}, b_i)$ with vertex with label $(b_1, \ldots, b_{i-1})$ at level $i - 1$. A two-dimensional $m \times n$ mesh of trees, where m and n are powers of 2, is a graph $MT_{m,n}$ obtained from an $m \times n$ mesh $M_{m,n}$ by: 1) removing all mesh edges; 2) building a height-h_1, $h_1 = \log n$, complete binary tree along each row, using the row-vertices as leafs of the tree; 3) building a height-h_2, $h_2 = \log m$, complete binary tree along each column, using the column-vertices as leafs of the tree.

In the meshes of trees, the trees along the rows and columns are called *row-trees* and *column-trees*, respectively. The definition is illustrated in Figure 1 A). Two-dimensional meshes of trees are non-regular as the mesh and root vertices have degree 2 while all other vertices have degree 3. $MT_{m,n}$ has $3 \cdot m \cdot n - m - n$ vertices and $4 \cdot m \cdot n - 2 \cdot m - 2 \cdot n$ edges.

In [11] relation (1) was used to provide the following:

Lemma 2. A minimum feedback vertex set of a two-dimensional mesh of trees $MT_{m,n}$ has size at least $m \cdot n/2 - m/2 - n/2 + 1$.

Moreover, in [11] Luccio presented a feedback vertex set V' of size at most $\frac{9}{16} \cdot m \cdot n$. This bound differs from the lower bound in Lemma 2 for a multiplicative constant $9/8$ in the term of highest order. We now improve this upper bound obtaining an optimal bound. To do so we propose two new vertex-removal strategies and we consider their performance. The simplest is the *mesh-removal*

B)

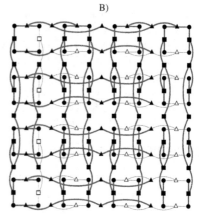

A)

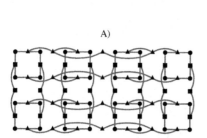

Fig. 1. A) $MT_{4,8}$. The circles give the mesh vertices, the triangles the vertices of the row-trees and the squares the vertices of the column-trees. B) Minimum feedback vertex set for $MT_{8,8}$. The removed vertices are colored white. The edges incident upon removed vertices are drawn with thin lines.

strategy, where V' consists of all mesh vertices. In the *coordinate-axis-removal strategy* V' consists of all vertices of the row-trees at one level above the leafs (alternatively one can take all vertices of the column-trees at one level above the leafs). For the proof of the following lemma refer to [12].

Lemma 3. *Applying the mesh-removal strategy to $MT_{m,n}$, $|V'| = m \cdot n$. Applying the coordinate-axis-removal strategy, $|V'| = m \cdot n/2$. Either strategy gives a feedback vertex.*

The mesh-removal strategy removes twice as many vertices as necessary, but the coordinate-axis-removal strategy is asymptotically optimal. We now consider how to further improve it. The set V' is constructed along the lines of the coordinate-axis-removal strategy with three modifications: 1) None of the vertices in row 0 is added to V'. 2) None of the leftmost vertices at level $\log n - 1$ of the row-trees (the vertices between column 0 and column 1) is added to V'. 3) All vertices in column 1 at level $\log m - 1$ of the column-tree are added to V'. The last two of these modifications were suggested by an anonymous referee of Sirocco 2004, which is gratefully acknowledged. The modified construction is illustrated in Figure 1 B), and for the proof of the following lemma refer to [12].

Theorem 1. *$MT_{m,n}$ has a vertex feedback set V' with $|V'| = m \cdot n/2 - m/2 - n/2 + 1$.*

2.2   Higher-Dimensional Meshes of Trees

Definition 3. *A d-dimensional $n_1 \times \cdots \times n_d$ mesh of trees, denoted $MT_{n_1,\ldots,n_d}$, is obtained from a d-dimensional $n_1 \times \cdots \times n_d$ cube by: 1) removing all mesh edges; 2) building a height-$\log n_j$ complete binary tree along each of the $\prod_{1 \leq i \neq j \leq d} n_i$ one-dimensional subarrays of length n_j running along coordinate axis j.*

A d-dimensional $n_1 \times \cdots \times n_{d-1} \times n_d$ mesh of trees can also be obtained by creating n_d copies of $d - 1$-dimensional $n_1 \times \cdots \times n_{d-1}$ meshes of trees and connecting these at the bottom level by $\prod_{1 \leq i \leq d-1} n_i$ complete binary trees of height $\log n_d$. For $d \geq 3$, the mesh vertices have maximum degree, their degree being d. All other vertices have degree 3 except for the tree roots with degree 2. In the following we are not interested in exactly matching the lower bounds, so there is no need for exact estimates of the number of vertices and edges either. For $MT_{n_1,\ldots,n_d}$ we use the following estimates:

Lemma 4.

$$|V| = (d+1) \cdot \prod_{1 \leq i \leq d} n_i - O\left(\sum_{1 \leq j \leq d} \prod_{1 \leq i \neq j \leq d} n_i \right), \tag{2}$$

$$|E| = 2 \cdot d \cdot \prod_{1 \leq i \leq d} n_i - O\left(\sum_{1 \leq j \leq d} \prod_{1 \leq i \neq j \leq d} n_i \right). \tag{3}$$

Proof. These estimates can be proven using induction from the recursive construction: The lemma is correct for $d = 2$. For larger d, the leading constant in (3) increases by 2 when going from dimension $d - 1$, because each of the new connecting trees contributes its $2 \cdot n_d - 2$ edges. The leading constant in (2) increases only by 1, because n_d of the $2 \cdot n_d - 1$ vertices of the added trees are grid vertices, which are identified with the vertices of the other trees. For $n_1 = \cdots = n_d$, the results of this lemma can be found in [8].   □

Substituting these values in (1) together with $r = d - 1$, we get:

Corollary 1. *For a d-dimensional $n_1 \times \cdots \times n_d$ mesh of trees with $d \geq 3$, a vertex feedback set has size at least $\prod_{1 \leq i \leq d} n_i - O(\sum_{1 \leq j \leq d} \prod_{1 \leq i \neq j \leq d} n_i)$.*

The two vertex-removal strategies can also be generalized: the mesh-removal strategy now amounts to removing all vertices from the d-dimensional $n_1 \times \cdots \times n_d$ mesh, the coordinate-axis-removal strategy amounts to removing all vertices along the $d-1$ coordinate axes at one level above the leafs. For example, V' can be built by taking all at level $\log n_j - 1$ from all meshes of trees running along coordinate axis j, for all $2 \leq j \leq d$. Now we obtain the following generalization of Lemma 3:

Theorem 2. *For a d-dimensional $n_1 \times \cdots \times n_d$ mesh of trees with $d \geq 2$, the set V' constructed with either the mesh-removal or the coordinate-axis-removal strategy is a feedback vertex set for $MT_{n_1,\ldots,n_d}$. With the first strategy $|V'| = \prod_{1 \leq i \leq d} n_i$, with the second strategy $|V'| = (d-1)/2 \cdot \prod_{1 \leq i \leq d} n_i$.*

Proof. The correctness is proven as before: removing the mesh results in a set of trees whose depth has been reduced by one. Leaving out all at one level above the leafs gives us a set of trees of depth two smaller than the original trees plus the isolated mesh vertices. Reinserting the vertices along one coordinate axis connects the mesh vertices with one tree, not creating any cycles. In total the coordinate-axis-removal strategy thus removes $d - 1$ sets of each $(\prod_{1 \leq i \leq d} n_i)/2$ vertices.   □

Both strategies are correct for all $d \geq 2$, but they are not equally good: for $d = 2$, only the coordinate-axis-removal strategy is asymptotically optimal, whereas for $d \geq 4$, we must use the mesh removal strategy to achieve this. Only for $d = 3$ we can choose.

3   Trees of Meshes

Definition 4. [9, 13] *For a power of two n, the two-dimensional $n \times n$ tree of meshes, TM_n, is an undirected graph (V_n, E_n) obtained by replacing each vertex of a complete binary tree with a mesh and each edge by several edges which link the meshes together. More precisely, the root is an $n \times n$ mesh and its two children are $n \times n/2$ meshes, their children are $n/2 \times n/2$ meshes and so on, until the leafs are replaced by 1×1 meshes. An $n' \times n'$ mesh in this tree is linked to its two $n' \times n'/2$ children by connecting the vertices in the rightmost (leftmost) column to the vertices in the leftmost (rightmost) column in the mesh corresponding to the right (left) child. An $n' \times n'/2$ mesh in this tree is linked to its two $n'/2 \times n'/2$ children by connecting the vertices in the topmost (bottommost) row to the vertices in the bottommost (topmost) row in the mesh corresponding to the right (left) child. These connections preserve the column and row order of the vertices and the planarity of the graph.*

For TM_n is a non-regular graph, with vertices of degree 1, 3 and 4. $|V_n| = 2 \cdot n^2 \cdot \log n + n^2$ and $|E_n| = 4 \cdot n^2 \cdot \log n - n^2 + n$. The definition is illustrated in Figure 2.

Using (1), we get the following lower bound on the number of vertices to remove: $|\overline{V}_n| \geq \lceil ((4 \cdot n^2 \cdot \log n - n^2 + n) - (2 \cdot n^2 \cdot \log n + n^2) + 1)/3 \rceil = \lceil (2 \cdot n^2 \cdot (\log n - 1) + n + 1)/3 \rceil$. For $n = 2, 4, 8, 16$, this gives 1, 13, 89, 518, respectively.

This lower-bound result can be sharpened by separately considering the numerous meshes embedded in a tree of meshes in combination with Lemma 1. Consider TM_n with $n = 2^k$. It is composed of four copies of $TM_{n/2}$ connected to an $n \times (2 \cdot n)$ mesh. For $n = 2^k$, it turns out that $\lceil ((n-1) \cdot (2 \cdot n - 1) + 1)/3 \rceil = (2 \cdot n^2 + 1)/3 - n + 1$. So we obtain the following recurrence relation:

$$|\overline{V}_1| = 0, \qquad |\overline{V}_n| \geq 4 \cdot |\overline{V}_{n/2}| + (2 \cdot n^2 - 3 \cdot n + 4)/3.$$

For $n = 2, 4, 8, 16$, this gives $2, 16, 100, 556$, respectively. This inhomogeneous recurrence can be made homogeneous, which results in a characteristic polynomial of degree 4. The roots can be guessed easily. Using the smallest values, the coefficients can be determined. This gives:

$$|\overline{V}_n| \geq 2/3 \cdot n^2 \cdot \log n - 5/9 \cdot n^2 + n - 4/9. \tag{4}$$

An upper bound can be found analogously. Again we use that TM_n consists of four copies of $TM_{n/2}$ connected to an $n \times (2 \cdot n)$ mesh. The feedback vertex set V'_n of TM_n is recursively constructed as follows: 1) V'_1 is the empty set. 2) V'_n is composed of the four subsets $V'_{n/2}$ which are taken to be identical in each of

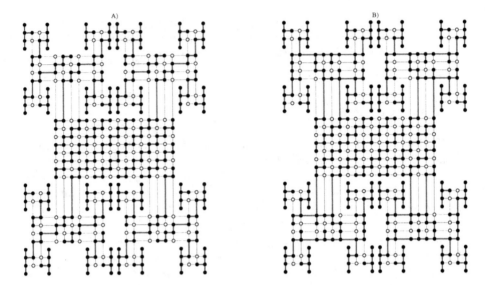

Fig. 2. A) A two-dimensional TM_8. The white vertices give a feedback vertex set of size 114, 14 more than the lower bound. B) Feedback vertex set for TM_8 of size 106, 6 more than the lower bound.

the four copies of $TM_{n/2}$ plus a suitable subset of vertices from the connecting $n \times (2 \cdot n)$ mesh. This mesh subset consists of every third diagonal. These diagonals run from upper-right to lower-left. The first is starting in position $(0, 2)$. This construction is illustrated in Figure 2 A). For its correctness refer to [12].

Let us now consider $|V_n'|$. $(2 \cdot n^2 - 2)/3$ vertices are added from the $n \times (2 \cdot n)$ mesh. So, we get the following recurrence relation:

$$|V_1'| = 0, \qquad |V_n'| = 4 \cdot |V_{n/2}'| + (2 \cdot n^2 - 2)/3.$$

For $n = 2, 4, 8, 16$, this gives 2, 18, 114, 626, respectively. There are considerable deviations from the lower bound, but these are limited to the lower-order terms:

Lemma 5. *The constructed set V_n' is a feedback vertex set for TM_n. $|V_n'| = 2/3 \cdot n^2 \cdot \log n - 2/9 \cdot n^2 + 2/9$.*

For the correctness of this lemma refer to [12]. It is not hard to improve this result by giving improved constructions for the smallest TM_n. For example, using a feedback vertex set for TM_4 with 16 instead of 18 vertices, saves 2 vertices for each of the $n^2/16$ disjoint copies of TM_4 in TM_n, saving $n^2/8$ vertices in total. Such an improved construction for TM_4 can be used in the construction of a feedback vertex set for TM_n without creating cycles as shown in Figure 2 B). This reduced-size feedback vertex set is denoted V_n''.

Theorem 3. *The constructed set V_n'' is a feedback vertex set for TM_n. $|V_n''| = 2/3 \cdot n^2 \cdot \log n - 25/72 \cdot n^2 + 2/9$.*

Comparison with (4) shows that this result is tight to within $5/24 \cdot n^2$. Going on like this the gap between upper and lower bound can be further reduced.

4   Pyramid Networks

Definition 5. [2, 14] *A pyramid network of height n is a graph $P_n = (V_n, E_n)$ with vertex set $V_n = \{(i, x, y)|0 \le i \le n, 0 \le x, y < 2^i\}$, and a set of undirected edges given by $E_n = \{((i, x, y), (i, x, y + 1))|1 \le i \le n, 0 \le x < 2^i, 0 \le y < 2^i - 1\} \cup \{((i, x, y), (i, x + 1, y))|1 \le i \le n, 0 \le x < 2^i - 1, 0 \le y < 2^i\} \cup \{((i, x, y), (i - 1, \lfloor x/2 \rfloor, \lfloor y/2 \rfloor))|1 \le i \le n, 0 \le x < 2^i, 0 \le y < 2^i\}$.*

Informally, a pyramid network is composed of a collection of meshes of increasing sizes. Level i, $0 \le i \le n$, consist of a $2^i \times 2^i$ mesh. The 1×1 mesh at level 0 constitutes the top of the pyramid. In addition to the mesh connections, each vertex is connected to one vertex in the level above it and four vertices in the level below it. $|V_n| = \sum_{i=0}^{n} 4^i = (4^{n+1} - 1)/3$; $|E_n| = \sum_{i=1}^{n} (2^i \cdot (2^i - 1) + (2^i - 1) \cdot 2^i + 4^i) = 4^{n+1} - 2^{n+2}$. A pyramid network is a non-regular graph with vertices that may have degree 3, 4, 5, 7, 8 and 9. The definition is illustrated in Figure 3 A).

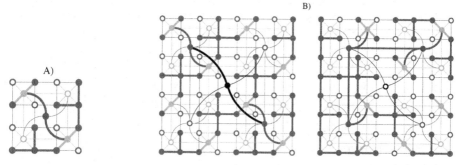

Fig. 3. A) P_2. The white vertices give a feedback vertex set of size 8, matching the lower bound. B) P_3. The white vertices give feedback vertex sets of size 34, matching the lower bound.

Small Pyramids. Consider the two-level pyramid P_1. Removing only the top vertex, the cycle in the 2×2 mesh at level 1 remains. Removing a single vertex at level 1, there remains a cycle through the other four vertices. So, at least two vertices must be removed from P_1. All cycles in P_1 can be eliminated by either removing the top vertex and an arbitrary vertex at level 1 or by removing two diagonally opposite vertices at level 1. Both constructions are used in Figure 3A). Thus, $|\overline{V}_1| = 2$.

Now consider P_2. It contains four copies of P_1. Therefore, $|\overline{V}_2| \ge 4 \cdot |\overline{V}_1| = 8$. As we can see from Figure 3 A) it also suffices to remove 8 vertices from P_2 in order to obtain a cycle-free graph. Thus, $|\overline{V}_2| = 8$.

Finally consider P_3. It contains four copies of P_2. Therefore, $|\overline{V}_3| \ge 4 \cdot |\overline{V}_2| = 32$. However, we can prove more, observing that P_3 contains in total 17 copies of P_1: 16 at the two lowest levels and 1 at the two highest levels. Thus, $|\overline{V}_3| \ge$

$17 \cdot |\overline{V}_1| = 34$. As we can see from Figure 3 B) it also suffices to remove 34 vertices from P_3 in order to obtain a cycle-free graph. Thus, $|\overline{V}_3| = 34$.

For larger P_n we do not know the exact value of $|\overline{V}_n|$. Particularly, for P_4 the best lower bound we know is $|\overline{V}_4| \geq 4 \cdot |\overline{V}_3| = 136$. If there would not have been a vertex feedback set V' for P_3 with $|V'| = 34$ containing the top of the pyramid we could have proven a lower bound of 138, but the right picture in Figure 3 B) shows that such a set V' exists. A feedback vertex set for P_4 of size 142 is shown in Figure 4 A). This is the best we were able to construct.

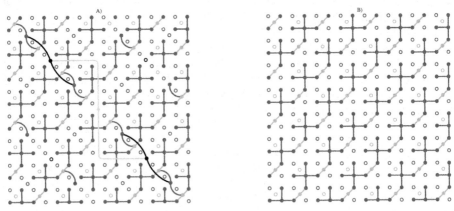

Fig. 4. A) P_4. The white vertices give a feedback vertex set of size 142, 6 more than the lower bound. Edges incident upon removed vertices are not drawn. B) An extendible two-level pattern.

A general lower bound on $|\overline{V}_n|$ is the following:

Lemma 6. *[11] For a pyramid network P_n of height n, we have $|\overline{V}_n| \geq 2/15 \cdot 4^{n+1} - 5$.*

In [11] Luccio presented a feedback vertex set V' of size at most $7/45 \cdot 4^{n+1} - 2^{n+2}/9 + 5/12 \cdot n + 1$. This bound differs from the lower bound in Lemma 6 for a multiplicative constant $7/6$ in the term of highest order.

Elementary Patterns. First observe that the following bound trivially holds: $|\overline{V}_n| \geq 4 \cdot |\overline{V}_{n-1}|$. At first one might hope that a vertex feedback set for P_n can be obtained by taking suitable feedback vertex sets for the four copies of P_{n-1} and removing the top. This would give $|\overline{V}_n| \leq 4 \cdot |\overline{V}_{n-1}| + 1$. However, the case $n = 3$ shows that this does not work. The reason is that when building P_n from four copies of P_{n-1} and a new top vertex, cycles do not only arise by the connections added at the top, but particularly also by the connections added at the lower levels. To cope with this problem we introduce the notion of an *extendible pattern* (see also Figure 4 B) and Figure 5):

Definition 6. *Let $P_{k,l}$, $1 \leq l \leq k+1$, be the graph with the structure of the l lowest levels of P_k. An extendible l-level pattern consists of a vertex feedback set*

V' for $P_{k,l}$ with the special property that taking 4^j, for any $j > 0$, copies of $P_{k,l}$ with the vertices in V' and their incident edges removed, no cycles arise when gluing these together in the way they would appear in $P_{k+j,l}$.

Lemma 7. There is an extendible two-level pattern removing 32 of the 80 vertices in $P_{3,2}$.

Proof. Figure 4 B) gives the two-level pattern. In this picture 4 copies of the pattern for $P_{3,2}$ are glued together. The result is a pattern of connected but cycle-free structures spanning three diagonals, separated by diagonals which have been entirely removed. From the $64 + 16 = 80$ vertices of $P_{3,2}$, $24 + 8 = 32$ vertices are removed. □

Assume that we know an extendible l-level pattern, then a feedback vertex set of P_n can be constructed by, starting from the bottom, alternatively removing the vertices in the pattern in l levels, and then removing all vertices in one level. More precisely, the levels $n + 1 - j \cdot (l + 1)$, $1 \leq j \leq \lfloor (n + 1)/(l + 1) \rfloor$, are removed entirely, while on the intermediate levels the removal is dictated by the extendible pattern. Let $V'_{n,2}$ be the feedback vertex set constructed for P_n with help of the pattern in Figure 4 B).

Lemma 8. $|V'_{n,2}| = 9/63 \cdot 4^{n+1} + O(1)$.

Proof. We do no longer consider every individual contribution. Rather we notice that from any disjoint copy of $P_{3,3}$ contained in P_n, $32+4 = 36$ out off $80+4 = 84$ vertices are removed. Only at the top there may be one partial copy of $P_{3,3}$. So, $|V'_{n,2}| = 36/84 \cdot |V_n| \pm O(1)$. □

Comparing with the results from Lemma 6, we see that this construction leaves a factor $15/14$ between upper and lower bound. Further improvements can be obtained by using higher-level patterns. The construction of Figure 5 gives the following:

Lemma 9. There is an extendible three-level pattern removing 140 of the 336 vertices in $P_{4,3}$.

Let $V'_{n,3}$ be the feedback vertex set constructed for P_n with help of the pattern in Figure 5. The following is proven analogously to Lemma 8:

Lemma 10. $|V'_{n,3}| = 12/85 \cdot 4^{n+1} + O(1)$.

Best Construction. The given feedback vertex set based on the three-level pattern can be further improved by making the construction of the pattern explicit and generalizing it for more levels. In this way we obtain a feedback set which, without adding new ideas, cannot be further improved.

Let us consider the extendible two- and three-level patterns as given in Figure 4 B) and Figure 5 in more detail. In this discussion the levels will be counted from the bottom, so level 0 now refers to the bottom level. In both patterns 6

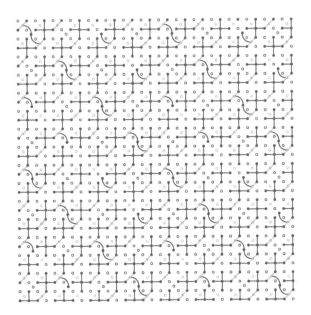

Fig. 5. An extendible three-level pattern.

out off any 16 level-zero vertices are removed. The remaining vertices together
with the edges running between them constitute the pattern consisting of shifted
crosses. In the two-level pattern every second level-one vertex is removed. This
suffices to eliminate all grid edges at this level and connects the crosses in a diagonal way. In the three-level pattern we could do the same, but then we could
have left only very few level-two vertices. Therefore we remove some extra level-
one vertices. By removing 36 out off any 64 level-one vertices, the diagonals are
cut in pieces which allows to keep every second level-two vertex. These level-two
vertices connect the diagonal pieces into diagonal bundles.

This idea can be generalized. If we want to construct an extendible four-
level pattern, we start with four copies of the three-level pattern. This gives
the situation in Figure 5. In order to be able to keep a substantial number of
level-three vertices, we must remove some extra level-two vertices. As in the
transition from the two-level to the three-level pattern, by removing 36 out off
any 64 level-two vertices, the diagonals are cut in pieces. Adding every second
level-three vertex connects these into diagonal bundles again. The lowest four
levels of vertices in Figure 6 A) illustrate this construction.

Arguing on, we see that an extendible $(k+1)$-level pattern can be constructed
by taking four copies of an extendible k-level pattern removing an additional
$1/16$ of the level-$(k-1)$ vertices and every second level-k vertex. This idea can
be applied for constructing an even better feedback vertex set for P_n, for any
$n \geq 3$: for the lower $n-2$ levels we use the extendible $(n-2)$-level pattern.
For the highest three levels we use the pattern depicted in Figure 6 B). Let V'_n
denote the thus constructed set of vertices to remove.

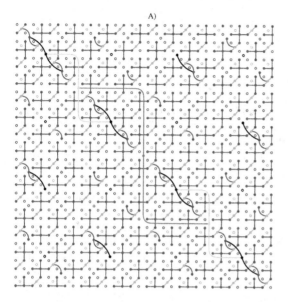

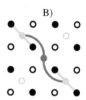

Fig. 6. A) P_5. The white vertices give a feedback vertex set of size 574, 29 more than the lower bound. B) The removal pattern at the top of P_n for $n \geq 3$.

Theorem 4. *For all $n \geq 3$, V'_n is a feedback vertex set for P_n. $|V'_n| = 9/64 \cdot 4^{n+1} - 2$.*

Proof. At the lower $n-2$ levels the extendible pattern connects the vertices into eight diagonal bundles spanning $2^{n-2} - 1$ diagonals each. By the construction at the top these bundles are interconnected without creating cycles. The top vertex is not removed, from level 1 and level 2 half of the vertices are removed. From the following levels 9/16 of the vertices are removed, except for level n where only 3/8 is removed. This gives: $|V'_n| = 0 \cdot 1 + 1/2 \cdot 4 + 1/2 \cdot 16 + 9/16 \cdot 64 + \cdots + 9/16 \cdot 4^{n-1} + 3/8 \cdot 4^n = 9/16 \cdot \sum_{i=0}^{n} 4^n - 9/16 \cdot 1 - 1/16 \cdot 4 - 1/16 \cdot 16 - 3/16 \cdot 4^n = 9/64 \cdot 4^{n+1} - 2$. $\qquad\square$

Comparing with the results from Lemma 6, we see that this construction leaves a factor $135/128 \simeq 1.055$ between upper and lower bound.

5   Conclusion

In this paper we have considered the minimum feedback vertex set problem in graphs. We have provided lower and upper bounds for the problem in the two- and higher-dimensional mesh of trees networks, in the tree of meshes and pyramid networks. Further study is needed to narrow the gap between lower and upper bounds for the pyramid networks. A possible approach is to prove a higher lower bound for P_4, as this would immediately lead to higher lower bounds for all P_n with $n > 4$. Such a proof might be given with help of a refined branch-and-bound algorithm. A good branch-and-bound algorithm would probably also allow to reduce the gap between upper and lower bound for the trees of meshes to less than $\epsilon \cdot n^2$, for any $\epsilon > 0$.

References

1. I. Caragiannis and C. Kaklamanis and P. Kanellopoulos. New bounds on the size of the minimum feedback vertex set in meshes and butterflies. Information Processing Letters. **83** (5) (2002) 275–280.
2. A. Dingle and H. Barada. Optimum embeddings of end-around meshes into pyramid networks. In: Proc. 5th IEEE International Parallel Processing Symposium. (1991) 445–451.
3. G. Fertin and E. Godard and A. Raspaud. Minimum feedback vertex set and acyclic coloring. Information Processing Letters. **84** (3), (2002) 131–139.
4. P. Festa and P.M. Pardolos and M.G.C. Resende. Feedback set problems. Encyclopedia of Optimization, Kluwer Academic Press. Vol. 2, (2001) 94–106.
5. R. Focardi and F.L. Luccio and D. Peleg. Feedback vertex set in hypercubes. Information Processing Letters. **76** (1-2), (2000) 1–5.
6. M.R. Garey and D.S. Johnson. Computers and intractability: A guide to the theory of NP-completeness, Freeman, San Francisco, (1979).
7. R. Královič and P. Ružička. Minimum feedback vertex sets in shuffle-based interconnection networks. Information Processing Letters. **86** (4), (2003) 191–196.
8. F.T. Leighton. Introduction to parallel algorithms and architectures: array, trees, hypercubes. Morgan Kaufmann Publishers Inc., San Mateo, CA, USA, (1992).
9. F.T. Leighton. New lower bound techniques for VLSI. Mathematical System Theory. **17**, (1984) 47–70.
10. F.L. Luccio. Almost exact minimum feedback vertex set in meshes and butterflies. Information Processing Letters. **66** (2), (1998) 59–64.
11. F.L. Luccio. Minimum feedback vertex set in pyramid and mesh of trees networks. Proc. 10th International Colloquium on Structural Information and Communication Complexity, Carleton Scientific. (2003) 237–250.
12. F.L. Luccio and J.F. Sibeyn. Minimum Feedback Vertex Set in Mesh-based Networks. Reports on Computer Science 04.11, Fachbereich Mathematik und Informatik, Martin-Luther-Universität, Halle, Germany, (2004).
13. B.M. Maggs and E.J. Schwabe. Real-time emulations of bounded-degree networks. Information Processing Letters. **66** (5), (1998) 269–276.
14. S. Öhring and S.K. Das. Incomplete hypercubes: Embeddings of tree-related networks. Journal of Parallel and Distributed Computing. **26**, (1995) 36–47.
15. A.L. Rosenberg. Product-shuffle networks: toward reconciling shuffles and butterflies. Discrete Applied Mathematics. **37/38**, (1992) 465–488.
16. P. Salinger and P. Tvrdík. Optimal broadcasting and gossiping in one-port meshes of trees with distance-insensitive routing. Parallel Computing. **28** (4), (2002) 627–647.

Perfect Token Distribution on Trees[*]

Luciano Margara, Alessandro Pistocchi, and Marco Vassura

Computer Science Department, University of Bologna
{margara,pistocch,vassura}@cs.unibo.it

Abstract. Load balancing on a multi-processor system consists of re-distributing tasks among processors so that all processors end up with roughly the same amount of work to perform. The *token distribution problem* is a variant of the load balancing problem where each task has unit-size and it represents an *atomic* element of work. We present an algorithm for computing a perfect token distribution (each processor has either $\lceil T/N \rceil$ or $\lfloor T/N \rfloor$ tasks, where N is the number of processors and T is the number of tasks scattered among processors) on distributed tree-connected networks having worst-case running time $O(TD)$ (D denotes the diameter of the tree). The number of token exchanges exceeds the optimum by at most $O(D \min\{T, N\})$.

In order to compute a perfect token distribution each node v must be able to store $\Theta(d_v(\log T + \log N))$ bits, where d_v is the degree (number of adjacent nodes) of v. This is the first fully decentralized algorithm for computing perfect token distributions on arbitrary tree-connected networks which does not receive as input any kind of aggregate information about the network (e.g., number of nodes or total number of tokens).

1 Introduction

The performance of a distributed network crucially depends on dividing up work effectively among its processing elements [8]. This type of *load balancing* problem has been studied in many different models. The basic idea in all of the models is to evenly redistribute initial job load among processors (static balancing) and to keep load distribution as balanced as possible during time (dynamic balancing). Many variants of the load balancing problem have been proposed and widely investigated in the literature [1–4, 6, 9–11].

In this paper we consider a basic variant of the static load balancing problem, called *token distribution* problem, restricted to tree-connected networks. A tree-connected network is represented by a tree with N nodes. Each node represents a processing element of the network and possesses a number of jobs of unit-size (tokens) to be processed. The total number of tokens is denoted by T. In a single message a token can be moved from any node to any other adjacent node in the tree. No token is created or destroyed during the redistribution process. The goal is to redistribute tokens across the tree so that each node ends up (being

---

[*] This work was partially supported by the Future & Emerging Technologies unit of the European Commission through Project BISON (IST-2001-38923).

R. Královič and O. Sýkora (Eds.): SIROCCO 2004, LNCS 3104, pp. 221–232, 2004.
© Springer-Verlag Berlin Heidelberg 2004

aware of that) with either $\lceil T/N \rceil$ or $\lfloor T/N \rfloor$ tokens (*perfect* token distribution). We adopt asynchronous single-port communication model with uni-directional communication links. Nodes are *anonymous*, i.e. they do not have identification labels. Synchronization between nodes (executing the same code in parallel) is achieved by exchanging messages.

Previous Results. For general networks Ghosh et al. [4] analyze two algorithms which reduce the maximum difference in tokens between any two nodes (called "discrepancy") to at most $O((d^2 \log N)/\alpha)$, where d is the maximum degree of the nodes of the network, N is the number of nodes in the network, and α is the edge expansion of the network. Many results have been carried out on specific network topologies. For ring networks Gehrke, Plaxton, and Rajaraman [3] give an algorithm having an asymptotically optimal message complexity which converges to a perfectly balanced state.

For meshes and torus Houle et al. [7] give an algorithm that reduces the discrepancy to the minimum degree of the nodes of the network and that runs in worst-case optimal time. The same algorithm used for complete binary trees obtains in the worst case a discrepancy equal to the height of the tree [6].

For arbitrary trees Houle, Symvonis, and Wood [5] give an algorithm for computing a perfect token distribution assuming that each node at the beginning of the computation knows the number of nodes in the tree.

Our Results. We present a fully decentralized algorithm for tree-connected networks which computes a perfect token distribution in time $O(TD)$ without receiving as input any additional (aggregate) information about the network such as the total number of nodes or the total number of tokens. The number of token exchanges made by our algorithm exceeds the optimum by at most $O(D \min\{T, N\})$. In addition, at the end of the redistribution process, all the nodes reach a distinguished final state and they are ready to start a new activity.

Our technique is based on a three phase approach.

– *Phase 1.* We compute (in a completely distributed way) the number of nodes of the tree and the number of tokens scattered across the tree in $\Theta(D\,d)$, where d is the maximum degree and D is the diameter of the tree[1].

– *Phase 2.* Let v be any node of the tree and T_i, $1 \le i \le k$, be the k sub-trees rooted at nodes adjacent to v. After Phase 1 v knows the number of nodes and the number of tokens contained in each T_i. Taking advantage of this information each node having more than $\lfloor T/N \rfloor$ tokens is able to decide where to send its extra tokens. Nodes that have less than $\lfloor T/N \rfloor$ tokens wait until they receive enough tokens from their neighbors. After Phase 2 we get an "almost perfect" token distribution in which at least $N-1$ nodes have exactly $\lfloor T/N \rfloor$ tokens. Note that almost perfect token distributions have discrepancy at most $(T \bmod N)$.

– *Phase 3.* We refine the distribution obtained after Phase 2 in order to make it perfect. To this extent, the node containing more than $\lceil T/N \rceil$ tokens (at most one node has this property) sends them to its neighbors. This procedure is then executed recursively by all the neighbors of v.

---

[1] In multi-port models such information can be computed in $\Theta(D)$.

The rest of the paper is organized as follows. In Section 2 we define some communication primitives that will be used in our algorithm. In Section 3 we present our algorithm while in Section 4 and 5 we sketch the proof of its correctness and we discuss its computational cost in terms of time, space, and number of token exchanges, respectively. Section 6 contains conclusions and a brief description of possible open problems.

2   Communication Primitives

We define the following communication primitives:

-$reserve(c)$ reserves channel c for communication and returns *busy* if c is already reserved;

-$release(c)$ releases channel c reserved for communication;

-$send(c,m)$ sends message m through the already reserved channel c;

-$receive(c)$ receives a message from channel c if a message was sent on channel c (channel needs not to be reserved). If no message is arriving from c the message returned by *receive* is *NULL*. If two or more messages are sent on a channel then they are received one by one in the order they were sent.

Using these primitives we build functions:

-$waitany(c)$ interrogates every channel until a message is received, then returns the message received, saving the channel through which it arrived in the parameter c;

-$wait(c)$ interrogates channel c until a message is received, then returns the message received;

-$safeSend(c,m)$ waits until c is not *busy*, reserves it, sends m through c, and then releases c;

-$receiveToken(c)$ receives a token from c, if any. If a message is received updates local variables containing the number of tokens;

-$receiveTokensFromAll()$ receives, using $receiveToken(c)$, tokens sent from neighbors, if any;

-$sendToken(c)$ sends, using $safeSend(c,m)$ a token from the current node through c and updates the local variables containing the number of tokens.

3   The Algorithm

Our algorithm consists of three phases: first it collects information about the tree structure and the initial token distribution, then uses this information for obtaining an almost perfect token distribution, and finally it refines the solution by redistributing a small number of tokens.

In what follows we assume that each node v can store $\Theta(d_v(\log T + \log N))$ bits of information.

3.1   Phase 1

Each node v has an internal state defined according to the values of two local variables, namely S and R which contain the number of messages sent and received by v so far. Let deg be the degree d_v of v. Both S and R are set to 0 at the beginning of the computation. According to the values of S and R we define 5 distinct states:

$S1.1$: $[\mathsf{S} = 0$ and $\mathsf{R} < (\mathbf{deg} -1)]$. All the nodes having degree greater than 1 start in this state. Nodes in this state are waiting for messages from their neighbors. Any node with k neighbors waits for $k-1$ messages and then changes state (from $S1.1$ to $S1.2$).

$S1.2$: $[\mathsf{S} = 0$ and $\mathsf{R} = (\mathbf{deg} -1)]$. Leaves start in this state. Each node with k neighbors reaches this state after receiving $k - 1$ messages. A node in this state tries to send information about its subtree to the only neighbor from which it did not receive any message yet. If succeding the node changes state (from $S1.2$ to $S1.3$). Otherwise, if the channel is busy then the last neighbor is sending subtree information. The node receives it and changes state (from $S1.2$ to $S1.4$).

$S1.3$: $[\mathsf{S} = 1$ and $\mathsf{R} = (\mathbf{deg} -1)]$. Nodes in this state are waiting for global information from the neighbor to which they sent their local information.

$S1.4$: $[\mathsf{S} = 0$ and $\mathsf{R} = \mathbf{deg}]$. Only one node reaches this state: the one computing global information. In this state it sends such information to all its neighbors and then ends.

$S1.5$: $[\mathsf{S} = 1$ and $\mathsf{R} = \mathbf{deg}]$. In this state a node has already received global information and forwards it to all its neighbors and then ends.

In phase 1 each message msg consists of two integer numbers, referred to as msg.T and msg.N representing the number of tokens and of nodes of the entire subtree rooted at the sender of msg. We assume that channels at each node v are numbered from 1 to deg.

3.2   Phase 2

In phase 2 we start moving tokens.

From now on we will refer to the node which received local information from all its neighbors as the root of the tree. Every node has two arrays, namely subTreeN and subTreeT, containing the total number of nodes and the total number of tokens in the subtrees rooted at its neighbors. Each node, knowing total number of tokens T and of nodes N in the tree, computes the average number of tokens per node $\mathsf{Avg} = \lfloor T/N \rfloor$. All nodes send or wait for tokens until they and the nodes in their subtrees contain exactly Avg tokens, except for the root of the tree which might contain extra tokens. Function $subTreeAvg(\mathsf{i})$ is a simple function computing the average number of tokens in the subtree rooted at v_i: it returns subTreeT[i]/subTreeN[i]. Each node v in Phase 2 has an internal state represented by

- a boolean variable subTreeBalanced (*true* if all the subtrees rooted at children of v are balanced, *false* otherwise);
- an integer variable T_v storing the number of tokens at v;
- an integer variable parent which is set to 0 if v is the root.

```
PHASE 1
INITIALIZATION
deg ← d_v                                  //neighbors of v
T_v ← tokens at v                          //tokens in subtree rooted at v
N_v ← 1                                    //nodes in subtree rooted at v
R ← 0                                      //messages received by v
S ← 0                                      //messages sent by v
parent ← 0                                 //the parent of the node
subTreeT array of deg NULL elements        //tokens in neighbor subtrees
subTreeN array of deg NULL elements        //nodes in neighbor subtrees

while(S < deg or R < deg) do
  case (S,R) of
  S = 0, R < (deg − 1):
      while(R < (deg − 1)) do   //receive local info from all neighbors but one
      | msg←waitany(i)                      //waits for a message from any neighbor
      | T_v ←T_v +msg.T                     //tokens in subtree rooted at v
      | N_v ←N_v +msg.N                     //nodes in subtree rooted at v
      | subTreeT[i]←msg.T                   //tokens in subtree rooted at v_i
      | subTreeN[i]←msg.N                   //nodes in subtree rooted at v_i
      | R←R+1                               //v received a message
      endwhile
  S = 0, R = (deg − 1):
      parent ← channel j for which subTreeT[j]=NULL
                                   //the channel from which v did not received msg
        if (reserve(parent) ≠busy) then    //check that v_i is not using the channel
        | msg←receive(parent)               //check if received the message
        | if (msg=NULL) then                //message not received
        | | msg.T←T_v                       //nodes in subtree rooted at v
        | | msg.N←N_v                       //tokens in subtree rooted at v
        | | send(parent,msg)                //send values of subtree rooted at v to v_parent
        | | release(parent)                 //release channel parent
        | | S←S+1                           //v sent a message
        | else
        | | release(parent)                 //release channel parent
        | | T_v ←T_v +msg.T                 //tokens in subtree rooted at v
        | | N_v ←N_v +msg.N                 //nodes in subtree rooted at v
        | | subTreeT[parent]←msg.T          //tokens in subtree rooted at v_parent
        | | subTreeN[parent]←msg.N          //nodes in subtree rooted at v_parent
        | | parent← 0                       //v is root, parent = 0
        | | R←R+1                           //v received a message
        | endif
        else
        | msg←wait(parent)                  //v_parent using the channel : wait for the message
        | T_v ←T_v +msg.T                   //tokens in subtree rooted at v
        | N_v ←N_v +msg.N                   //nodes in subtree rooted at v
        | subTreeT[parent]←msg.T            //tokens in subtree rooted at v_parent
        | subTreeN[parent]←msg.N            //nodes in subtree rooted at v_parent
        | parent← 0                         //v is root, parent = 0
        | R←R+1                             //v received a message
        endif
  S = 1, R = (deg − 1):
      msg←wait(parent)                      //waits for global message
      subTreeT[parent]←msg.T−T_v            //tokens in subtree rooted at v_parent
      subTreeN[parent]←msg.N−N_v            //nodes in subtree rooted at v_parent
      T_v ←msg.T                            //total tokens of the tree
      N_v ←msg.N                            //total nodes of the tree
      R←R+1                                 //v received a message
  S = 0, R = deg:
      msg.T←T_v                             //total tokens of the tree
      msg.N←N_v                             //total nodes of the tree
      for i← 1 to deg do                    //for all neighbors v_i
      | safeSend(i,msg)                     //send global values to v_i
      endfor
      S←deg                                 //v sent deg messages
  S = 1, R = deg:
      msg.T←T_v                             //total tokens of the tree
      msg.N←N_v                             //total nodes of the tree
      for i← 1 to deg do                    //has to send global values to all neighbors
      | if (i ≠ parent) then                //except v_parent
      | | safeSend(i,msg)                   //send global values to v_i
      | endif
      endfor
      S←deg                                 //v sent deg − 1 messages
  endcase
endwhile
```

PHASE 2
INITIALIZATION
Avg $\leftarrow \lfloor T_v/N_v \rfloor$
$T_v \leftarrow$ tokens at v
$R \leftarrow 0$
bigger $\leftarrow \{i | subTreeAvg(i) > Avg, \ i \neq parent\}$
smaller $\leftarrow \{i | subTreeAvg(i) < Avg, \ i \neq parent\}$
subTreeBalanced $\leftarrow$ (bigger$= \emptyset$ and smaller$= \emptyset$)

while(not((subTreeBalanced=true and T_v=Avg)
    or (subTreeBalanced=true and parent= 0)) do
  case (subTreeBalanced, T_v, parent) of
    subTreeBalanced $= false$, $T_v \leq$ **Avg, parent any value**
      receiveTokensFromAll()   *//check for incoming tokens and update data*
      bigger $\leftarrow \{i | subTreeAvg(i) > Avg, \ i \neq parent\}$
      subTreeBalanced $\leftarrow$ (bigger$= \emptyset$ and smaller$= \emptyset$)
    subTreeBalanced $= true$, $T_v <$ **Avg, parent** > 0
      receiveToken(parent)   *//all sons have finished,*
                               //checking tokens coming from parent
    subTreeBalanced $= false$, $T_v >$ **Avg, parent any value**
      receiveTokensFromAll()   *//check for incoming tokens and update data*
      bigger $\leftarrow \{i | subTreeAvg(i) > Avg, \ i \neq parent\}$
      if (smaller$\neq \emptyset$)
        j $\leftarrow$ an element $\in$ smaller
        sendToken(j)   *//send token to v_j and update data*
        smaller $\leftarrow \{i | subTreeAvg(i) < Avg, \ i \neq parent\}$
      endif
      subTreeBalanced $\leftarrow$ (bigger$= \emptyset$ and smaller$= \emptyset$)
    subTreeBalanced $= true$, $T_v >$ **Avg, parent** > 0
      sendToken(parent)   *//send token to v_{parent} and update data*
  endcase
endwhile

$S2.1$: [**subTreeBalanced** $= false$ and $T_v \leq$ **Avg**]. Nodes remain in this state receiving tokens and updating local variables until they have more than Avg tokens and then change state. (from $S2.1$ to $S2.3$).

$S2.2$: [**subTreeBalanced** $= true$ and $T_v <$ **Avg, parent** > 0]. Nodes in this state need not to exchange tokens with children, since all such subtrees have Avg number of tokens. They wait for tokens from their parent (a node in this state is not the root).

$S2.3$: [**subTreeBalanced** $= false$ and $T_v >$ **Avg**]. Nodes in this state need to balance their descending subtrees: they receive tokens from any node and then send tokens to subtrees with less than Avg number of tokens.

$S2.4$: [**subTreeBalanced** $= true$ and $T_v >$ **Avg, parent** > 0]. Nodes in this state have already balanced subtrees rooted at their children and send extra tokens towards their parents.

As we already mentioned, a node stops exchanging tokens if it has balanced its subtrees (subTreeBalanced $= true$ and $T_v = $ Avg) (Fig. 1 part (b)) or if it has balanced the subtrees of all its children (subTreeBalanced$= true$) and it is the root (parent $= 0$)) (Fig. 1 part (c)).

3.3 Phase 3

In this phase tokens keep on moving until a *Finished* message is sent.

In order to distinguish between tokens and finished messages we use two internal variables, namely msg.Token and msg.Finished. At any time if msg.Finished=

$true$ then msg.Token = $NULL$, else if msg.Finished = $false$ then msg.Token contains a token.

In phase 3 extra tokens sent to the root during phase 2 are redistributed down the tree. The root (and all the other nodes recursively) sends its extra tokens to subtrees which can accept them[2]. In this phase each node can be in one of the two following states:

$S3.1$: **[T_v > Avg+1 and parent = 0].** Nodes in this state are root of a subtree and send their extra tokens to subtrees which can accept them.

$S3.2$: **[T_v any value and parent > 0].** Nodes in this state receive messages from their parent. If they receive tokens they update local variables. If they receive a $Finished$ message they set parent to 0 and become a root.

When a node is root and has at most Avg+1 number of tokens it sends a $Finished$ message to all its children. As proven in Section 4, eventually all nodes become a root and receive a $Finished$ message (Fig. 1 part (d)).

```
PHASE 3
INITIALIZATION
p ← parent                              //p is the parent of v

while(not(T_v ≤Avg+1 and parent= 0)) do
 case (T_v, parent) of
   T_v > Avg+1, parent = 0
       i ← a neighbor for which subTreeAvg(j) < Avg + 1,  i ≠p
       msg.Token ← a token at v
       msg.Finished ← false
       safeSend(i,msg)                  //send token to v_i
       T_v ← T_v − 1                    //tokens at v
       subTreeT[i] ← subTreeT[i]+1      //tokens in subtree rooted at v_i
   T_v any value, parent > 0
       msg ← receive(parent)
       if(msg≠NULL)                     //if a message msg received
         if(msg.Finished=true)          //if message received was Finished
         | parent ← 0                   //the parent finished: become root
         else                           //else a token was received
         | T_v ← T_v +1
         | subTreeT[parent] ← subTreeT[parent]−1
       endif
     endif
 endcase
endwhile
for i← 1 to deg do                      //for all neighbors v_i
   | msg.Token ← NULL
   | msg.Finished ← true
   | if(i≠p)                            //except the parent
   | | safeSend(i,msg)                  //send Finished to v_i
   | endif
endfor
```

4   Proof of Correctness

In this Section we prove the correctness of our algorithm.

---

[2] Eventually, each node ends up with either Avg or Avg+1 tokens, then each subtree must contain less than Avg+1 tokens.

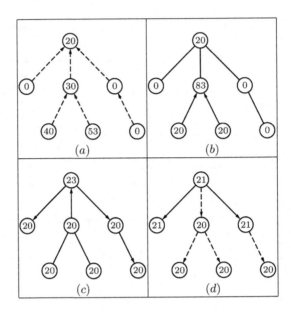

Fig. 1. Example of execution of our algorithm. Dashed arrows represent information messages, solid arrows represent token exchanges. (a) Phase 1: local information is sent from leaves to a single node which becomes the root of the tree. (b) Phase 2: token are redistributed in subtrees until each subtree has **Avg** number of tokens. (c) End of Phase 2: each node has **Avg** tokens but the root that may contain some extra token. (d) End of Phase 3: a perfect token distribution is achieved and each node receives a *Finished* message.

4.1   Phase 1

Let $S1.final$ be the final state of Phase 1 defined by [S =deg and R =deg].

Theorem 1. *At the end of phase 1, $\forall v \in V$ we have that $T_v = T$ and $N_v = N$. In addition, exactly one node (the root of the tree) has **parent** = 0 while each other node has **parent** = i where i is the number of the channel which links the node to its parent in the rooted tree.*

Sketch of the proof. At the beginning of the computation, each node of degree 1 is in state $S1.2$ while all the other nodes are in state $S1.1$. Then, each node in state $S1.2$ sends its subtree information to nodes in state $S1.1$ and moves to state $S1.3$. Eventually each node in state $S1.1$ moves to state $S1.2$ and decides which node is its parent. When state $S1.2$ contains only two nodes, one of them moves to state $S1.4$. All the other nodes go to state $S1.3$ and wait for global information. The node in state $S1.4$ is the root, computes global information and sends it to its neighbors. Neighbors receive such information, identify the node which sent them as they parent, move to state $S1.5$, and then forward it to their children. At the end of this process all the nodes receive global information and set $T_v = T$ and $N_v = N$. Eventually each node moves to state $S1.final$.   □

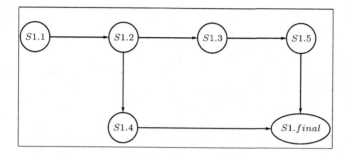

Fig. 2. State transitions diagram of phase 1.

4.2 Phase 2

During Phase 1 we implicitly elect a leader, namely the only node with parent=0. We refer to this node as the root of the tree. Links to parent nodes are provided by local variables parent.

Lemma 1. *Let v_j be a node different from the root of the tree and such that $subTreeAvg(j)^3 \geq$ Avg. Let v be any node of the subtree rooted at v_j. Then at the end of phase 2 v contains exactly Avg tokens.*

Proof. We make the proof by induction on the height of the subtree:

Base Case. If the subtree has height 0 (only one node) subTreeBalanced $= true$. If the node has Avg tokens it will do nothing in Phase 2 and the lemma is true, if it has more it will send tokens to its parent (state $S2.4$) until it remains Avg tokens.

Induction. If the subtree has height $n + 1$ it has subsubtrees of height n. Since the algorithm behaves exactly in the same way for every node in the tree except the root we can apply induction hypothesis on them. When subTreeBalanced $= true$ the root of the subtree has at least Avg tokens and sends extra tokens toward the root of the tree as in the base case (state $S2.4$) and every other node finishes with Avg tokens for induction hypothesis, so the lemma holds. If subTreeBalanced $= false$ for induction hypothesis all subsubtrees with $subTreeAvg \geq$ Avg will have Avg tokens in each node and send extra tokens to the root of the subtree. The root will receive such tokens (states $S2.1$ and $S2.3$) and, since the total number of tokens of the subtree is at least subTreeN[j] $\cdot$ Avg, will eventually have at least Avg tokens. If there are subsub-Trees with $subTreeAvg <$ Avg their root can not be in state $S2.4$ and so will receive tokens from their parent. Their parent, the root of the subTree, has more than Avg tokens and sends some of them to such subsubtrees (state $S2.3$). Reasoning again on number of tokens the root will make such subsubtrees to have $subTreeAvg =$ Avg remaining at least Avg tokens. At this point subTreeBalanced will become $true$ and the lemma holds, as explained above.   □

---

[3] Each neighbor of v_j has a different value for $subTreeAvg(j)$ but, being the tree rooted, we refer the value owned by the neighbor v_{parent} which is the parent of v_j.

Theorem 2. *At the end of phase 2 the root of the tree has at least Avg tokens while all the other nodes have exactly Avg tokens.*

Proof. Consider subTreeBalanced $= true$ at the root of the tree. Then we apply lemma 1 to all the children v_j of the root: all nodes in their subtrees have exactly Avg tokens. Then, for definition of Avg, the root has at least Avg tokens and the theorem holds. Consider subTreeBalanced $= false$ at the root of the tree. Then we apply lemma 1 to all the children v_j of the root having $subTreeAvg(j) \geq$ Avg. During phase 2 the root receives tokens (states $S2.1$ and $S2.3$) from subtrees having $subTreeAvg(j) >$ Avg. Since the total number of tokens of the tree is at least $N \cdot$ Avg, when such subtrees reach balancing the root has at least Avg tokens. If there are subTrees having $subTreeAvg(j) <$ Avg the root will have more than Avg tokens and will send some of them to such subsubtrees (state $S2.3$). The root v_j of such subtrees can not be in state $S2.4$ and then receives tokens sent from the root. Reasoning again on number of tokens the root will make such subsubtrees to have $subTreeAvg(j) =$ Avg remaining at least Avg tokens. At this point subTreeBalanced will become $true$. We apply lemma 1 to all the children v_j of the root and the theorem holds.   □

4.3   Phase 3

Theorem 3. *After phase 3 each node in the tree has Avg or Avg +1 tokens.*

Proof. We make the proof by induction on the height of the tree:

Base Case. If the tree has height 0 (only one node) being the total number of tokens at most Avg +1 the theorem is trivially true.

Induction. If the subtree has height $n + 1$ it has subsubtrees of height n. If the root has at most Avg +1 tokens the theorem is proved for theorem 2. If it has more tokens it sends them to its sons until it remains Avg +1 tokens, due to the total number of tokens this will happen before that all subtrees have $subTreeAvg =$ Avg +1. At this time the root will send $Finished$ to all its sons. On the other hand each son receives tokens without sending them and sets parent to 0 upon receiving the $Finished$ message. Now the algorithm repeats recursively on each subtree rooted at each son of the root. Since the average number of tokens for each subtree is at most Avg +1 the total number of tokens in the subtree is such that its root can send all extra tokens to sons before their subtrees have more than Avg +1 average number of tokens. We can then apply induction hypothesis to every subtree proving the theorem.   □

5   Complexity

We call d the maximum degree of the tree, T the total number of tokens, D the diameter of the tree and N the number of nodes. Phase 1 runs in (worst case) $\Theta(Dd)$ in single-port model[4]. Phase 2 runs in (worst case) $O(TD)$, while phase 3

---

[4] $\Theta(D)$ in multi-port model.

runs in (worst case) $O(D(T \bmod N))$. Since phase 2 has the greatest complexity the overall complexity of the algorithm is $O(TD)$.

The space needed to store local information at each node is that for subtrees load and number of nodes, plus a constant number of local data of size less or equal to the elements of subtrees array, so size complexity at single node is $\Theta(d(\log T + \log N))$.

In phase 2 all tokens travelling through a link go in the same direction and, if $\mathsf{Avg} = T/N$, algorithm achieves perfect token distribution. This means that the number of token exchanges is optimum. If $\mathsf{Avg} < T/N$ then at most $T \bmod N$ tokens generate extra token exchanges. The longest path done by such tokens is from a leaf to the root. So the number of token exchanges is at most $O(D(T \bmod N))$ greater than optimum. In phase 3, unless extra tokens were already at the root, we have already done more token exchanges than the optimum, so all the $O(D(T \bmod N))$ worst case token exchanges are beyond the optimum. The total number of token exchanges is then no more than $O(D \min\{T, N\})$ greater than optimum.

6   Conclusions and Further Work

We presented a distributed algorithm which computes a perfect token distribution on anonymous tree connected networks without taking as input any global information about the tree structure or about the initial distribution of tokens.

Some other questions remain open and deserve further investigation. Among them we wish to point out the following two that are strictly related to the results presented in this paper.

(1) Each node in our algorithm has a local memory of size $\Theta(d(\log T + \log N))$. Is it possible to compute perfect token distributions on trees assuming that each node has only constant size local storing capacity?

(2) Is it possible to compute perfect token distribution in a fully decentralized way minimizing the number of token exchanges?

References

1. J. E. Boillat. Load balancing and Poisson equation in a graph. Concurrency: Practice and Experience, 2(4):289–311, 1990.
2. G. Cybenko. Dynamic load balancing for distributed memory multiprocessors. Journal of Parallel and Distributed Computing, 7:279–301, 1989.
3. Johannes E. Gehrke, C. Greg Plaxton, and Rajmohan Rajaraman. Rapid convergence of a local load balancing algorithm for asynchronous rings. Theoretical Computer Science, 220:247–265, 1999.
4. B. Ghosh, F. T. Leighton, B. M. Maggs, S. Muthukrishnan, C. G. Plaxton, R. Rajaraman, A. W. Richa, R. E. Tarjan, and D. Zuckerman. Tight analyses of two local load balancing algorithms. In Proc. of the 27th Annual ACM Symposium on Theory of computing, 548–558, May 1995.

5. M. E. Houle, A. Symvonis, and D. R. Wood. Dimension-Exchange Algorithms for Token Distribution on Tree-Connected Architectures. Journal of Parallel and Distributed Computing, to appear. Also in Proc. of 9th International Colloquium on Structural Information and Communication Complexity (SIROCCO '02), 181–196, Carleton Scientific, 2002.

6. M. E. Houle, E. Tempero, and G. Turner. Optimal dimension-exchange token distribution on complete binary trees. Theoretical Computer Science, 220(2):363–376, 1999.

7. M. E. Houle and G. Turner. Dimension-exchange token distribution on the mesh and the torus. Parallel Computing, 24(2):247–265, 1998.

8. B. Shirazi, A. Hurson, and K. Kavi. Scheduling and local balancing in parallel and distributed systems. IEEE computer society press, 1995.

9. J. Song. A partially asynchronous and iterative algorithm for distributed load balancing. Parallel Computing, 20:853–868, 1994.

10. R. Subramanian and I. D. Scherson. An analysis of diffusive load-balancing. ACM Symposium on Parallel Architectures and Algorithms, 220–225, 1994.

11. A. N. Tantawi and D. Towsley. Optimal static load balancing in distributed computer systems. Journal of the ACM, 32:445–465, 1985.

Approximation Algorithm
for Hotlink Assignment in the Greedy Model

Rachel Matichin and David Peleg*

The Weizmann Institute of Science, Rehovot, 76100, Israel
david.peleg@weizmann.ac.il

Abstract. Link-based information structures such as the web can be enhanced through the addition of hotlinks. Assume that each node in the information structure is associated with a weight representing the access frequency of the node by users. In order to access a particular node, the user has to follow a path leading to it from the root node. By adding new hotlinks to the tree, it may be possible to reduce the access cost of the system, namely, the expected number of steps needed to reach a leaf from the root, assuming the user can decide which hotlinks to follow in each step. The *hotlink assignment* problem involves finding a set of hotlinks maximizing the *gain* in the expected cost. The paper addresses this problem in the more realistic *greedy* user model recently introduced in [3], and presents a polynomial time 2-approximation algorithm for the hotlink assignment problem on trees.

1 Introduction

1.1 Motivation

Large databases containing diverse information types are often organized on the basis of a hierarchical classification index. The Web, for example, is provided with this type of organizational scheme in Yahoo [6] and the Open Directory Service [7]. A user who searches for an information item in a hierarchically structured database has to follow a path from the root to the desired node in the classification tree. The degree of this tree is often relatively low and its average depth is often high, causing the search path to be long. Another factor contributing to the length of the search path is that the classification usually does not take into account the access frequency of the items by users. This implies that the depth of certain frequently visited items in the tree may be high. Hence the access cost of the tree, defined as the expected number of steps needed to reach an item from the root, may be high.

An ad-hoc solution used in practice in the Web is to augment the tree organization with "*hotlinks*" added to various nodes. These hotlinks lead directly to the most frequently accessed items. The hotlinks to be added should be selected based on the access probability of the various items in the database.

---

* Supported in part by a grant from the Israel Science Foundation.

R. Královič and O. Sýkora (Eds.): SIROCCO 2004, LNCS 3104, pp. 233–244, 2004.
© Springer-Verlag Berlin Heidelberg 2004

When searching for an item in the original tree structure, the user starts from the root and advances along tree edges along the unique path to the desired destination leaf. An implicit assumption underlying the common hierarchical approach is that at any node along the search in the tree, the user is able to select the correct link leading towards the desired leaf. Evidently, the user does not necessarily know the tree topology. However, the user usually has some general knowledge about the domain, and the links followed at any node are selected on the basis of the natural partitioning of the domain. Thus, when the user sees several tree edges and hotlinks at a node, the user is capable of selecting the right link downwards in the tree.

With hotlinks added, the situation is more involved, since the resulting hotlink-enhanced index structure now becomes a directed acyclic graph (DAG), with possibly more than one path to some destinations. Again, it is assumed that when faced with a hotlink in the current page, the user will be able to tell whether or not this hotlink may lead it to a closer point on the path to the desired destination. However, the hotlink selection process is not known to the user. Subsequently, while at a particular node, the user knows only the hotlinks emanating from that node (or possibly also hotlinks emanating from nodes previously visited by it on the way from the root to the current node). In particular, the user cannot know whether any hotlinks emanate from descendants of the current node, or where such hotlinks may lead.

This implies that any approach taken for designing a hotlink-enhanced index structure must take into account certain assumptions regarding the user's search policy. Two models have been considered concerning user capabilities. One natural model, referred to as the "clairvoyant" user model, captures situations where the user somehow knows the entire topology of the enhanced structure. This model is based on the following assumption.

The clairvoyant user model: At each node in the enhanced structure, the user can infer from the link labels which of the tree edges or hotlinks available at the current page is on a shortest path (in the enhanced structure) to the desired destination. The user always chooses that link.

This model appears to be too strong to be considered realistic. An alternative model recently proposed in [3] is based on the assumption that the user does not have this knowledge. This forces the user to deploy a *greedy* strategy.

The greedy user model: At each node in the enhanced structure, the user can infer from the link labels which of the tree edges or hotlinks available at the current page leads to a page that is closest *in the original tree structure* to the desired destination. The user always chooses that link.

In this paper we address the optimization problem of constructing a set of hotlinks that achieves a maximum improvement in the access cost. The problem has been given an exact algorithm on trees in the greedy user model in [3], but the complexity of that algorithm might be exponential in the depth of the tree, hence it yields a polynomial time solution only for trees of logarithmic depth. In the clairvoyant user model, the problem has been given a polynomial time 2-approximation algorithm on arbitrary trees [4]. Unfortunately, this algorithm does not appear to extend to the more realistic greedy user model. Our goal here is thus to develop an approximation algorithm for the problem on arbitrary trees in the greedy model.

1.2   Formal Definitions

Formally, given a rooted directed tree T with root r storing the database, a *hotlink* is a new directed edge which is not part of the tree. The hotlink *starts* at some node v and *ends* at (or *leads* to) one of its descendants u. To each node v we may assign at most one hotlink. Each node x of T has a *weight* $p(x)$, representing its access frequency, namely, the fraction of the user visits to that node, compared with the total number of user visits to tree nodes. Normalizing the weights allows us to interpret $p(x)$ as the probability that a user wants to access node x. For simplicity, we assume that for each nonleaf node x, $p(x) = 0$. Denote by $\Gamma(x)$ all x immediate descendants in the tree, and define $P(x) = \sum_{y \in \Gamma(x)} P(y)$ where a leaf x, will have value $P(x) = p(x)$.

Let S be a set of hotlinks constructed on the tree T (obeying the bound of K outgoing hotlinks per node). Denote by $T \oplus S$ the enhanced structure obtained by augmenting the tree T with the hotlinks of S. Denote the length of a path L (in edges) by $|L|$.

In the greedy model, let $D_G(T, v)$ and $D_G(T \oplus S, v)$ denote, respectively, the greedy path from the root to the node v in T (i.e., without using the hotlinks) and the greedy path to v in $T \oplus S$ (i.e., including the hotlinks). The gain from a hotlink assignment S is defined as

$$g_G(T, p, S) = \sum_{v \in Leaves(T)} \left(|D_G(T, v)| - |D_G(T \oplus S, v)| \right) \cdot p(v) .$$

Two optimization variants can be considered. The *hotlink enhancement* problem requires finding a set of hotlinks S optimizing the expected cost. The *hotlink assignment* problem requires finding a set of hotlinks S which optimizes $g_G(T, p, S)$ and achieves the optimal gain, namely,

$$g_G^*(T, p) = \max_S \{g_G(T, p, S)\}.$$

A set S attaining this optimum is termed an *optimal* set of hotlinks.

In the clairvoyant model, denote by $D_C(T, v)$ and $D_C(T \oplus S, v)$, respectively, the clairvoyant (shortest) path that does not include hotlinks and the clairvoyant path including hotlinks from the root to node v. Subsequently, define the gain $g_C(T, p, S)$ from a hotlink assignment S and the optimal gain $g_C^*(T, p)$ as in the greedy model, except for replacing the subscript G by C.

Note that while the optimization problems are equivalent (in the sense that a set of hotlinks S is an optimal solution to one if and only if it is an optimal solution to the other), the corresponding approximation versions may be very different, namely, the same set of hotlinks S may well-approximate one but not the other.

1.3   Related Work

Most past discussions on the problem concentrated on trees and DAG's. The NP-completeness of the hotlink enhancement problem on DAGS is proven in [1] by a reduction from the problem of Exact Cover by 3-Sets. (As this problem is equivalent to the hotlink assignment problem, the latter is NP-complete as well.) That article also discusses several distribution functions on the leaves, including the uniform, geometric and Zipf distributions, but restricts the discussion to full binary trees. An interesting analogy is presented therein between the hotlink enhancement problem on trees and coding theory. A classification tree can be interpreted as a coding of words (associating a move down to the ith child with the letter 'i'). Under this interpretation, every leaf corresponds to a codeword. The addition of a hotlink adds a letter to the alphabet. This provides a lower bound for the problem based on Shannon's theorem. By this lower bound, denoting the entropy of the probability distribution p by $H(p)$ and the maximal degree of the tree by Δ, the expected access cost, $cost(T \oplus S, p) = \mathbb{E}[\sum_{v \in Leaves(T)} |D_C(T \oplus S, v)| \cdot p(v)]$, is at least $\frac{H(p)}{\log \Delta + 1}$ [1, 2].

Based on these bounds, an approximation algorithm for the hotlink enhancement problem on bounded degree trees in the clairvoyant user model is presented in [2]. This algorithm approximates the expected access cost, $cost(T \oplus S, p)$. The access cost guaranteed by this algorithm is no more than $\frac{H(p)}{\log(\Delta+1) - (\Delta \log \Delta)/(\Delta+1)}$ $+ \frac{\Delta+1}{\Delta}$, hence the approximation ratio achieved by the algorithm is $\frac{(\Delta+1) \log(\Delta+1)}{\Delta \cdot H(p)}$ $+ \frac{\log(\Delta+1)}{\log(\Delta+1) - (\Delta \log \Delta)/(\Delta+1)}$, which is in general at least $\log(\Delta + 1)$.

Another recent article [5] discusses an interesting application of hotlink assignments in asymmetric communication protocols for achieving better performance bounds.

The greedy user model was presented in [3]. The underlying assumption is that the user has limited a-priori knowledge regarding the structure of the classification tree. The paper proposes an exact algorithm for solving the hotlink enhancement problem in that model on trees. The time complexity of that algorithm is polynomial over trees of logarithmic depth. The solution is also generalized to situations where more than one hotlink per node is allowed. For the case in which the distribution on the leaves is unknown, the paper gives an algorithm guaranteeing (an optimal) logarithmic upper bound on the access cost.

The hotlink assignment problem (namely, optimizing the gain) is introduced in [4]. The paper presents a polynomial time approximation algorithm for the hotlink assignment problem in the clairvoyant model. The algorithm achieves a worst case approximation ratio of 2, and works on general graphs. Unfortunately, this algorithm does not apply in the greedy user model, and it is not clear how to extend it to that model. Note that the algorithms of [2] and [4] cannot be directly compared as the approximation applies to different measures.

1.4   Our Results

In this paper we present a new polynomial time approximation algorithm for the hotlink assignment problem on rooted trees. The algorithm applies to both the clairvoyant model and the greedy model, and achieves an approximation ratio of 2. The algorithm is based on restricting attention to hotlinks of length 2, namely, hotlinks from a node to one of its grandchildren, and finding the best length two hotlink assignment for the given tree structure. The approximation is based on the crucial observation that the best length two hotlink assignment for a given tree T yields at least half the gain of the optimal (unrestricted) hotlink assignment for T. Moreover, we use this fact to show that the solutions generated by our algorithm achieve an approximation ratio of 2 in both the clairvoyant model and the greedy model, implying that the ratio between the optimal solutions in the two models is at most 2.

2   Basic Properties and Preliminaries

Denote by T_v the subtree of T rooted at node v. Define the height of a tree T, denoted $height(T)$, as the length of the longest path from the root of T to some leaf u. For each node v denote its set of children by $C(v)$.

Since any path from the root to a node u in the greedy model cannot be shorter than the shortest possible path to u, which is chosen in the clairvoyant model, we have:

Lemma 1. *For any tree T, and for any assignment S of hotlinks,*

$$g_G(S, T, p) \leq g_C(S, T, p) \ .$$

Let us represent a hotlink from node u to node v as a pair of parentheses, with u marking the left parenthesis and v marking the right one. The hotlinks placed along any path from the root can now be viewed as a sequence of parentheses. We say that a set of hotlinks S is *well-formed* if for any path P from the root to a leaf over the tree, the parentheses sequence of S on this path is balanced, namely, hotlinks do not cross each other.

Lemma 2. *For a well-formed hotlink assignment S, the gains in the greedy user model and in the clairvoyant model are the same, i.e., $g_G(S, T, p) = g_C(S, T, p)$.*

Proof. For a well-formed set of hotlinks S forming a hotlink enhanced index structure $T \oplus S$, the greedy path from the root to any leaf v coincides with the shortest path, i.e., $D_G(T \oplus S, v) = D_C(T \oplus S, v)$.   □

Based on these observations, the following lemma is proved in [3].

Lemma 3. [3] *For every index tree T, there exists an optimal solution to the hotlink assignment problem in the greedy user model which is well-formed.*

The *length* of a hotlink is defined as the number of edges in the path connecting its end points in the tree. A set of hotlinks S is called a *length two* assignment if all of its hotlinks are of length 2. Denote by *Length two Hotlink Assignment (L2HA)* the variant of the hotlink assignment problem in which it is required to find the set of hotlinks S which optimizes $g_C(T, p, S)$ under the constraint that the hotlink set S is a length two assignment. Denote this optimal gain by $g^*_{C_{L2}}(T, p) = \max_S\{g_C(T, p, S) \mid S \text{ is a length two assignment}\}$.

In the same manner, define $LkHA$ as the variant of the hotlink assignment problem restricted to hotlinks of length exactly k.

Lemma 4. *For every index tree T, there exists an optimal solution to the L2HA problem in the* clairvoyant *user model which is well-formed. Furthermore, any length two hotlink assignment that is not well-formed can be manipulated into well-formed assignment with the same clairvoyant model gain.*

Proof. Consider a length two hotlink assignment S which is not well-formed. Since the length of the hotlinks is bounded by 2, the only possibility is that there exists a path of four nodes v_1, v_2, v_3, v_4 such that v_{i+1} is the child of v_i and that v_1 has a hotlink to v_3 and v_2 has a hotlink to v_4 (see Figure 1). In such a case suppose that several hotlinks were added that changed the depth of v_1 and v_2, and that their depth after the hotlink assignment is d_1 and d_2 respectively. If $d_1 \geq d_2$ then the hotlink from v_1 to v_3 can be removed since the path from v_1 to v_3 can be replaced by travelling to v_2 and then directly to v_3 by the connecting edge. In case $d_1 < d_2$, the hotlink from v_2 to v_4 can be removed and v_4 can be reached through v_3. In either case, the assignment has lower cost and one less violation of the well-formed requirement.   □

Fig. 1. An example for assignment which is not well-formed in $L2HA$.

Corollary 1. *To solve L2HA it suffices to consider only well-formed length two assignments.*

Lemma 5. *For any $k \geq 2$, to solve LkHA it suffices to consider only well-formed length k assignments.*

Proof. Assume, to the contrary, that in any optimal solution there are crossing hotlinks. Observe the first two hotlinks that cross each other (the highest in the tree). The first starts at v and ends k nodes lower, the second starts at node u which is d nodes lower than v and ends in k nodes lower than u at node w. Then, since these are the highest crossing hotlinks there are no incoming hotlinks to the nodes between v and u and thus travelling to w through the second hotlink takes at least $d + 1$ steps, plus the number of steps it takes to travel to v. By removing this second hotlink the distance to w does not increase since we can travel to v plus take the first hotlink and then travel d steps down to w, and that will also be $d + 1$ steps.   □

Lemma 6. *For a set of well-formed length two hotlinks $S = S_1 \bigcup S_2$, where S_1 and S_2 are disjoint, the gain from S is the sum of the gains from S_1 and S_2, namely,*

$$g_C(T, p, S) = g_C(T, p, S_1) + g_C(T, p, S_2) ,$$

$$g_G(T, p, S) = g_G(T, p, S_1) + g_G(T, p, S_2) .$$

Proof. Consider the greedy (or clairvoyant) path to some node u in $T \oplus S_1$. It includes some subset S_1^u of the hotlinks from S_1. Likewise, the greedy path to u in $T \oplus S_2$ includes some subset S_2^u of the hotlinks from S_2. Since $S = S_1 \bigcup S_2$ is well-formed, no two hotlinks cross each other, and since it is a length two assignment they cannot overlap each other. Hence in an assignment containing all the hotlinks of S, a user travelling from the root r to u can take both sets of hotlinks S_1^u and S_2^u, and hence the gain achieved is the sum of the two.   □

Corollary 2. *For a set S of well-formed length two hotlinks,*

$$g_C(T, p, S) = \sum_{s \in S} g_C(T, p, s) .$$

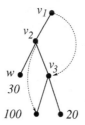

Fig. 2. An example tree T.

This corollary does not hold in case of non-well-formed assignments (even in length two assignments). For example, in Figure 2, the hotlink from v_2 has gain $100 \cdot 1$, and the hotlink from v_1 has gain $100 \cdot 1 + 20 \cdot 1 = 120$. The sum of the gains of these two hotlinks is thus 220, but when assigning both of them to T, the gain is only $100 \cdot 1 + 20 \cdot 1 = 120$. A better choice would be to assign a hotlink from v_1 to w with separate gain $30 \cdot 1 = 30$ and a hotlink from v_2 as before. The sum of gains for the two hotlinks is $100 + 30 = 130 < 220$ but the actual gain achieved by adding both of them to the tree is 130, which is higher.

3   A Polynomial Time Algorithm for *L2HA*

The number of possible length two assignments for a single node is bounded by Δ^2 where Δ is the maximum degree of the tree T. Therefore, based on Corollary 2, a dynamic programming algorithm can be devised to solve the problem. The algorithm is based on filling a table M of length n with a value for every node and a $n \times \Delta$ table M' with a value for every node-child pair. The value of $M(v)$ corresponds to the maximum gain possible in the subtree T_v, while $M'(v, \overline{u})$ for a vertex v and its child $\overline{u}$ is the greatest gain possible in the subtree T_v without hotlinks directed from v to one of $\overline{u}$'s descendants.

The calculations proceed from the leaves up. The base of the recursion is when the tree rooted at v has height 2. In such a case the best possible length two hotlink assignment is to the heaviest grandchild of v and the value of $M(v)$ is set to be the weight of that grandchild. The value of $M'(v, \overline{u})$ is set to be the maximum weight grandchild of v which is not a descendant of $\overline{u}$. If the tree is of $height(T_v) < 2$ then the value of both $M(v)$ and $M'(v, \overline{u})$ is set to 0.

$M(v)$ is calculated by breaking the problem to subproblems. For every child $\widetilde{u}$ of v, calculate the gain if the hotlink would be directed to one of $\widetilde{u}$'s children. This gain is the sum of all gains from the other children of v plus the gain from a hotlink directed to some child $\widetilde{w}$ of $\widetilde{u}$ (which is $1 \cdot P(\widetilde{w})$). In addition we need to take into account the possible gain from hotlinks directed from $\widetilde{u}$, but restricting $\widetilde{u}$ from choosing any hotlink to $\widetilde{w}$'s descendants. The grandchild $\widetilde{w}$ is chosen from all children of $\widetilde{u}$ to maximize the gain of the assignment.

$M(v, \overline{u})$ is also calculated similarly. Formally, the values of $M(v)$ and $M'(v, \overline{u})$ are defined as follows (see Figure 3).

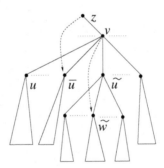

Fig. 3. Calculating $M(v)$ and $M'(v, \overline{u})$.

$$M(v) = \max_{\widetilde{u} \in \Gamma(v)} \left(\sum_{u \in \Gamma(v), u \neq \widetilde{u}} M(u) + \max_{\widetilde{w} \in \Gamma(\widetilde{u})} \left[P(\widetilde{w}) + M'(\widetilde{u}, \widetilde{w}) \right] \right),$$

$$M'(v, \overline{u}) = \max_{\widetilde{u} \in \Gamma(v), u \neq \overline{u}} \left(\sum_{u \in \Gamma(v), u \neq \widetilde{u}} M(u) + \max_{\widetilde{w} \in \Gamma(\widetilde{u})} \left[P(\widetilde{w}) + M'(\widetilde{u}, \widetilde{w}) \right] \right).$$

The algorithm generates all possible length two well-formed assignments. Hence by Corollary 1 the optimal gain for $L2HA$ is achieved. The time bound is the number of table entries multiplied by the time to calculate each entry, which is polynomial.

4   A 2 Approximation for Hotlink Assignment

4.1   The Approximation

We now prove that in the clairvoyant model the optimal solution for $L2HA$ yields a 2 approximation for the hotlink assignment problem, namely, $2 \cdot g^*_{C_{L2}}(T,p) \geq g^*_C(T,p)$.

Intuitively, every hotlink that shortens k steps to a certain leaf can be broken into a chain of $\lfloor (k+1)/2 \rfloor$ hotlinks of length 2, henceforth referred to as a *worm* of hotlinks, and the total gain achieved will be at least $k/2$. For example, in Figure 4(a) the direct hotlink from r has a gain of $5 \cdot 100 = 500$, and the worm gain is $3 \cdot 100 = 300$. in Figure 4(b) the direct hotlink from r has a gain of $4 \cdot 100 = 400$, and the worm gain is $2 \cdot 100 = 200$.

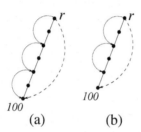

(a)   (b)

Fig. 4. A worm example.

Lemma 7. *For every tree T, the gain of an optimal solution to the $L2HA$ problem is at least half of the optimal gain using arbitrary length hotlinks in the clairvoyant user model, namely,*

$$2 \cdot g^*_{C_{L2}}(T,p) \geq g^*_C(T,p) .$$

Proof. By induction on the tree height h. (For simplicity the proof will be presented for binary trees, but it applies to all trees.)

For a tree of height 2 with root v, the value of $M(v)$ is the weight of the heaviest grandchild. This gain is achieved by choosing a hotlink from v to that leaf, which is also the best possible choice of hotlink assignment in the clairvoyant model, and thus the gain of the hotlink assignment is the same. In particular, $2 \cdot g^*_{C_{L2}}(T,p) \geq g^*_C(T,p)$.

Assume the claim is correct for any tree of height at most h, and consider a tree of height $h + 1$. Consider the optimal assignment of an arbitrary length hotlink from the root v to some node w. Without loss of generality, this hotlink leads to the left subtree, T_{v_L}. By the inductive hypothesis, the solution for the right subtree T_{v_R} can be replaced by an $L2HA$ assignment with at least half the gain of the optimal assignment on that subtree, i.e., $G^*_{C_{L2}}(T_{v_R}) \geq \frac{1}{2}G^*_C(T_{v_R})$.

The subtree T_{v_L} is also of height at most h, hence its entire hotlink assignment, S_{HA}, can be replaced by a length two hotlink assignment S_{L2HA}, still attaining at least half the gain.

The only remaining modification required is to discard the hotlink from v while still retaining at least half of the gain achieved by this hotlink, without decreasing the gain of the other length two hotlinks assignments. This is done by creating a worm $W(v, w)$ from v to w. This worm may "collide" with other hotlinks that were already assigned by S_{L2HA}, in the sense that it may require setting a hotlink from some node u along its path, where such a hotlink has already been selected. When such a case occurs, we remove the previous hotlink from S_{L2HA} and add the one required by the worm. Denote this assignment by S_{L2HA_w}. It is clear that such a worm achieves half the gain of a direct hotlink from v to w. Hence to complete the proof, it remains to show that the worm $W(u, w)$ does not decrease the gain achieved by the set of hotlinks S_{L2HA}.

Consider the structure of the worm $W(v, w)$. It has a length two hotlink emanating from every second node on the path from v to w. Denote the depth of nodes on the path as their distance from the node v (see Figure 5). The nodes that are assigned hotlinks by the worm are all even depth nodes (see Figure 5), so the only possible decrease in the gain achieved can be caused by previous hotlinks from S_{L2HA} that were removed due to the addition of the worm. The gain achieved by any assignment is the summation over all the nodes, of the shortening of nodes cost multiplied by its relative weight. Therefore, if for every node u the shortening achieved by S_{L2HA} is achieved also after adding the worm then the gain remains the same.

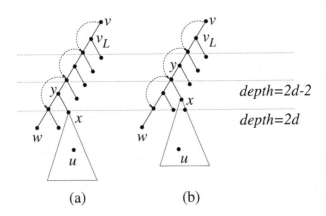

(a)   (b)

Fig. 5. The two possibilities for node u.

Consider a node u in T_{v_L} and let y be the lowest common ancestor of u and w (see Figure 5). Observe that at most one length two hotlink on the path leading from v to u could have been deleted by the worm. This hotlink must have originated at some even depth node, specifically, either y or its parent (see cases (a) and (b) in Figure 5), and aimed at some grandchild node x which is u's ancestor and is not on the path to w (specifically either y's child or grandchild on the path to u). From this node x and onward the path to u remains untouched. Thus we must show that the travelling path to x is of same length as in the original S_{L2HA} assignment.

Suppose x is of depth $2d$ from v (as explained it must be an even depth). In the S_{L2HA} assignment, the path to x first travels from v to v_L and then follows the shortest path to u, consisting of tree edges and possibly some hotlinks. Since x is at depth $2d$, the optimal choice of hotlinks for x is a direct worm $W(v_L, x)$ of length $d - 1$ (since the length of the path from v_L to x is $2d - 1$ the worm will end at the parent of x) and the total minimal length of the path from v_L to x will be $(d - 1) + 1$ and from v to x a total of $|D_C(T \oplus S_{L2HA}, x)| \geq d + 1$ steps to reach x. In the S_{L2HA_w} assignment the hotlink to x was removed, but there was a direct worm from v up until the last hotlink. The total length of the new path to x is thus $|D_C(T \oplus S_{L2HA_w}, x)| \leq \frac{d-2}{2} + 2 = d + 1$. $\qquad\square$

Theorem 1. *The hotlink assignment problem has a 2 approximation algorithm both in the clairvoyant and the greedy model, namely, $2 \cdot g^*_{C_{L2}}(T, p) \geq g^*_G(T, p)$.*

Proof. The $L2HA$ optimal solution, S^*_{L2HA}, is well-formed and thus by Lemma 2 its value is the same under the greedy model. Namely, $g_G(S^*_{L2HA}, T, p) = g_C(S^*_{L2HA}, T, p)$. Hence

$$\begin{aligned} 2 \cdot g_G(S^*_{L2HA}, T, p) &= 2 \cdot g_C(S^*_{L2HA}, T, p) \\ &= 2 \cdot g^*_{C_{L2}}(T, p) \geq g^*_C(T, p) \geq g^*_G(T, p) \ , \end{aligned}$$

hence S^*_{L2HA} is a solution for the greedy model with approximation ratio 2. $\quad\square$

We conclude with a bound on the gap between the clairvoyant and greedy models.

Corollary 3. *For any tree T, $g^*_G(T, p) \leq g^*_C(T, p) \leq 2 \cdot g^*_G(T, p)$.*

Proof. By the theorem, there exists a well-formed solution S^*_{L2HA} using length two hotlinks for the tree T, satisfying $g^*_G(T, p) \geq g_G(S^*_{L2HA}, T, p) \geq g^*_C(T, p)/2$. Recalling also Fact 1, the claim follows. $\qquad\square$

4.2 Generalization to K Hotlinks in Each Node

The same ratio of 2 can be proven when K hotlinks are allowed per node. In fact, when adding the K worms from v to $w_{1 \leq i \leq K}$, we can remove all K hotlinks previously assigned to a node by S_{L2HA} when collision occurs, since the argument above will apply to all of the hotlinks destinations. Hence we are left with $K - 1$

additional hotlinks we can now assign to reduce farther the cost (in some cases it will be possible to use it in others not). But it seems that generally the value of $L2HA$ relative to the general hotlink assignment problem changes for the better as K increases.

4.3   Example of Tightness

There are instances where the best gain achieved by length two hotlinks is exactly half of that possible by general hotlinks. An example is given in Figure 6. For this example, the optimal length two hotlink assignment shown in Figure 6(a) yields gain $10 + 2 \cdot 20 = 50$, and the optimal arbitrary length hotlink assignment shown in Figure 6(b) yields gain $2 \cdot 10 + 4 \cdot 20 = 100$.

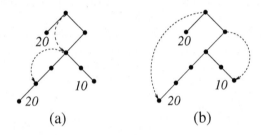

(a) (b)

Fig. 6. Tight example.

References

1. P. Bose, J. Czyzowicz, L. Gasieniec, E. Kranakis, D. Krizanc, A. Pelc and M.V. Martin, Strategies for hotlink assignments, In Proc. 11th Symp. on Algorithms and Computation, 2000, pp. 23–34.
2. E. Kranakis, D. Krizanc and S. Shende, Approximating hotlink assignments, In Proc. 12th Symp. on Algorithms and Computation, 2001, pp. 756–767.
3. O. Gerstel, S. Kutten, R. Matichin and D. Peleg, Hotlink enhancement algorithms for web directories, In Proc. 14th Symp. on Algorithms and Computation, 2003, pp. 68–77.
4. R. Matichin and D. Peleg, Approximation algorithm for hotlink assignments in web directories, In Proc. Workshop on Algorithms and Data Structures, 2003, pp. 271–280.
5. P. Bose, D. Krizanc, S. Langerman and P. Morin, Asymmetric communication protocols via hotlink assignments, In Proc. 9th Colloq. on Structural Information and Communication Complexity, 2002, pp. 33–39.
6. http://www.yahoo.com/
7. http://www.dmoz.org/

Optimal Decision Strategies
in Byzantine Environments

Michel Paquette and Andrzej Pelc

Département d'informatique, Université du Québec en Outaouais,
Gatineau, Québec J8X 3X7, Canada
michel.paquette@polymtl.ca, andrzej.pelc@uqo.ca

Abstract. A Boolean value of given *a priori* probability distribution
is transmitted to a deciding agent by several processes. Each process
fails independently with given probability, and faulty processes behave
in a Byzantine way. A deciding agent has to make a decision concerning
the transmitted value on the basis of messages obtained by processes.
We construct a deterministic decision strategy which has the provably
highest probability of correctness. It computes the decision in time linear
in the number of processes.

Decision optimality may be alternatively approached from a local, rather
than global, point of view. Instead of maximizing the total probability
of correctness of a decision strategy, we may try to find, for every set of
values conveyed by processes, the conditionally most probable original
value that could yield this set. We call such a strategy *locally optimal*, as
it locally optimizes the probability of a decision, given a set of relayed
values, disregarding the impact of such a choice on the overall probability of correctness. We construct a locally optimal decision strategy which
again computes the decision value in time linear in the number of processes. We establish the surprising fact that, in general, local probability
maximization may lead to a decision strategy which does not have the
highest probability of correctness. However, if the probability distribution of the Boolean value to be conveyed is uniform, and all processes
have the same failure probability smaller than $1/2$, this anomaly does
not occur.

1   Introduction

In faulty environments, decisions must often be made on the basis of erroneous
or corrupted data. This is the situation, e.g., in system-level diagnosis [10] where
processors have to be diagnosed on the basis of tests conducted by other processors, and faulty testers are unreliable. In network communication, messages
travel to their destinations by multiple paths, some of which contain faulty components and may relay corrupted information. In nuclear reactors, temperature
of the core is monitored by multiple sensors, some of which may be dysfunc-
tional. In all these situations it is desirable to achieve fault tolerance, i.e., the
ability to make correct decisions with high probability, in spite of faults occur-
ring in the system. The fundamental mechanism used to achieve this capacity is

R. Královič and O. Sýkora (Eds.): SIROCCO 2004, LNCS 3104, pp. 245–254, 2004.
© Springer-Verlag Berlin Heidelberg 2004

based on redundancy. The same information (diagnostic decision, message in a network, temperature value) is conveyed by multiple relayers (processors, communication paths, sensors). A deciding agent makes a decision concerning the transmitted information based on data obtained through these relayers, using some kind of voting mechanism which maps each set of obtained values to a single decision. It is important to design strategies of the deciding agent that have a high probability of correctness.

In this paper we consider a very general model of a fault-prone system in which a decision has to be made on the basis of unreliable information. We assume that a Boolean value is conveyed to the deciding agent by several *processes*. The a priori probability distribution of this value can be arbitrary and is known to the agent. If the agent does not have any information on the distribution, it may assume that it is uniform. Relaying processes (which model processors in a system, communication paths in a network, sensors in a monitoring system) are assumed to fail independently (and independently of the stochastic choice of the value to be conveyed), with known (but arbitrary and not necessarily equal) probabilities. Fault-free processes relay the correct value but faulty ones may behave arbitrarily, in a Byzantine way (they may relay the value correctly, corrupt it, or relay no value at all). The deciding agent receives the vector of relayed values and must make a decision concerning the original value. The aim is to design a deterministic decision strategy which has the highest possible probability of correctness.

1.1 Our Results

We first define precisely the probability of correctness of a (deterministic) decision strategy. This is not an obvious notion, due to the Byzantine behaviour of processes: our model involves a mixture of stochastic and adversarial ingredients (on the one hand random distributions of the decision values and faulty processes, and on the other hand the Byzantine nature of faults) and both these ingredients have to be accounted for in the definition of correctness probability.

Given this definition we design a decision strategy and prove that it has the highest possible probability of correctness among all (deterministic) decision strategies. This strategy is simple to implement and the decision value is computed in time linear in the number of processes. In fact, it is computed in time linear in the number of processes whose failure probability is less than $1/2$, if such processors exist, and in *constant* time if they do not exist. Indeed, it turns out that the optimal decision strategy is based only on data conveyed by processes whose failure probability is less than $1/2$, if such processes exist. Otherwise, the a priori more probable Boolean value is chosen.

Decision optimality may be alternatively approached from a local, rather than global, point of view. Instead of maximizing the total probability of correctness of a decision strategy, we may try to find, for every set of values conveyed by processes, the conditionally most probable original value that could yield this set. We call such a strategy *locally optimal*, as it locally optimizes the probability of a decision, given a set of relayed values, disregarding the impact of such a choice on the overall probability of correctness. We construct a locally optimal decision strategy which again computes the decision value in time linear in the number of processes. It is natural to ask if these two ways of optimizing the decision are equivalent. The surprising answer is no. We show that, in general, local probability maximization may lead to a decision strategy which has not the highest probability of correctness. However, if the probability distribution of the Boolean value to be conveyed is uniform, and all processes have the same failure probability smaller than 1/2, this anomaly does not occur.

1.2   Related Work

Voting mechanisms have been proposed as a fundamental way to achieve fault tolerance. Coteries, defined in [7] and further studied, e.g., in [1, 2, 6, 8, 11], are considered to provide the most general framework for such schemes. An important way of using voting mechanisms, considered also in the present paper, is to produce a single value on the basis of many unreliable reported values. A similar approach was adopted in [3, 4], however, unlike in our case, the authors considered a purely stochastic model in which not only the a priori distribution of the transmitted value and distribution of faulty processes but also reports of faulty processes (and even – in one of the models – reports of fault-free processes) are random. This purely stochastic model facilitates the definition of correctness of a decision strategy but has limited applicability, as the probability distribution of faulty process reports is often unknown, and in some cases this behaviour may not be random but controlled by a malicious adversary. Our approach of treating faulty processes as Byzantine makes it possible to deal with this very general kind of faults. Moreover, in the purely stochastic model used in [3, 4], computing the value of the optimal decision strategy given a set of data conveyed by processes was a task taking time exponential in the number of processes, and hence intractable for larger systems. This should be contrasted with our framework which, in addition to treating a more general type of faults, permits to compute the decision value in time linear in the number of processes.

Decisions based on data subject to stochastically distributed faults of Byzantine type have been also considered in a different framework, that of system-level diagnosis of multiprocessor systems (cf. [10]). In particular, in [5, 9], the authors considered the problem of finding a diagnosis scheme having the highest probability of correctness, when faulty processors are distributed randomly but behave in a Byzantine way when testing other processors. While the setting in both cases is different, some of the definitions and techniques have the same spirit.

2   Terminology, Definitions and Preliminary Results

Let b denote the Boolean value to be conveyed. Assume that the a priori probability of 0 is p and of 1 is $1 - p$. Without loss of generality assume that $p \geq 1 - p$ (the other case is symmetric). For any $v \in \{0, 1\}$, we let $q(v) = p$ if $v = 0$, and $q(v) = 1 - p$ otherwise. Let $\{1, ..., n\}$ denote the set of processes and let p_i

denote the failure probability of process i. We assume that processes fail independently. A set of values conveyed by processes, called a *report*, is any function $R : \{1, ..., n\} \longrightarrow \{0, 1\}$. $R(i)$ denotes the value reported by process i. Fault-free processes report the correct value b, and faulty processes may report any value or no value at all. If process i does not report any value (this may be the case only for a faulty process), we set $R(i) = 0$ by default. The set of all reports is denoted by $\mathcal{R}$.

We consider only deterministic strategies, although we evaluate them probabilistically, due to the stochastic ingredient in the environment. Hence a decision strategy is a function assigning a Boolean decision to any report. More precisely, a *decision strategy* is any function $D : \mathcal{R} \longrightarrow \{0, 1\}$. We now formally define the probability of correctness of any decision strategy. The sample space is the set $\Omega = \{(G, v) : G \subseteq \{1, ..., n\}, v \in \{0, 1\}\}$. Intuitively, the elementary event (G, v) means that fault-free processes are exactly those in G (G will be called a *good set*) and that v is the value to be conveyed, i.e., $v = b$. For any $G \subseteq \{1, ..., n\}$, we denote $G' = \{1, ..., n\} \setminus G$. The probability distribution in this sample space is as follows: $P(G, v) = q(v) \cdot \prod_{i \in G}(1 - p_i) \cdot \prod_{i \in G'} p_i$

This definition captures the assumption about stochastic independence of all failures and of the a priori Boolean value distribution. The probability function P is extended to all events (subsets of Ω) in the following standard way: $P(\mathcal{E}) = \sum_{(G,v) \in \mathcal{E}} P(G, v)$.

A report R is *compatible* with an elementary event (G, v), if it can occur when the value to convey is v and the set of fault-free processes is G. This simply means that $R(i) = v$, for all $i \in G$ (recall that faulty processes may report arbitrary values). Let $\rho(G, v)$ denote the set of reports compatible with (G, v). Hence $\rho(G, v) = \{R \in \mathcal{R} : \forall i \in G \quad R(i) = v\}$.

If D is a decision strategy, the event that D is correct, denoted by $\mathrm{Cor}(D)$, consists of elementary events (G, v), such that the decision strategy outputs v on any report compatible with (G, v), i.e., of those elementary events for which the strategy outputs the correct value *regardless* of the behaviour of faulty processes. Formally, $\mathrm{Cor}(D) = \{(G, v) : \forall R \in \rho(G, v) \quad D(R) = v\}$. The reliability of the decision strategy D is defined as the probability of its correctness: $\mathrm{Rel}(D) = P(\mathrm{Cor}(D))$. A decision strategy is *optimal* if $\mathrm{Rel}(D) \geq \mathrm{Rel}(D')$, for any decision strategy D'.

An event $\mathcal{E} \subseteq \Omega$ is an *intersecting family* if $G_1 \cap G_2 \neq \emptyset$ for all $(G_1, 0), (G_2, 1) \in \mathcal{E}$. An event $\mathcal{E} \subseteq \Omega$ is a *filter* if it is an intersecting family and if, for all $(G_1, v) \in \mathcal{E}$ and $G_1 \subseteq G_2$, we have $(G_2, v) \in \mathcal{E}$. For any intersecting family $\mathcal{E}$ we define $\mathcal{F}(\mathcal{E}) = \{(A, v) \in \Omega : \exists (B, v) \in \mathcal{E} \quad B \subseteq A\}$. Clearly, for any intersecting family $\mathcal{E}$, the event $\mathcal{F}(\mathcal{E})$ is a filter.

Lemma 1. *For any decision strategy D, $Cor(D)$ is an intersecting family.*

The following lemma shows that the probability of a filter is decreasing with respect to processor failure probabilities.

Lemma 2. *Let $\mathcal{F}$ be a filter and let $\mathcal{P} = (p_1, ..., p_n)$, $\mathcal{P}' = (p'_1, ..., p'_n)$ be two sets of failure probabilities of processes $i = 1, ..., n$, such that $p_i \leq p'_i$ for all*

$i = 1, ..., n$. Let P and P' be probability functions on Ω corresponding to the same distribution $(p, 1 - p)$ of Boolean value b and to process failure probabilities $\mathcal{P}$ and $\mathcal{P}'$, respectively. Then $P(\mathcal{F}) \geq P'(\mathcal{F})$.

Proof. It is enough to prove the lemma for $p_{i_0} \leq p'_{i_0}$ and $p_i = p'_i$, for all $i \neq i_0$. Denote $r = p_{i_0}$ and $r' = p'_{i_0}$. Consider any $C \subseteq \{1, ..., n\} \setminus \{i_0\}$. Let $v \in \{0, 1\}$. Denote $X = \prod_{i \in C}(1 - p_i) \cdot \prod_{i \in C' \setminus \{i_0\}} p_i$. Then $P(C, v) = q(v)rX$, $P'(C, v) = q(v)r'X$, $P(C \cup \{i_0\}, v) = q(v)(1 - r)X$, and $P'(C \cup \{i_0\}, v) = q(v)(1 - r')X$. It follows that $P'(C, v) - P(C, v) = q(v)X(r' - r)$ and $P'(C \cup \{i_0\}, v) - P(C \cup \{i_0\}, v) = q(v)X(r - r')$. Consequently, any gain of probability on an event (C, v) is compensated by an equal loss of probability on the event $(C \cup \{i_0\}, v)$. By definition of a filter, $(C, v) \in \mathcal{F}$ implies $(C \cup \{i_0\}, v) \in \mathcal{F}$. This concludes the proof. (The possibility of sharp inequality $P(\mathcal{F}) < P'(\mathcal{F})$ comes from the fact that there may exist sets $C \subseteq \{1, ..., n\} \setminus \{i_0\}$ such that $(C, v) \notin \mathcal{F}$ but $(C \cup \{i_0\}, v) \in \mathcal{F}$.) $\square$

3 Description and Analysis of an Optimal Decision Strategy

In this section we describe a decision strategy which is then proved optimal. Call a process i *heavy* if $p_i < 1/2$, and call it *light* otherwise. Denote by H the set of all heavy processes and denote by L the set of all light processes. The strategy is called Ignore-the-Suspects (IS, for short), as it completely ignores report values of light (suspect) processes. The definition depends on whether heavy processes exist. For any set $G \subseteq \{1, ..., n\}$, define $I(G) = \prod_{i \in G \cap H}(1 - p_i) \cdot \prod_{i \in H \setminus G} p_i$. For example, $I(\emptyset) = \prod_{i \in H} p_i$, and $I(\{1, ..., n\}) = I(H) = \prod_{i \in H}(1 - p_i)$. The following lemma follows from the definition of heavy processes.

Lemma 3. $I(G_1) \leq I(G_2)$ whenever $G_1 \subseteq G_2$.

For any report R and any value $v \in \{0, 1\}$, denote by $Z(R, v)$ the set of processes that "vote" for v in report R. More precisely, $Z(R, v) = \{i \in \{1, ..., n\} : R(i) = v\}$.

Case 1. The set H of heavy processes is nonempty.
For any report R,

$$\text{IS}(R) = \begin{cases} 0 \text{ if } p \cdot I(Z(R, 0)) \geq (1 - p) \cdot I(Z(R, 1)) \\ 1 \text{ otherwise.} \end{cases} \tag{1}$$

In this case, the strategy compares two events, for $v = 0, 1$: that the value to be conveyed is v, all heavy processes voting for v are fault free and all heavy processes voting against v are faulty. Then it picks the value v yielding the more probable of these events, breaking ties arbitrarily (we choose to break ties in favour of 0 but this decision does not influence optimality).

Case 2. The set H of heavy processes is empty.

$$\text{IS}(R) = 0 . \tag{2}$$

In this case, the strategy disregards the report and picks the a priori more probable value: recall that we assumed $p \geq 1 - p$.

This concludes the description of strategy IS. Notice that, for any report R, in order to compute $IS(R)$, $O(|H|)$ arithmetic operations have to be performed in Case 1 and no operations in Case 2. Hence the value of the strategy on any report can be computed in time $O(|H|)$, when $H \neq \emptyset$, and in constant time otherwise. This implies

Theorem 1. *The value of strategy IS on any report can be computed in time linear in the number of processes.*

The rest of this section is devoted to the proof of optimality of strategy IS. We first prove a lemma describing the set Cor(IS).

Lemma 4. *If the set of heavy processes is nonempty then*

$$
\begin{aligned}
Cor(IS) = &\{(G, v) \in \Omega : q(v) \cdot I(G) > q(1 - v) \cdot I(G')\} \\
&\cup \{(G, 0) \in \Omega : p \cdot I(G) = (1 - p) \cdot I(G')\} ;
\end{aligned}
\tag{3}
$$

otherwise $Cor(IS) = \{(G, 0) : G \subseteq \{1, ..., n\}\}$.

Proof. First suppose that the set H of heavy processes is nonempty. Let $(G, v) \in \Omega$ and consider four cases.

Case 1. $q(v) \cdot I(G) > q(1 - v) \cdot I(G')$.

Let $R \in \rho(G, v)$. Hence, for all $i \in G$, we have $R(i) = v$. Consequently, $G \subseteq Z(R, v)$ and hence $I(G) \leq I(Z(R, v))$ and $I(G') \geq I(Z(R, 1 - v))$, in view of Lemma 3. This implies $q(v) \cdot I(Z(R, v)) > q(1 - v) \cdot I(Z(R, 1 - v))$, and hence $IS(R) = v$. Thus $IS(R) = v$ for all $R \in \rho(G, v)$. This implies $(G, v) \in Cor(IS)$.

The arguments in the three other cases:

Case 2. $q(v) \cdot I(G) = q(1 - v) \cdot I(G')$ and $v = 0$.
Case 3. $q(v) \cdot I(G) = q(1 - v) \cdot I(G')$ and $v = 1$.
Case 4. $q(v) \cdot I(G) < q(1 - v) \cdot I(G')$.
are similar.

This concludes the proof when H is nonempty. If the set H of heavy processes is empty, the lemma follows immediately from the definition of the strategy IS. □

Lemma 5. *Rel(IS) does not depend on failure probabilities of light processes.*

Proof. If the set H of heavy processes is empty, Lemma 4 implies that $Rel(IS) = p$, which implies independence of $Rel(IS)$ of any p_i. If the set of heavy processes is nonempty, Lemma 4 implies the following equivalence: for any set S of heavy processes, any set T of light processes, and any $v \in \{0, 1\}$, we have $(S, v) \in Cor(IS)$ if and only if $(S \cup T, v) \in Cor(IS)$. This implies that $Rel(IS)$ does not depend on p_i, for any light process i. □

Theorem 2. *The decision strategy IS is optimal.*

Proof. We will show that $\text{Rel(IS)} \geq P(\mathcal{I})$, for any intersecting family $\mathcal{I} \subseteq \Omega$. By Lemma 1 this implies $\text{Rel(IS)} \geq \text{Rel}(D)$, for any decision strategy D. First suppose that $p_i = 1/2$ for all light processes i. Fix an intersecting family $\mathcal{I}$. Consider two cases.

Case 1. The set H of heavy processes is nonempty.

Without loss of generality suppose that $1 \in H$. Call a set $S \subseteq H$ *dominating* if either $I(S) > I(H \setminus S)$ or $I(S) = I(H \setminus S)$ and $1 \in S$. For any dominating set S consider the family $\mathcal{C}(S)$ of all events $(S \cup T, 0)$, $(H \setminus S \cup T, 0)$, $(S \cup T, 1)$, $(H \setminus S \cup T, 1)$, for all $T \subseteq L$. In order to prove $\text{Rel(IS)} \geq P(\mathcal{I})$, it is enough to show that $P(\text{Cor(IS)} \cap \mathcal{C}(S)) \geq P(\mathcal{I} \cap \mathcal{C}(S))$, for all dominating sets S. Fix a dominating set S and consider two cases.

Subcase 1.1. $p \cdot I(H \setminus S) \geq (1 - p) \cdot I(H)$.

By definition of a dominating set we have $p \cdot I(S) \geq (1 - p) \cdot I(H \setminus S)$ and by the assumption of Subcase 1.1 we have $p \cdot I(H \setminus S) \geq (1 - p) \cdot I(H)$. Hence, by Lemma 4 we have $\text{Cor(IS)} \cap \mathcal{C}(S) = \{(S \cup T, 0), (H \setminus S \cup T, 0) : T \subseteq L\}$.

Now consider $\mathcal{I} \cap \mathcal{C}(S)$. If $(S \cup T, 1) \in \mathcal{I}$, for some $T \subseteq L$, then $(H \setminus S \cup L \setminus T, 0) \notin \mathcal{I}$, by definition of an intersecting family. Similarly, if $(H \setminus S \cup T, 1) \in \mathcal{I}$, for some $T \subseteq L$, then $(S \cup L \setminus T, 0) \notin \mathcal{I}$. However $P((H \setminus S \cup L \setminus T, 0)) \geq P((S \cup T, 1))$ and $P((S \cup L \setminus T, 0)) \geq P((H \setminus S \cup T, 1))$. (Here we use the assumption that $p_i = 1/2$ for all light processes i.) Consequently, in $\mathcal{I} \cap \mathcal{C}(S)$, an event (G, v) from $\text{Cor(IS)} \cap \mathcal{C}(S)$ can only be replaced by an event of probability not larger than $P((G, v))$. This implies $P(\text{Cor(IS)} \cap \mathcal{C}(S)) \geq P(\mathcal{I} \cap \mathcal{C}(S))$ in Subcase 1.1.

Subcase 1.2. $p \cdot I(H \setminus S) < (1 - p) \cdot I(H)$.

Now we have $p \cdot I(S) \geq (1 - p) \cdot I(H \setminus S)$ and $p \cdot I(H \setminus S) < (1 - p) \cdot I(H)$. Hence $\text{Cor(IS)} \cap \mathcal{C}(S) = \{(S \cup T, 0), (S \cup T, 1) : T \subseteq L\}$. Similarly as in Subcase 1.1, we conclude that an event (G, v) from $\text{Cor(IS)} \cap \mathcal{C}(S)$ can only be replaced by an event of probability not larger than $P((G, v))$. This implies $P(\text{Cor(IS)} \cap \mathcal{C}(S)) \geq P(\mathcal{I} \cap \mathcal{C}(S))$ in Subcase 1.2 as well.

Case 2. The set H of heavy processes is empty.

By our assumption, all processes have failure probability $1/2$. Hence we have $P(G, 0) = p/2^n$ and $P(G, 1) = (1 - p)/2^n$, for any $G \subseteq \{1, ..., n\}$. In Case 2, $\text{Cor(IS)} = \{(G, 0) : G \subseteq \{1, ..., n\}\}$. If $(G, 1) \in \mathcal{I}$, for some $G \subseteq \{1, ..., n\}$ then $(G', 0) \notin \mathcal{I}$, by definition of an intersecting family. Consequently, an event of probability $p/2^n$ from Cor(IS) can only be replaced in $\mathcal{I}$ by an event of smaller or equal probability $(1 - p)/2^n$. This implies $\text{Rel(IS)} \geq P(\mathcal{I})$ in Case 2.

It remains to remove the assumption that $p_i = 1/2$ for all light processes i. Let $\mathcal{I}$ be any intersecting family. Let P denote, as before, the probability function yielded by process failure probabilities p_i, where $p_i = 1/2$ for all light processes i. Let P' denote the probability function yielded by process failure probabilities p'_i, where $p'_i = p_i$ for all heavy processes i, and p'_i for light processes i are arbitrary reals from the interval $[1/2, 1]$. By Lemma 5 we have $P(\text{Cor(IS)}) = P'(\text{Cor(IS)})$. Since $\mathcal{F}(\mathcal{I})$ is an intersecting family, the first part of this proof implies $P(\mathcal{F}(\mathcal{I})) \leq P(\text{Cor(IS)})$. Since $\mathcal{F}(\mathcal{I})$ is also a filter, Lemma 2 implies $P'(\mathcal{F}(\mathcal{I})) \leq P(\mathcal{F}(\mathcal{I}))$.

Finally, since $\mathcal{I} \subseteq \mathcal{F}(\mathcal{I})$, we have $P'(\mathcal{I}) \leq P'(\mathcal{F}(\mathcal{I}))$. Hence we get:

$$P'(\mathcal{I}) \leq P'(\mathcal{F}(\mathcal{I})) \leq P(\mathcal{F}(\mathcal{I})) \leq P(\text{Cor(IS)}) = P'(\text{Cor(IS)}) . \tag{4}$$

This concludes the proof for arbitrary values of process failure probabilities. □

4   Locally Optimal Decision Strategies

An alternative way of optimizing the quality of a decision is adopting a local rather than a global point of view. Instead of maximizing the total probability of correctness of a decision strategy, as we did in the previous section, we may want to find the conditionally most probable value, given a report R. It is natural to call such a strategy *locally optimal*, as it locally optimizes the probability of a decision, given a report obtained from processes, disregarding the impact of such a choice on the overall probability of correctness. The phrase "conditionally most probable value, given a report R" has to be formalized very carefully. Its meaning is not obvious, for the same reason as the meaning of probability of correctness of a strategy, examined in Sect. 3, was not straightforward. Since the behaviour of faulty processes is Byzantine rather than stochastic, the event corresponding to the occurrence of a given report R is not immediately clear. Nevertheless, we will define the notion of the probability of a value given any fixed report; a locally optimal decision strategy should pick, for any report R, the value for which this probability is larger. We will then define a locally optimal decision strategy which computes the decision value in time linear in the number of processes. It is natural to ask if these two ways of optimizing the decision (the one presented in Sect. 3 which could be called *global optimization* and the local optimization discussed in this section) are equivalent. The surprising answer is no. We show that, in general, local maximization of probability may lead to a decision strategy which does not have the highest probability of correctness.

Fix any report R. Recall that $Z(R, v) = \{i \in \{1, ..., n\} : R(i) = v\}$. Given the report R, the probability that value v was to be transmitted, is the probability of the following event: $b = v$ and all fault-free processes reported v. More formally: $P(v|R) = P(\{(G, v) : G \subseteq Z(R, v)\})$. Hence we adopt the following definition. A decision strategy $D : \mathcal{R} \longrightarrow \{0, 1\}$ is called *locally optimal*, if

$$D(R) = \begin{cases} 0 \text{ if } P(\{(G, v) : G \subseteq Z(R, 0)\}) \geq P(\{(G, v) : G \subseteq Z(R, 1)\}) \\ 1 \text{ otherwise.} \end{cases} \tag{5}$$

Since $P(\{(G, v) : G \subseteq Z(R, v)\}) = q(v) \cdot \prod_{i \in Z(R, 1-v)} p_i$, the (unique) locally optimal strategy Loc: $\mathcal{R} \longrightarrow \{0, 1\}$ can be defined as follows:

$$\text{Loc}(R) = \begin{cases} 0 \text{ if } p \cdot \prod_{i \in Z(R, 1)} p_i \geq (1 - p) \cdot \prod_{i \in Z(R, 0)} p_i , \\ 1 \text{ otherwise.} \end{cases} \tag{6}$$

The following result follows from the definition of strategy Loc.

Theorem 3. *The decision strategy Loc is locally optimal. Its value on any report can be computed in time linear in the number of processes.*

We now present an example witnessing to the (somewhat surprising) fact that optimality and local optimality of decision strategies are not equivalent notions.

Example 1. Consider a system in which the probability distribution of the Boolean value to be conveyed is uniform, i.e., $p = 1/2$, and there are three processes, 1, 2, and 3, with failure probabilities $p_1 = 1/4$ and $p_2 = p_3 = 2/5$, respectively. Consider the report R, such that $R(1) = 1$ and $R(2) = R(3) = 0$. Since $p = 1 - p$ and

$$I(Z(R,1)) = \frac{3}{4} \cdot \frac{2}{5} \cdot \frac{2}{5} > \frac{3}{5} \cdot \frac{3}{5} \cdot \frac{1}{4} = I(Z(R,0)) , \tag{7}$$

the optimal strategy IS gives value 1 on report R. On the other hand, since

$$P(\{(G,0) : G \subseteq Z(R,0)\}) = \frac{1}{4} > \frac{2}{5} \cdot \frac{2}{5} = P(\{(G,1) : G \subseteq Z(R,1)\}) , \tag{8}$$

the locally optimal strategy Loc gives value 0 on report R. This shows that an optimal decision strategy need not be locally optimal.

The converse is not true either. For the above described system we have

$$\text{Cor(IS)} = \{(\{1,2,3\}, v), (\{1\}, v), (\{1,2\}, v), (\{1,3\}, v) : v \in \{0,1\}\} , \tag{9}$$

hence Rel(IS)=0.75. On the other hand,

$$\text{Cor(Loc)} = \{(\{1,2,3\}, v), (\{2,3\}, v), (\{1,2\}, v), (\{1,3\}, v) : v \in \{0,1\}\} , \tag{10}$$

hence Rel(Loc)=0.72. This shows that the locally optimal decision strategy need not be optimal.

We conclude the paper with the observation that the above anomaly does not occur in the special case when the probability distribution of the Boolean value to be conveyed is uniform and all processes have *the same* failure probability less than $1/2$.

Proposition 1. *If $p = 1/2$ and all process failure probabilities are equal and smaller than $1/2$, then the optimal strategy IS and the locally optimal strategy Loc are equal.*

Proof. Let $p = 1/2$ and let $q < 1/2$ be the failure probability of all processes. Fix any report R. Let $z = |Z(R,0)|$ and hence $n - z = |Z(R,1)|$. We have $I(Z(R,0)) = (1 - q)^z q^{n-z}$ and $I(Z(R,1)) = (1 - q)^{n-z} q^z$. It follows that $I(Z(R,0)) \geq I(Z(R,1))$ if and only if $z \geq n - z$. Consequently, $\text{IS}(R) = 0$ if and only if $z \geq n - z$. On the other hand, by definition of Loc, $\text{Loc}(R) = 0$ if and only if $q^{n-z} \geq q^z$, which is also equivalent to $z \geq n - z$. This concludes the proof.   □

5   Conclusion

We considered decision strategies based on reports of fault-prone processes, assuming that fault probability distribution among processes is known, and that faulty processes behave in a Byzantine way. We constructed a deterministic decision strategy which has the highest possible probability of correctness. We also showed that a locally optimal decision strategy (that finds, for every set of results conveyed by processes, the conditionally most probable value that could yield these results) need not have the highest probability of correctness. Both our strategies, the globally optimal and the locally optimal, compute the decision value for any report of the processes in time linear in the number of processes.

Our strategies work for arbitrary fault probability distributions among processes, and for arbitrary a priori distribution of transmitted values, but we assume that these values are Boolean. It would be interesting to extend our results to the case of arbitrary finite sets of possible transmitted values, with arbitrary a priori probabilities.

Acknowledgments. The research of Michel Paquette was partially supported by the CALDI M.Sc. Scholarship from the Research Chair in Distributed Computing at the Université du Québec en Outaouais. The research of Andrzej Pelc was partially supported by NSERC grant OGP 0008136 and by the Research Chair in Distributed Computing at the Université du Québec en Outaouais.

References

1. D. Barbara and H. Garcia-Molina, The Vulnerability of Vote Assignments, ACM Transactions on Computer Systems 4 (1986), 187-213.
2. D. Barbara and H. Garcia-Molina, The Reliability of Voting Mechanisms, IEEE Transactions on Computers 36 (1987), 1197-1208.
3. D. Blough and G. Sullivan, A comparison of voting strategies for fault-tolerant distributed systems, Proc. 9th Symp. on Reliable Distr. Systems (1990), 136-145.
4. D. Blough and G. Sullivan, Voting using predispositions, IEEE Trans. on Reliability 43 (1994), 604-616.
5. K. Diks, A. Pelc, Globally optimal diagnosis in systems with random faults, IEEE Transactions on Computers 46 (1997), 200-204.
6. K. Diks, E. Kranakis, D. Krizanc, B. Mans, A.Pelc, Optimal coteries and voting schemes, Information Processing Letters 51 (1994), 1-6.
7. H. Garcia-Molina and D. Barbara, How to assign Votes in a distributed system, Journal of the ACM 32 (1985), 841-860.
8. H. Kakugawa, S. Fujita, M. Yamashita and T. Ae, Availability of k-Coterie, IEEE Transactions on Computers, 42 (1993), 553-558.
9. A. Pelc, Optimal diagnosis of heterogeneous systems with random faults, IEEE Transactions on Computers 47 (1998), 298-304.
10. F. Preparata, G. Metze and R. Chien, On the connection assignment problem of diagnosable systems, IEEE Trans. on Electron. Computers 16 (1967), 848-854.
11. M. Spasojevic and P. Berman, Voting as the optimal static pessimistic scheme for managing replicated data, IEEE Transactions on Parallel and Distributed Systems 5 (1994), 64-73.

Sharing the Cost of Multicast Transmissions in Wireless Networks*

Paolo Penna and Carmine Ventre

Dipartimento di Informatica ed Applicazioni "R.M. Capocelli",
Università di Salerno, via S. Allende 2, I-84081 Baronissi (SA), Italy
{penna,ventre}@dia.unisa.it

Abstract. We investigate the problem of sharing the cost of a multi-cast transmission in a wireless network where each node (radio station) of the network corresponds to (a set of) user(s) potentially interested in receiving the transmission. As in the model considered by Feigenbaum *et al* [2001], users may act *selfishly* and report a false "level of interest" in receiving the transmission trying to be charged less by the system. We consider the issue of designing a so called *truthful mechanisms* for the problem of maximizing the *net worth* (i.e., the overall "happiness" of the users minus the cost of the transmission) for the case of *wireless* networks. Intuitively, truthful mechanism guarantee that no user has an incentive in reporting a false valuation of the transmission. Unlike the "wired" network case, here the cost of a set of connections implementing a multicast tree is *not* the sum of the single edge costs, thus introducing a complicating factor in the problem. We provide both positive and negative results on the existence of optimal algorithms for the problem and their use to obtain VCG truthful mechanisms achieving the same performances.

1 Introduction

One of the main benefits of ad-hoc wireless networks relies in the possibility of communicating without any fixed infrastructure. Indeed, each station is a radio transmitter/receiver and communication between two stations that are not within their respective transmission ranges can be achieved by *multi-hop* transmissions: a set of intermediate stations forwards the message till its destination.

In this work, we consider the problem of sharing the cost of a multicast transmission in such wireless networks. A set of radio stations implements a *directed communication graph* which can be used to broadcast a (set of) messages from a given source node s to any subset of *users*. In particular, each user j is sitting close to some station i and she can receive the transmission only if i does. In addition, user j benefits from receiving the transmission some amount specified by a value v_j (say, how much j valuates the transmission).

---

* Work supported by the European Project IST-2001-33135, Critical Resource Sharing for Cooperation in Complex Systems (CRESCCO).

R. Královič and O. Sýkora (Eds.): SIROCCO 2004, LNCS 3104, pp. 255–266, 2004.

Since transmissions along the edges (i.e., links) of the communication graph require some costs (i.e., the power consumption of the station forwarding the messages), one would like to select a suitable set of nodes, to which the transmission is sent to, so that (i) the users share the cost of the transmission, and (ii) the overall *net worth* is maximized: the net worth is defined as the sum of the v_js of all users receiving the transmission minus the overall cost due to the used links.

Although the costs of the links are a property of the network (thus known to the "protocol"), the valuation v_j is clearly a property of user j. Thus, each v_j is a *private* piece of information (a part of the input) held by user j only. So, a user may act *selfishly* and report a different value b_j trying to receive the transmission at a lower price. We thus need to design so called *truthful mechanism* for our problem, that is, a suitable combination of an algorithm A and payments to the users which guarantee that (i) no user j has an incentive in reporting $b_j \neq v_j$, and (ii) the algorithm A, once provided with the correct v_js, returns an optimal solution.

This problem has been previously considered in the context of "classic" wired networks in [15]. In this work we consider a different model, that is, the wireless network one. The main difference between the two models relies on the different cost functions, which turns out to be a key point for solving the problem above.

Cost of Wireless Connections. Consider a *directed weighted communication graph* $\mathcal{G} = (\mathcal{S}, \mathcal{E}, w)$ defined as follows: $\mathcal{S}$ is the set of stations, and $\mathcal{G}$ contains a *directed edge* $(i, j) \in \mathcal{E}$ if and only if the direct transmission from i to j is feasible; in this case the weight $w(i, j)$ is the minimum power required for station i to directly transmit to station j. For instance, in the empty space $w(i, j) = \mathsf{d}(i, j)^2$, where $\mathsf{d}(i, j)$ is the Euclidean distance between i and j.

Each station is a radio transmitter/receiver and a station i is able to *directly* transmit a message to station j if and only if the power P_i used by station i satisfies $P_i \geq w(i, j)$. Stations use *omnidirectional antennas*, and a message sent by station i to j can be also received by every other station j' for which $w(i, j') \leq w(i, j)$. In order to reduce the power consumption, every station i can adjust its transmission power P_i, thus implementing the set of connections $\{(i, j) \mid w(i, j) \leq P_i\}$. Hence, given a set of connections $C \subseteq \mathcal{E}$, its cost is defined as follows:

$$\mathsf{Cost}(C) = \sum_{i \in \mathcal{S}} \max_{j:(i,j) \in C} w(i, j), \tag{1}$$

that is, the *overall power consumption* required to implement all these connections.

Net Worth of a Multicast Transmission. We are interested in sets $C \subseteq \mathcal{E}$ which guarantee that, given a distinguished *source node* s, the set C connects s to a suitable set $D(C) \subseteq \mathcal{S}$ of *destination nodes*. Consider a set U of users, each of them located close to some of the nodes in $\mathcal{S}$. The source s can send some kind of transmission (say a movie or a sport event) to a user j only if j is close to some of the destination nodes $D(C)$. In addition, every user j has a *valuation* v_j of the transmission representing how much she would benefit from receiving it (i.e., how much she would pay for it). As in the model of [15], we consider the situation in which each user j is sitting close to one station, say i; the latter represents the router of the network at distance one hop from user j. So, user j can receive the transmission only if node i does. Observe that, we can always reduce the case of several users located close to the same node to the case of (at most) one user close to one node (consider each user as a node with no outgoing edges and one ingoing edge of cost 0). Given a solution $T \subseteq \mathcal{E}$, its *net worth* is defined as

$$\mathsf{NW}(T) = \mathsf{Worth}(T) - \mathsf{Cost}(T), \tag{2}$$

where $\mathsf{Worth}(T) = \sum_{i \in D(T)} v_i$.

The *cost sharing problem* asks for a $T \subseteq \mathcal{E}$ that, for a given source s, maximizes the net worth function above.

Selfish Users and Economical Constraints. Associated to each node there is a *selfish agent* reporting some (not necessarily true) valuation b_i; the true value v_i is *privately known* to agent i. Based on the reported values $b = (b_1, b_2, \ldots, b_n)$ a *mechanism* $M = (A, P)$ constructs a multicast tree T using algorithm A (i.e. $T = A(b)$) and charges, to each agent i, an amount of money to pay for receiving the transmission equal to $P^i(b)$, with $P = (P^1, P^2, \ldots, P^n)$.

There is a number of natural constraints/goals that we would like a mechanism $M = (A, P)$ to satisfy/meet:

1. Truthfulness (or Strategyproofness)[1]. For every i, let $b_{-i} := (b_1, b_2, \ldots, b_{i-1}, b_{i+1}, \ldots, b_n)$ and $(b_i, b_{-i}) := b$. The *utility* of agent i when she reports b_i, and the other agents report b_{-i}, is equal to

$$u_i(b_i, b_{-i}) := \begin{cases} v_i - P^i(b_i, b_{-i}) & \text{if } T = A(b_i, b_{-i}) \text{ and } i \in D(T), \\ 0 & \text{otherwise.} \end{cases}$$

We require that, for every i, for every b_{-i}, and for every $b_i \neq v_i$, it holds that $u_i(v_i, b_{-i}) \geq u_i(b_i, b_{-i})$. In other words, whatever strategy the other agents follow, agent i has no incentive to lie about her true valuation v_i. A mechanism satisfying this property is called *truthful*.

2. Efficiency. The *net worth* $\mathsf{NW}(T)$ yielded by the computed solution T is maximum, that is, $\mathsf{NW}(T) = \max_{C \subseteq \mathcal{E}} \{\mathsf{NW}(C)\}$.

3. No Positive Transfer (NPT). No user receives money from the mechanism, i.e., $P^i(\cdot) \geq 0$.

4. Voluntary Participation (VP). We never charge a user an amount of money grater than her *reported* valuation, that is, $\forall b_i, \forall b_{-i} \quad b_i \geq P^i(b_i, b_{-i})$. In particular, a user has always the option to not paying for a transmission for which she is not interested.

5. Consumer Sovereignty (CS). Every user is guaranteed to receive the transmission if she reports a high enough valuation.

6. Budget Balance (BB). $\sum_i P^i(b) = \mathsf{Cost}(A(b))$.

7. Cost Optimality (CO). The set of connections T is optimal w.r.t. the set of receivers $D(T)$, that is, $\mathsf{Cost}(T) = \min_{C \subseteq \mathcal{E}, D(T) = D(C)} \{\mathsf{Cost}(C)\}$.

---

[1] In Sect. 2 we provide a more general definition of truthfulness which applies to a wide class of problems involving selfish agents.

Clearly, the Efficiency requirement implies the Cost Optimality. Unfortunately, in some cases it is impossible to achieve efficiency, so we will relax it to *r-efficiency*, that is, $r \cdot \mathsf{NW}(T) \geq \max_{C \subseteq \mathcal{E}} \{\mathsf{NW}(C)\}$. In these cases, we will take into account CO and *r*-CO, i.e., $\mathsf{Cost}(T) \leq r \cdot \min_{\substack{C \subseteq \mathcal{E}, \\ D(T)=D(C)}} \{\mathsf{Cost}(C)\}$.

1.1   Previous Work

Power Consumption and Range Assignment Problems. The problem of computing a broadcast tree of minimal cost for wireless networks has been investigated in [19, 13, 9, 30]. In particular, in [19] the authors proved that the problem is NP-hard to approximate within logarithmic factors, while it remains NP-hard even when considering geometric 2-dimensional networks [9]. Several variants of this problem have been considered in [23, 12, 13, 9, 30, 5, 6, 1] (see also [10] for a survey). However, to our knowledge, no algorithmic solution for optimizing the net worth has been given so far.

Recently, the design of truthful mechanisms for the range assignment problem in presence of "selfish transmitters" (i.e., selfish agents that want to minimize the energy their station has to use) has been investigated in [2] for the strongly connectivity problem, and in [3] for point-to-point transmissions, respectively.

Mechanism Design and Cost-Sharing Mechanisms in Wired Networks. The theory of mechanism design dates back to the seminal papers by Vickrey [29], Clarke [8] and Groves [18], and recently found a natural application to (algorithmic) questions related to the Internet [25] (see also [16] and [26]). VCG mechanisms guarantee the truthfulness under the hypothesis that the mechanism is able to compute the optimum and the optimization function is *utilitarian* (see Sect. 2 for a formal definition of utilitarian problem).

This technique is employed in [15] (and in this work) where the authors consider the wired networks case. They indeed provide a *distributed* optimal algorithm for the case in which the communication graph is a directed tree. This yields a *distributed mechanism*[2] which, for this problem version, satisfies all requirements mentioned above (truthfulness, efficiency, etc.) except for budget balance.

Noticeably, a classical result in game theory [17, 28] implies that, for this model, budget balance and efficiency are mutually exclusive. Additionally, in [14] (see also Theorem 5 in [7]) it is shown that no α-efficiency and β-efficiency can be guaranteed simultaneously, for any two $\alpha, \beta > 1$. So, the choice is to either optimize the efficiency (as in [15]) or to meet budget balance (as in [20, 21, 7]). In the latter case, it is also possible to obtain so called *group strategyproofness*, a stronger notion of truthfulness which can also deal with *coalitions*. On the other hand, if we insist on efficiency, NPT, VP, and CS, then there is essentially only one such mechanism: the marginal-cost mechanism [24], which belongs to the VCG family.

All such negative results apply also to our problem (i.e., wireless networks). Indeed, a simple observation is that every instance of the "wired" case can be

---

[2] The mechanism is able to compute both the solution and the payments in distributed fashion.

reduced to the wireless one using the following trick: replace every edge (i, j), with two edges $(i, x(i, j))$ and $(x(i, j), j)$ with $w(i, x(i, j)) = 0$ and $w(x(i, j), j) = w(i, j)$. So, also for our problem we have to choose between either budget balance of efficiency.

1.2   Our Results

We consider the problem of designing mechanisms that satisfy truthfulness, efficiency, NPT, VP, CS, and CO in the case of wireless networks. We first show that, even though the problem is not utilitarian, it is possible to adapt the VCG technique so as to obtain truthful mechanisms based on exact algorithms (Sect. 2).

Unfortunately, the problem is NP-hard, thus preventing from a straightforward use of the VCG result to obtain *polynomial-time* truthful mechanisms. Motivated by this negative result, we first consider the problem restricted to communication graphs that are trees (this is the analogous of the result for wired networks in [15]). We prove that, in this case, the optimal net worth can be computed via a polynomial time *distributed* algorithm (Sect. 3.1). The importance of this result[3] is threefold:

- It shows that the hardness of the problem is confined in the choice of a "good" multicast tree, and not in its use: if an "oracle" provides us with a tree containing an optimal multicast tree, then we can compute the optimum in polynomial-time.
- It is used to obtain a truthful *distributed* polynomial-time mechanism satisfying NPT, VP, CS, and efficiency when the given communication graph is a tree. In this case, both the solution and the payments can be computed in distributed fashion using $O(1)$ messages per link.
- It can be used to approximate the problem in some situations for which a good "universal" tree exists, i.e., a tree containing a set of connections of cost not much larger than the optimal solution and reaching the same set of nodes. This approach is similar to that of several wireless multicast protocols[4] which construct a multicast tree by pruning a broadcast tree T (e.g., MIP, MLU and MLiMST in [13]). In all such cases, one can assume that the communication graph G is the tree T.

Moreover, we show that a shortest-path tree can be used as universal tree so to obtain a polynomial-time mechanism satisfying truthfulness, NPT, VP, CS, and $O(n)$-CO, for the case of any communication graph G. In addition, our mechanism guarantees $|D(T^*)|$-efficiency, for all instances that admit an optimal solution T^* satisfying $|D(T^*)| \leq \gamma \frac{\mathsf{Worth}(T^*)}{\mathsf{Cost}(T^*)}$, for some constant $\gamma < 1$ (Sect. 3.2). We also prove that, in general, for any $R > 0$, no polynomial-time algorithm can guarantee R-efficiency, unless $\mathsf{P} = \mathsf{NP}$. Notice that, this result rules out the possibility of having polynomial-time mechanisms satisfying $O(n)$-efficiency.

---

[3] Independently from this work, Bilò *et al* [4] also provide polynomial-time truthful mechanisms for trees in the case of wireless networks.

[4] In this case the set of destination nodes (termed multicast group) is given in input.

We then extend our positive result to a class of graphs denoted as trees with *metric free edges* (see Sect. 3.3). Our technical contribution here is a non-trivial algorithm extending the technique and the results for trees.

Finally, we turn our attention to the Euclidean versions of the problem, that is, the case in which points are located on a d-dimensional Euclidean space and $w(i,j) = \mathsf{d}(i,j)^\alpha$, for a fixed constant $\alpha \geq 1$. We first show that the problem remains NP-hard even when $d = 2$ and for any $\alpha > 1$ (Sect. 3.4). For the case $d = 1$ we provide a polynomial-time mechanism satisfying truthfulness, efficiency, NPT, VP, CS and CO, with the additional property of ensuring multicast trees of depth at most h, for any $1 \leq h \leq n-1$ given in input. This result exploits the broadcasting algorithm in [11]. For the case $d = 2$, we present a solution based on the construction of so called *Light Approximate Shortest-path Trees* (LASTs) given in [22]. This achieves a better performance w.r.t. MST-based solutions in several cases.

Due to lack of space some of the proofs are omitted. We refer the interested reader to the full version of this work [27].

2   Optimal Algorithms Yield Truthful Mechanisms

For the sake of completeness, we first recall the classical technique to obtain truthful mechanisms for utilitarian problems known as VCG-mechanism [29, 8, 18]. We then show how to adapt this technique to our (non-utilitarian) problem.

Let us first consider a more general situation in which each agent i has a certain type t_i. The *valuation* of agent i of a solution X is represented by a function $\mathsf{VAL}_i(X, t_i)$, where the function $\mathsf{VAL}_i(\cdot, \cdot)$ is known to the mechanism. A *maximization* problem is *utilitarian* if its objective function $g(\cdot)$, which depends on the agents type vector $t = (t_1, t_2, \ldots, t_n)$, satisfies

$$g(X, t) = \sum_{j=1}^{n} \mathsf{VAL}_j(X, t_j), \tag{3}$$

for any solution X. Each agent i can declare a different type b_i to the mechanism. We have the following result on the VCG mechanism $M = (\mathsf{ALG}, P_{VCG})$:

Theorem 1. *[18] If* ALG *is an optimal algorithm for a utilitarian problem* Π, *then the mechanism* $M = (\mathsf{ALG}, P_{VCG})$ *is truthful for* Π.

Let $\sigma_i(T) = 1$ if $i \in D(T)$, and 0 otherwise. Notice that, simply setting $\mathsf{VAL}_i(T, v_i) := \sigma_i(T) \cdot v_i$ does not satisfy the definition of utilitarian problem since $\mathsf{NW}(T) \neq \sum_{i=1}^{n} \mathsf{VAL}_i(T, v_i) = \mathsf{Worth}(T)$. Nevertheless, the next results states that the VCG technique can be adapted to our problem. The main idea is to initially charge each node by the cost of its ingoing edge in the tree (computed as in the wireless network case) so to "reduce" our problem to a utilitarian one (see [27] for the details).

Theorem 2. *Let* A *be a (polynomial-time) exact algorithm for maximizing the* $\mathsf{NW}(\cdot)$ *function. Then the cost sharing problem on wireless networks admits a (polynomial-time) mechanism* $M = (A, P_A)$ *satisfying truthfulness, efficiency, NPT, VP, CS and CO.*

3   Special Communication Graphs

Motivated by the result of the previous section, we focus on the existence of *polynomial-time exact* algorithms for the cost sharing problem on wireless networks. Since the problem is NP-hard, also for Euclidean 2-dimensional instances (Sect. 3.4), in the following we will focus on restrictions for which such algorithms exist.

3.1   Trees

We proceed similarly to [15] and assume that the communication graph is a directed tree $\mathcal{T} = (\mathcal{S}, \mathcal{E})$.

Definition 1. *For every $i \in \mathcal{T}$, let $p(i)$ denote its parent in $\mathcal{T}$. Also let $c_i = w(p(i), i)$ and $c_s = 0$. We denote by $\mathcal{T}_i$ the subtree rooted at i, and by $\mathsf{ch}(i)$ the set of i's children. Finally, $\mathsf{NW_{opt}}(i)$ denotes the optimal net worth of $\mathcal{T}_i$, that is, the optimum for the instance in which i is the source node and the universal tree is $\mathcal{T}_i$.*

Algorithm Wireless_Trees at node i

1. After receiving a message μ^j from each child $j \in \mathsf{ch}(i)$ do
   (a) $Add(j) := -c_j + \sum_{k \in \mathsf{ch}(i), c_k \leq c_j} \mu^j$;
   (b) $T(i) := \emptyset$;
   (c) if $\max_{j \in \mathsf{ch}(i)} Add(j) < 0$ then $\mu^i := v_i$
   (d) else do
       i. $Add := \max_{j \in \mathsf{ch}(i)} Add(j)$;
       ii. $\mu^i := v_i + Add$;
       iii. $j^* := \arg\max\{j \in \mathsf{ch}(i) | Add(j) = Add\}$;
       iv. $T(i) := \{(i, j) | w(i, j) \leq w(i, j^*)\}$;
2. send μ^i to parent $p(i)$;

Fig. 1. The distributed algorithm for trees computing an optimal solution in bottom-up fashion.

Lemma 1. *For every node i, it holds that*

$$\mathsf{NW_{opt}}(i) = v_i + \max\{0, \max_{j \in \mathsf{ch}(i)} \{-c_j + \sum_{k \in \mathsf{ch}(i), c_k \leq c_j} \mathsf{NW_{opt}}(k)\}\}. \qquad (4)$$

Proof. The proof is by induction on the hight h of $\mathcal{T}_i$. Obviously, for $h = 1$, since i is a leaf node, then $\mathsf{NW_{opt}}(i) = v_i$. Let us now assume that the lemma holds for any $h' \leq h - 1$, and let us prove it for h. Let T_i^* denote an optimal solution for $\mathcal{T}_i$. Let us first observe that, if $\mathsf{NW_{opt}}(i) = \mathsf{NW}(T_i^*) > v_i$, then T_i^* must contain at least one outgoing edge from node i. Let (i, j) be the longest such edge in

T_i^*. For any node $k \in \mathcal{T}_i$, let $T_{i,k}^*$ denote the subtree of T_i^* rooted at k. Since $(i,j) \in T_i^*$, then $k \in D(T_i^*)$, for all $k \in \text{ch}(i)$ such that $c_k \leq c_j$. So,

$$\text{Worth}(T_i^*) = v_i + \sum_{k \in \text{ch}(i), c_k \leq c_j} \text{Worth}(T_{i,k}^*), \tag{5}$$

$$\text{Cost}(T_i^*) = c_j + \sum_{k \in \text{ch}(i), c_k \leq c_j} \text{Cost}(T_{i,k}^*). \tag{6}$$

Let us now suppose, by contradiction, that there exists a $l \in \text{ch}(i)$, with $c_l \leq c_j$ and $\text{NW}(T_{i,l}^*) < \text{NW}_{\text{opt}}(l)$. Let T_l' denote the subtree of $\mathcal{T}_l$ yielding optimal net worth w.r.t. $\mathcal{T}_l$, that is, $\text{NW}(T_l') = \text{NW}_{\text{opt}}(l)$. Let T_i' be the solution obtained by replacing, in T_i^*, $T_{i,l}^*$ with T_l'. Since T_i' still contains all edges (i,k), with $w(i,k) \leq w(i,j)$, we have that

$$\text{Worth}(T_i') \geq \text{Worth}(T_l') + v_i + \sum_{k \in \text{ch}(i), c_k \leq c_j, k \neq l} \text{Worth}(T_{i,k}^*). \tag{7}$$

Clearly, $\text{Cost}(T_i') = \text{Cost}(T_i^*) - \text{Cost}(T_{i,l}^*) + \text{Cost}(T_l')$. This, combined with Eq. 7, yields $\text{NW}(T_i') \geq \text{NW}(T_i^*) - \text{NW}(T_{i,l}^*) + \text{NW}(T_l')$. From the hypothesis $\text{NW}(T_{i,l}^*) < \text{NW}_{\text{opt}}(l) = \text{NW}(T_l')$, we obtain $\text{NW}(T_i') > \text{NW}(T_i^*)$, thus contradicting the optimality of T_i^*. So, for every $k \in \text{ch}(i)$ with $c_k \leq c_j$, it must hold $\text{NW}(T_{i,k}^*) = \text{NW}_{\text{opt}}(k)$. From Eq.s 6-5 we obtain

$$\text{NW}_{\text{opt}}(i) = \text{NW}(T_i^*) = v_i - c_j + \sum_{k \in \text{ch}(i), c_k \leq c_j} \text{NW}(T_{i,k}^*)$$

$$= v_i - c_j + \sum_{k \in \text{ch}(i), c_k \leq c_j} \text{NW}_{\text{opt}}(k). \tag{8}$$

The optimality of T_i^* implies that, if $\text{NW}_{\text{opt}}(i) > v_i$, then j must be the node in $\text{ch}(i)$ maximizing the right quantity in Eq. 8. The lemma thus follows from the fact that $\text{NW}_{\text{opt}}(i) \geq v_i$: taking no edges in $\mathcal{T}_i$ yield a net worth equal to v_i. This completes the proof.

Theorem 3. *For any communication graph $\mathcal{T}$ which is a tree, algorithm* Wireless_Trees *computes the optimal net worth in polynomial time, using $O(n)$ total messages and sending $O(1)$-messages per link.*

In Fig. 2 we show a distributed top-down algorithm for computing $P_A^j(\cdot)$ of Theorem 2 (see the proof in [27]). The code refers to a non-source node; the computation is initialized by s which, at the end of the bottom-up phase of Algorithm Wireless_Trees, has computed the value $\text{NW}_{\text{opt}} = \mu^s$ and it executes the instructions 1a-1e for every $j \in \text{ch}(s)$.

Using the algorithm in Fig. 2, from Theorem 3 and from Theorem 2 we obtain the following:

Corollary 1. *If the communication graph is a tree, then the cost sharing problem on wireless networks admits a distributed polynomial-time truthful mechanism $M = (A, P_A)$ satisfying efficiency, NPT, VP, CS and CO.*

Algorithm Wireless_Trees_Pay at node $i = p(j)$

1. After receiving the message $(\mathsf{NW}_{opt}, \lambda^i)$ from parent $p(i)$, for each child $j \in \mathsf{ch}(i)$ do
   - (a) $\lambda^{j=0} := \mathsf{NW}_{opt} - b_j$;
   - (b) $x := \max_{k \in \mathsf{ch}(i), k \neq j}\{-c_k + \sum_{l \in \mathsf{ch}(i), c_l \leq c_k} \mu^l\}$;
   - (c) $\lambda^{-j} := \mathsf{NW}_{opt} - \mu^i + v_i + \max\{0, x\}$;
   - (d) $\lambda^j := \lambda^{-j} - \lambda^{j=0}$;
   - (e) send $(\mathsf{NW}_{opt}, \lambda^j)$ to child j;

Fig. 2. The distributed algorithm for computing the payments in top-down fashion.

3.2   Do Good Universal Muticast Trees Exist?

In this section we propose an application of our optimal algorithm given in Sect. 3.1. In particular, we consider mechanisms that pre-compute *some* broadcast tree T (i.e., $D(T) = S$) and then solve the problem by computing an optimal subtree T of T. Let ALG_{un} denote the resulting algorithm. The following result is a simple generalization of Corollary 1.

Theorem 4. *There exists a payment function P_{un} such that $M_{un} = (\mathsf{ALG}_{un}, P_{un})$ satisfies truthfulness, NPT, VP and CS. Moreover, if ALG_{un} runs in polynomial time, then the payment functions P_{un} are computable in polynomial time as well.*

The next result provides an upper bound on the approximability of the cost sharing problem when the input communication graph is *not* a tree. The idea is to use a shortest-path tree as universal tree.

Theorem 5. *There exists a polynomial-time mechanism $M_{un} = (\mathsf{ALG}_{un}, P_{un})$ satisfying truthfulness, NPT, VP, CS and $O(l)$-CO, where $l = |D(T)|$ and T is the computed solution. Additionally, for any $\gamma < 1$ and for all instances that admit an optimal solution T^* satisfying $|D(T^*)| \leq \gamma \frac{\mathrm{Worth}(T^*)}{\mathrm{Cost}(T^*)}$, M_{un} guarantees also $(\frac{k}{1-\gamma})$-efficiency, with $k = |D(T^*)|$. Thus, in this case, M_{un} guarantees $O(n)$-efficiency.*

The above result guarantees $O(|D(T^*)|)$-efficiency only in some cases. The next theorem rules out the possibility of obtaining polynomial-time $O(|D(T^*)|)$-efficiency in general. Its proof is a simple adaptation of an analogous result for the wired case in [15].

Theorem 6. *For any $R > 0$, no polynomial-time R-approximation algorithm (mechanism) exists, unless $\mathsf{P} = \mathsf{NP}$.*

3.3   Trees with Metric Free Edges

We consider the case in which the set of edges of the communication graph $\mathcal{G} = (\mathcal{S}, \mathcal{E})$ is partitioned into two sets $\mathcal{T} \cup \mathcal{F}$. Similarly to the case considered in Sect. 3.1, $\mathcal{T}$ induces a spanning tree of $\mathcal{G}$, i.e., $D(\mathcal{T}) = \mathcal{S}$, and, for every node i, we must select a set of outgoing edges in the set $\mathcal{T}$. However, each of these edge also induces a set of "additional connections for free" in the set $\mathcal{F}$, that is, for every $(i, k) \in \mathcal{F}$, $w(i, k) = 0$. So, adding all such connections to the solution does not increase its cost. These connections are specified as follows. Let $w^*(i, j)$ denote the weight of the path in $\mathcal{T}$ connecting the node i to one of its descendants j. Let us define $\mathsf{FREE}(i, j) := \{(i, k) | \ w^*(i, k) \le w(i, j) \wedge (i, j) \in \mathcal{T} \wedge (i, k) \notin \mathcal{T}\}$. Moreover, for every $T \subseteq \mathcal{T}$, let $\mathcal{F}(T) := \bigcup_{(i,j) \in T} \mathsf{FREE}(i, j)$, and $\mathcal{F} := \mathcal{F}(\mathcal{T})$. We consider the restriction of the problem in which every feasible solution $C \subseteq \mathcal{T} \cup \mathcal{F}$ must fulfill the following property: C contains an edge $(i, k) \in \mathcal{F}$ if and only if it also contains an edge $(i, j) \in \mathcal{T}$ with $w^*(i, k) \le w(i, j)$. In other words, no edge in $\mathcal{F}$ can appear as the longest outgoing edge of a node in any solution C, thus implying $\mathsf{Cost}(C) = \mathsf{Cost}(C \cap \mathcal{T})$. Observe that this definition captures some restrictions of the 2-dimensional Euclidean case (see [27] for a discussion).

The following result allows us to apply Theorem 2 and obtain a truthful mechanism.

Theorem 7. *The optimal net worth, in the case of metric free edge, of any given tree $\mathcal{T}$ can be computed in polynomial time.*

Corollary 2. *The cost sharing problem on wireless networks, in the case of trees with metric free edges, admits a polynomial-time mechanism $M = (A, P_A)$ satisfying truthfulness, efficiency, NPT, VP, CS, CO.*

3.4   Geometric Euclidean Graphs

In this section we consider so called geometric communication graphs, that is, stations are located on the d-dimensional Euclidean space, the communication graph is a *complete graph* with $w(i, j) := \mathsf{d}(i, j)^\alpha$, where $\alpha \ge 1$ is a fixed constant and $\mathsf{d} : \mathcal{R}^d \to \mathcal{R}^+$ is the Euclidean distance.

Theorem 8. *The problem of maximizing $\mathsf{NW}(\cdot)$ is NP-hard, even when restricted to geometric wireless networks, for any $d \ge 2$ and any $\alpha > 1$.*

Definition 2. *Let $\mathsf{MST}(\mathcal{S})$ denote the minimum spanning tree of a set of points $\mathcal{S} \subseteq \mathcal{R}^d$. Given a source node $s \in \mathcal{S}$, $\mathsf{MST}_{\mathsf{brd}}(\mathcal{S}, s)$ denotes the directed spanning tree obtained by considering all edges of $\mathsf{MST}(\mathcal{S})$ downward directed from s. Let $\mathsf{opt}_{\mathsf{brd}}(\mathcal{S}, s)$ denote the minimum cost among all $T \subseteq \mathcal{S} \times \mathcal{S}$ such that $D(T) = \mathcal{S}$.*

The following result concerns the problem of constructing a tree T minimizing the cost for transmitting to *all* nodes in $\mathcal{S}$:

Theorem 9. *[9, 30] For any $d \ge 1$ and for every $\alpha \ge d$, there exists a constant c_α^d such that, for any $\mathcal{S} \in \mathcal{R}^d$, and for every $s \in \mathcal{S}$, $\mathsf{Cost}(\mathsf{MST}(\mathcal{S}, s)) \le c_\alpha^d \cdot \mathsf{opt}_{\mathsf{brd}}(\mathcal{S}, s)$. In particular, $c_2^2 = 12$ [30].*

Unfortunately, the same approximability result does not hold if the set of destinations is required to be a *subset* of $\mathcal{S}$ (e.g., stations that form a grid of size $\sqrt{n} \times \sqrt{n}$ and only one node with a strictly positive valuation).

A slightly better result can be obtained by using so called *Light Approximate Shortest-path Trees* (LASTs) introduced in [22]:

Theorem 10. *For any k, for any $\beta > 1$, for any $d \geq 2$, and for any $\alpha \geq d$, there exists a polynomial-time mechanism M satisfying truthfulness, NPT, VP, and CS. Additionally, M satisfies $O(1)$-CO whenever the computed solution T' satisfies $D(T') = S$ or $|D(T')| \leq k$.*

We now consider the problem restricted to *linear* networks, i.e., the Euclidean case with $d = 1$. By using the result in [11] it is possible to prove the following:

Theorem 11. *The optimal h net worth[5] of any given communication graph G corresponding to a linear network can be computed in polynomial time.*

Corollary 3. *The cost sharing problem on linear wireless networks admits a polynomial time mechanism $M = (A, P_A)$ satisfying truthfulness, efficiency, NPT, VP, CS, and CO. This holds also with the additional constraint of computing multicast trees of depth at most h.*

Very recently and independently form this work, Biló *et al* [4] also consider the Euclidean case and provide $O(1)$-BB and $O(1)$-CO mechanisms.

References

1. C. Ambuehl, A. Clementi, M. Di Ianni, N. Lev-Tov, A. Monti, D. Peleg, G. Rossi, and R. Silvestri. Efficient algorithms for low-energy bounded-hop broadcast in ad-hoc wireless networks. In Proc. of STACS, LNCS 2996, pages 418–427, 2004.
2. C. Ambuehl, A. Clementi, P. Penna, G. Rossi, and R. Silvestri. Energy Consumption in Radio Networks: Selfish Agents and Rewarding Mechanisms. In Proc. of SIROCCO, pages 1–16, 2003.
3. L. Anderegg and S. Eidenbenz. Ad hoc-VCG: A Truthful and Cost-Efficient Routing Protocol for Mobile Ad Hoc Networks with Selfish Agents. In Proc. of ACM MOBICOM, pages 245 – 259, 2003.
4. V. Biló, C. Di Francescomarino, M. Flammini, and G. Melideo. Sharing the Cost of Multicast Transmissions in Wireless Networks. In Proc. of SPAA, June 2004.
5. D.M. Blough, M. Leoncini, G. Resta, and P. Santi. On the symmetric range assignment problem in wireless ad hoc networks. In Proc. of IFIP-TCS, pages 71–82. Kluwer vol. 223, 2002.
6. G. Calinescu, I.I. Mandoiu, and A. Zelikovsky. Symmetric connectivity with minimum power consumption in radio networks. In Proc. of IFIP-TCS, pages 119–130. Kluwer vol. 223, 2002.
7. S. Chawla, D. Kitchin, U. Rajan, R. Ravi, and A. Sinha. Profit maximization mechanisms for the extended multicast game. Technical report, School of Computer Science, Carnegie Mellon University, 2002.
8. E.H. Clarke. Multipart Pricing of Public Goods. Public Choice, pages 17–33, 1971.

---

[5] With h net worth we denote the net worth that we obtain connecting, in at most h hops, the source s with a set of nodes X

9. A. Clementi, P. Crescenzi, P. Penna, G. Rossi, and P. Vocca. On the complexity of computing minimum energy consumption broadcast subgraphs. In Proc. of STACS, LNCS 2010, pages 121–131, 2001.
10. A. Clementi, G. Huiban, P. Penna, G. Rossi, and Y.C. Verhoeven. Some recent theoretical advances and open questions on energy consumption in ad-hoc wireless networks. In Proc. of Workshop on Approximation and Randomization Algorithms in Communication Networks (ARACNE), pages 23–38, 2001.
11. A. Clementi, M. Di Ianni, and R. Silvestri. The minimum broadcast range assignment problem on linear multi-hop wireless networks. Theoretical Computer Science, 299(1-3):751–761, 2003.
12. A. Clementi, P. Penna, and R. Silvestri. Hardness results for the power range assignment problem in packet radio networks. In Proc. of APPROX, LNCS 1671, pages 197–208, 1999.
13. A. Ephremides, G.D. Nguyen, and J.E. Wieselthier. On the Construction of Energy-Effficient Broadcast and Multicast Trees in Wireless Networks. In Proc. of INFOCOM, pages 585–594, 2000.
14. J. Feigenbaum, K. Krishnamurthy, R. Sami, and S. Shenker. Hardness results for multicast cost sharing. Theoretical Computer Science, 304:215–236, 2003.
15. J. Feigenbaum, C.H. Papadimitriou, and S. Shenker. Sharing the cost of multicast transmissions. Journal of Computer and System Sciences, 63(1):21–41, 2001.
16. J. Feigenbaum and S. Shenker. Distributed algorithmic mechanism design: Recent results and future directions. In Proc. of ACM DIALM, pages 1–13, 2002.
17. J. Green, E. Kohlberg, and J.J. Laffont. Partial equilibrium approach to the free rider problem. Journal of Public Econmics, 6:375–394, 1976.
18. T. Groves. Incentive in Teams. Econometrica, 41:617–631, 1973.
19. S. Guha and S. Khuller. Approximation algorithms for connected dominating sets. Algorithmica, 20(4):374–387, 1998.
20. K. Jain and V.V. Vazirani. Applications of approximation algorithms to cooperative games. In Proc. of STOC, 2001.
21. K. Jain and V.V. Vazirani. Equitable cost allocations via primal-dual-type algorithms. In Proc. of STOC, 2002.
22. S. Khuller, B. Raghavachari, and N. Young. Balancing minimum spanning trees and shortest-path trees. Algorithmica, 14(4):305–321, 1995.
23. L.M. Kirousis, E. Kranakis, D. Krizanc, and A. Pelc. Power consumption in packet radio networks. Theoretical Computer Science, 243:289–305, 2000.
24. H. Moulin and S. Shenker. Strategyproof sharing of submodular costs: Budget balance versus efficiency. Economic Theory, 1997.
25. N. Nisan and A. Ronen. Algorithmic Mechanism Design. Games and Economic Behavior, 35:166–196, 2001. Extended abstract in STOC'99.
26. C. H. Papadimitriou. Algorithms, Games, and the Internet. In Proc. of STOC, 2001.
27. P. Penna and C. Ventre. Sharing the Cost of Multicast Transmissions in Wireless Networks. Technical report, CRESCCO, www.ceid.upatras.gr/crescco/, 2003.
28. K. Roberts. The characterization of implementable choice rules. Aggregation and Revelation of Preferences, pages 321–348, 1979.
29. W. Vickrey. Counterspeculation, Auctions and Competitive Sealed Tenders. Journal of Finance, pages 8–37, 1961.
30. P.-J. Wan, G. Calinescu, X.-Y. Li, and O. Frieder. Minimum-energy broadcasting in static ad hoc wireless networks. Proc. of INFOCOM, 2001. Journal version in Wireless Networks, 8(6): 607-617, 2002.

NP-Completeness Results
for All-Shortest-Path Interval Routing

Rui Wang[1], Francis C.M. Lau[1,*], and Yan Yan Liu[2]

[1] Department of Computer Science and Information Systems
The University of Hong Kong
[2] Department of Computer Science
The Ocean University of China

Abstract. k-Interval Routing Scheme (k-IRS) is a compact routing method that allows up to k interval labels to be assigned to an arc. A fundamental problem is to characterize the networks that admit k-IRS. Many of the problems related to single-shortest-path k-IRS have already been shown to be NP-complete. For all-shortest-path k-IRS, the characterization problem remains open for $k \geqslant 1$. We investigate the time complexity of devising minimal-space all-shortest-path k-IRS and prove that it is NP-complete to decide whether a graph admits an all-shortest-path k-IRS, for every integer $k \geqslant 3$, as well as whether a graph admits an all-shortest-path k-strict IRS, for every integer $k \geqslant 4$. These are the first NP-completeness results for all-shortest-path k-IRS where k is a constant and the graph is unweighted. Moreover, the NP-completeness holds also for the linear case.

Keywords: interval routing; compact routing, NP-completeness.

1 Introduction

Interval Routing is a space-efficient routing method for communication networks [9]. The routing table stored at each node is made compact by grouping the set of destination addresses that use the same output port into intervals of consecutive addresses. Formally, the network is modeled as a finite graph $G = (V, E)$, where the set of vertices, V, represents the nodes of the network and the set of edges, E, represents the bidirectional links. Each edge between the nodes u and v is viewed as two opposite arcs, (u, v) and (v, u). A routing scheme for network G is to assign each arc (u, v) a subset $I(u, v) \subseteq V$, such that the union of the subsets assigned to the arcs emanating from u covers the set $V - \{u\}$. Routing is then performed according to the assignment I, such that at vertex u, a message will be sent on the arc (u, v) whose $I(u, v)$ contains the destination of the message. A good interval routing scheme would try to minimize the number of intervals in $I(u, v)$ over all the arcs by selecting an address mapping $L : V \rightarrow \{1, 2, \ldots, |V|\}$ and an assignment $I : E \rightarrow 2^V$ such that the maximum number of intervals over all arcs is minimized. If each $I(u, v)$ contains no more than k intervals under L,

---

* Correspondence: F.C.M. Lau, Department of Computer Science and Information Systems, The University of Hong Kong, Hong Kong / Email: fcmlau@csis.hku.hk

R. Králović and O. Sýkora (Eds.): SIROCCO 2004, LNCS 3104, pp. 267–278, 2004.

the routing scheme, denoted by $R = (L, I)$, is called a *k-Interval Routing Scheme* (*k*-IRS).

The standard definition of IRS assumes a single routing path between any two nodes. It therefore forces the arcs emanating from a node u to be assigned disjoint subsets, i.e., $I(u, v) \cap I(u, w) = \phi$ for $v \neq w$. The routing process with such an IRS is deterministic. A more flexible routing scheme, called *multipath* IRS or *non-deterministic* IRS [10, 7], allows multiple arcs of a node to point to a destination; the routing process can pick an arc randomly or according to dynamic trafic conditions.

There are two different models of networks. The weighted model associates each edge of the graph with a positive number to denote the cost of communication along that edge; the unweighted model assumes that the cost of every edge is one unit. The length of a path under both models is the sum of the costs of the edges in the path. In practice, routing along shortest paths is desirable. A *shortest-path* IRS always induces shortest paths. A *single-shortest-path* IRS offers a *unique* shortest path between any two vertices in the graph. An *all-shortest-path* IRS is a multipath IRS that gives exactly *all* the shortest paths between any pair of vertices in the graph.

1 and $|V|$ being considered consecutive, the interval $[a, b]$ with $a > b$ denotes the set $\{i | a \leqslant i \leqslant |V|\} \cup \{i | 1 \leqslant i \leqslant b\}$. An interval $[a, b]$ is linear if $a \leqslant b$, and circular otherwise. An IRS using only linear intervals is a *linear* IRS, LIRS in short. An IRS is *strict*, denoted SIRS, if every arc (u, v) satisfies $u \notin I(u, v)$. An IRS is donoted SLIRS if it is both strict and linear.

The space efficiency of an IRS is measured by its compactness, which is defined to be the maximum, over all the arcs (u, v), of the number of intervals in $I(u, v)$. The characterization of the networks that admit *shortest-path* interval routing scheme with compactness k (*k*-IRS for short) is a fundamental question in this field (note the stress on shortest path because if the shortest path requirement is relaxed, every graph supports a single path 1-IRS). The question has been a subject of extensive study for many variants of the basic interval routing method, and under different models. Succesful work has been done for many special classes of graphs, including trees, outerplannar graphs, hypercubes and meshes, r-partite graphs, interval graphs, unit-circular graphs, tori, 2-trees, chordal rings, and general graphs. A summary of these and other results on interval routing can be found in [7]. For general graphs, existing complexity results for various models and variants are summarized in Table 1, where an entry for a variant of single-path IRS refers to both the strict and non-strict versions of the problem; an entry for all-shortest-path IRS refers to both the linear and non-linear versions of the problem; NPC denotes an NP-complete problem.

All the problems related to IRS with a single shortest path are known to be NP-complete. The completeness for every constant $k \geqslant 3$ in the table needs an explanation. It follows from combining the result of [3] with that of [8], or with [11]. In [3], Flammini gave a polynomial time construction of graphs from biniary matrices such that there are at most k blocks of consecutive 1's in each column of the matrix under some row permutation if and only if there is a single-

Table 1. Complexity results on characterizations of IRS

Shortest paths represented	Compactness k	Variants	Graph model	
			Unweighted	Weighted
Single	fixed $k = 1$	IRS	NPC, [2]	NPC, [2]
		LIRS	NPC, [2]	NPC, [2]
	fixed $k = 2$	IRS	NPC, [3]	NPC, [3]
		LIRS	NPC, [3]	NPC, [3]
	fixed $k \geqslant 3$	IRS	NPC, [3, 8, 11]	NPC, [3, 8, 11]
		LIRS	NPC, [3, 8, 11]	NPC, [3, 8, 11]
	general k	IRS	NPC, [2]	NPC, [4]
		LIRS	NPC, [2]	NPC, [2]
All	fixed $k = 1$	IRS	?	?
		SIRS	P, [1, 5]	P, [1, 5]
	fixed $k = 2$	IRS	?	?
		SIRS	?	?
	fixed $k = 3$	IRS	NPC, this paper	NPC, this paper
		SIRS	?	?
	fixed $k \geqslant 4$	IRS	NPC, this paper	NPC, this paper
		SIRS	NPC, this paper	NPC, this paper
	general k	IRS	NPC, this paper	NPC, [4]
		SIRS	NPC, this paper	NPC, [4]

shortest-path $(k+1)$-IRS for the constructed graph; in [8], Goldberg et al. proved that for every constant $k \geqslant 2$, deciding whether a given binary matrix can be row permuted leading to each column has at most k blocks of consecutive 1's is NP-complete; in [11], we strengthened the result by showing that the same NP-completeness holds even if the problem is restricted to symmetric matrices.

For the all-shortest-path IRS, only partial answers (both positive and negative) have been given. 1-SIRS can be reduced to the consecutive ones property of binary matrices which can be solved in linear time [1]. Flammini et al. in [5] presented characterizations for 1-SLIRS and 1-LIRS. On the negative side, in [4], it was shown that the optimization problem of determining the minimal k such that a given weighted network belongs to the class of all-shortest-path k-IRS is NP-hard. For unweighted networks, their characterization remains open for all-shortest-path IRS of compactness $k \geqslant 1$ [7].

In this paper, we study the characterization question of all-shortest-path IRS under the most basic model – unweighted graph model. Specifically, we prove that the characterization of networks which admit all-shortest-path k-IRS (linear or non-linear) for every constant $k \geqslant 3$ is NP-complete. These results can be easily generalized to the weighted network model, and hence significantly extend the related NP-completeness results of [4].

The rest of the paper is organized as follows. The next section gives some formal definitions of the IRS models and their variants and the characterization problems. In Section 3 we prove the NP-completeness results. In the last section, we give some conclusive remarks about the implications of our results.

2   Preliminaries

The graphs we consider are connected, loopless, and do not contain multi-edges. The length of the path in the graph is the sum of the costs of the edges in the path (for the unweighted model this is equal to the number of the edges in the path). For an arc $e = (u, v)$, $S(u, v)$ denotes the set of those vertices which can be reached from vertex u through a shortest path passing through e; with this notation, $S(u, v) \neq S(v, u)$. We use $Adj(v)$ to denote the set of neighbours of vertex v in the graph.

The basic idea of interval routing is to label each vertex of the graph $G = (V, E)$ by a unique number from the set $\{1, 2, \ldots, |V|\}$, and then at each vertex v, to label each incident edge by a set of intervals over the set $\{1, 2, \ldots, |V|\}$, in such a way that the union of the intervals assigned to the edges emanating from v covers the set $\{1, 2, \ldots, |V|\}$. The routing is performed according to the labeling; in any vertex, a message will be sent on (one of) the edges whose label contains an interval including the destination of the message. Formally, we define an Interval Routing Scheme as follows.

Definition 1. *Let $G = (V, E)$ be a graph. An Interval routing Scheme (IRS) on G is a pair $\langle L, I \rangle$ defined by:*

(1) *L is a one to one vertex labeling, $L : V \to \{1, 2, \ldots, |V|\}$, that labels the vertices of V;*
(2) *I is an arc labeling, $I : E \to 2^V$, assigning a subset of V to each edge of E, such that at every vertex $u \in V$, $\bigcup_{(u,v) \in E} I(u, v) \bigcup \{u\} = V$;*
(3) *for every $x, y \in V$:*
      (3.1) *there exists a sequence of vertices $x = u_0, u_1, \ldots, u_s = y$ such that for $1 \leqslant i < s$, $y \in I(u_{i-1}, u_i) - \{u_{i-1}\}$; this sequence is called a routing path induced by $\langle L, I \rangle$;*
      (3.2) *any routing path induced by $\langle L, I \rangle$ between x and y is a simple path of G, i.e., $u_0, u_1, \ldots, u_s$ are mutually different vertices of V.*

To save space in its routing table, an IRS expresses $I(u, v)$, the subset of V assigned to an arc $e = (u, v)$, with intervals over $\{1, 2, \ldots, |V|\}$.

Definition 2. *An interval of $\{1, 2, \ldots, |V|\}$ is one of the following:*

(1) *A linear interval $[i, j] = \{i, i + 1, \ldots, j\}$, where $i, j \in \{1, 2, \ldots, |V|\}$ and $i \leqslant j$;*
(2) *a circular interval $[i, j] = \{i, \ldots, |V|, 1, \ldots, j\}$, where $i, j \in \{1, \ldots, |V|\}$ and $i > j$; or*
(3) *the null interval $[\ ]$ which is the empty set ϕ.*

For simplicity, we will not always strictly distinguish between a vertex v and its label $L(v)$, and will say that a vertex $v \in V$ is contained in an interval $[i, j]$ if $L(v) \in [i, j]$.

Definition 3. *Given $U \subseteq V$ and a labeling L of V, we denote by $N(L, U)$ the minimum number of disjoint intervals such that their union is equal to $\{L(v) | v \in U\}$.*

For example, if $L(v_i) = i$ then $N(L, \{v_5, v_6, v_7, v_9\}) = 2$, because $\{5, 6, 7, 9\} = [5, 7] \cup [9, 9]$. $N(L, \{v_1, v_2, v_5, v_6, v_7, v_9\})$, however, depends on whether the circular interval is allowed; it is 2 if yes, 3 otherwise. Apart from this use of the circular interval in expressing $I(u, v)$ as intervals, $I(u, v)$ itself may or may not be allowed to include the starting vertex u. These restrictions give rise to the following basic variants of IRS.

Definition 4. *Let $R = \langle L, I \rangle$ be an IRS on a graph $G = (V, E)$; we call R an*

(1) *Strict Interval Routing Scheme (SIRS) if for every arc $(u, v) \in E$, $u \notin I(u, v)$;*
(2) *Linear Interval Routing Scheme (LIRS) if for every arc $(u, v) \in E$ the intervals representing $I(u, v)$ are restricted to be linear;*
(3) *Strict Linear Interval Routing Scheme (SLIRS) if it is both an SIRS and an LIRS.*

When $u \in I(u, v)$, there is an interval on edge (u, v) that contains the starting vertex u. An interval on an arc leaving a vertex containing that vertex is called an *inner case interval*; otherwise, it is *strict*.

Definition 5. *Let $R = \langle L, I \rangle$ be an IRS (SIRS, LIRS, SLIRS, respectively) on a graph $G = (V, E)$. The compactness of R is the minimum integer k such that for every arc $(u, v) \in E$, $N(L, I(u, v)) \leqslant k$. We denote by k-IRS (k-SIRS, k-LIRS, k-SLIRS, respectively) every IRS (SIRS, LIRS, SLIRS, respectively) of compactness no more than k.*

In practice, we are interested in designing an IRS that induces only the shortest paths.

Definition 6. *Let $R = \langle L, I \rangle$ be an IRS (respectively SIRS, LIRS, SLIRS) on a graph $G = (V, E)$; we call R*

(1) *a single-shortest path IRS (respectively SIRS, LIRS, SLIRS) if it induces one and only one of the shortest paths between every pair $x, y \in V$; or*
(2) *an all-shortest-path IRS (respectively SIRS, LIRS, SLIRS) if it induces exactly the set of all possible shortest paths between every pair $x, y \in V$.*

By the definitions, for all-shortest-path IRS, the arc labeling I does not have much flexibility in assigning subsets to arcs. $I(u, v)$ is either $S(u, v)$ or $S(u, v) \cup \{u\}$. For all-shortest-path SIRS, the arc labeling I has no freedom but being identical with S, i.e., $I(u, v) = S(u, v)$ for every arc (u, v).

Given a graph G and an integer k, the problem of determining whether G supports an all-shortest-path k-IRS can be formally defined as follows.

The all-shortest-path k-IRS problem:
    Instance: A graph G, and a positive integer k.
    Question: Is there an all-shortest-path k-IRS for G?
The all-shortest-path k-SIRS problem:
    Instance: A graph G, and a positive integer k.
    Question: Is there an all-shortest-path k-SIRS for G?

The problems for all-shortest-path k-LIRS and k-SLIRS are defined similarly. Clearly, all of these problems are in the class of NP. In fact, given a graph G (with or without a cost function), an integer k, a vertex labeling L, and an arc labeling I, it can be verified in polynomial time whether $\langle L, I \rangle$ is an all-shortest-path k-IRS (respectively k-LIRS, k-SIRS, k-SLIRS) for G.

In this paper, we prove the above problems to be NP-complete; in particular, the all-shortest-path k-IRS and k-LIRS are NP-complete for every constant $k \geqslant 3$, and k-SIRS and k-SLIRS are NP-complete for every constant $k \geqslant 4$. The proof is based on a polynomial transformation to these problems from the following NP-complete problem.

Definition 7. *The Consecutive Block for Symmetric Matrices (k-CBS for short):*
Instance: *An $n \times n$ symmetric binary matrix M, and an integer $k > 0$.*
Question: *Is there a permutation of the rows of M such that for each column j the number of blocks of consecutive $1's$ (i.e., the number of entries such that $M[i, j] = 1$ and either $M[i + 1, j] = 0$ or $i = n$) is at most k?*

In [11], we proved that k-CBS is complete for every fixed $k \geqslant 2$.

In the next section, we show a construction of graphs from symmetric matrices satisfying that the matrices can be row permuted such that each column has no more than k blocks of consecutive 1's if and only if the constructed graph supports an all-shortest-path $(k + 1)$-IRS $((k + 2)$-SIRS). In the discussion section, we will point out that the tranformation can start from a similar NP-complete problem in [4] to prove the NP-completeness of all-shortest-path k-IRS for general integer k, but not constant k.

3   NP-Completeness Results

Starting with any instance of k-CBS, $\langle M_{n \times n}, k \rangle$, where M is a symmetric binary matrix, we will construct a graph $G = (V, E)$ such that there is a row permutation on M leading to each column having no more than k consecutive 1's blocks if and only if G supports an all-shortest-path $(k + 1)$-IRS. The construction is simple. For each row i of M, create a set $R_i = \{r_{i,1}, r_{i,2}, \ldots, r_{i,k+4}\}$ of $k + 4$ vertices in G, which we call row vertices; for each column j of M, create a set $C_j = \{c_{j,1}, c_{j,2}, \ldots, c_{j,2n(k+4)+1}\}$ of $2n(k + 4) + 1$ vertices in G, called column vertices; these two types of vertices induce a bipartite subgraph of G, and $R_i - C_j$ edges exist if and only if $M[i, j] = 1$. Finally, add a new vertex a to G to link all the other vertices so that the diameter of G is at most 2. Formally, $G = (V, E)$ is obtained as follows (refer to Figure 1, where for simplicity, a line between R_i and C_j represents an edge set, $\{(r_{i,l}, c_{j,h}) | r_{i,l} \in R_i, \ c_{j,h} \in C_j\}$).

$$V = R + C + \{a\}, \quad \text{where}$$

$$R = \bigcup_{1 \leqslant i \leqslant n} R_i \quad \text{and} \quad R_i = \{r_{i,l} | 1 \leqslant l \leqslant k + 4\}; \quad \text{and}$$

$$C = \bigcup_{1 \leqslant j \leqslant n} C_j \quad \text{and} \quad C_j = \{c_{j,h} | 1 \leqslant h \leqslant 2n(k + 4) + 1\}; \quad \text{and}$$

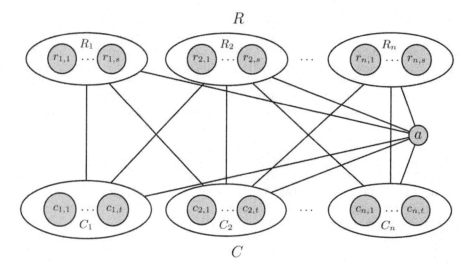

Fig. 1. The transformation graph ($s = k + 4$ and $t = 2n(k + 4) + 1$)

$$E = \{(r_{i,l}, c_{j,h}) | r_{i,l} \in R_i, \ c_{j,h} \in C_j, \ M[i,j] = 1\}$$
$$\cup \{(a, r_{i,l}) | r_{i,l} \in R\} \cup \{(a, c_{j,h}) | c_{j,h} \in C\}.$$

Note that any two vertices $r_{i,l}$ and $r_{i,l'}$ of R_i have identical neighbour sets $Adj(r_{i,l}) = Adj(r_{i,l'})$, and so do any two vertices $c_{j,h}$ and $c_{j,h'}$ of C_j. By $\Gamma(R_i)$ we denote the set of those column vertices that are neighbours of $r_{i,l} \in R_i$. Clearly, $\Gamma(R_i) = Adj(r_{i,l}) - \{a\} = \{c_{j,h} | (r_{i,l}, c_{j,h}) \in E\} = \bigcup_{M[i,j]=1} C_j$. Similarly, we use $\Gamma(C_j)$ to refer to the set of those row vertices that are linked with $c_{j,h} \in C_j$. Then, $\Gamma(C_j) = \bigcup_{M[i,j]=1} R_i$.

As the diameter of G is 2, for each arc $e = (u, v)$, $S(e)$, the optimally reachable vertices from u via e, is $Adj(v) - Adj(u) + \{v\} - \{u\}$. We summarize in the next proposition the $S(e)$ for various types of arcs e in G.

Proposition 1. *In the transformation graph $G = (V, E)$, for $1 \leqslant i, \ j \leqslant n$, $1 \leqslant l \leqslant k + 4$, $1 \leqslant h \leqslant 2n(k + 4) + 1$,*

$$S(a, \ r_{i,l}) = \{r_{i,l}\};$$
$$S(a, \ c_{j,h}) = \{c_{j,h}\};$$
$$S(r_{i,l}, \ a) = \{a\} + R - \{r_{i,l}\} + C - \Gamma(R_i);$$
$$S(r_{i,l}, c_{j,h}) = \{c_{j,h}\} + \Gamma(C_j) - \{r_{i,l}\};$$
$$S(c_{j,h}, \ a) = \{a\} + C - \{c_{j,h}\} + R - \Gamma(C_j);$$
$$S(c_{j,h}, r_{i,l}) = \{r_{i,l}\} + \Gamma(R_i) - \{c_{j,h}\}.$$

In the above, and in the following, we use $+$ and $-$ respectively for the union and the difference operation on sets.

Lemma 1. *There exists a permutation of the rows of the symmetric matrix $M_{n \times n}$ that would result in a matrix having at most k blocks of consecutive 1's per column if and only if the graph G obtained by the above transformation from $\langle M_{n \times n}, k \rangle$ supports an all-shortest-path $(k+1)$-IRS.*

Proof. Suppose first that there exists a permutation π of the rows of $M_{n \times n}$ that leads to at most k blocks per column. Without loss of generality, assume $\pi(i) = i$; then the $(k+1)$-IRS can be constructed as follows. The vertex labeling L is such that $L(r_{i,l}) = (i-1)(k+4) + l$ for each row vertex $r_{i,l} \in R$; $L(a) = n(k+4) + 1$; and $L(c_{j,h}) = n(k+4) + 1 + (j-1)(2n(k+4)+1) + h$ for each column vertex $c_{j,h} \in C$. That is, the vertices are ordered in such a way that each $R_i \subset R$ as well as R itself form one interval, and so does each $C_j \subset C$ and C; vertex a is in the middle and adjacent to the last vertex $r_{n,k+4}$ of R and the first vertex $c_{1,1}$ of C, as depicted below.

$$
\overbrace{r_{1,1} \cdots r_{1,s}}^{R_1} \overbrace{r_{2,1} \cdots , r_{2,s}}^{R_2} \cdots \overbrace{r_{n,1} \cdots r_{n,s}}^{R_n} \; a \; \overbrace{c_{1,1} \cdots c_{1,t}}^{C_1} \overbrace{c_{2,1} \cdots c_{2,t}}^{C_2} \cdots \overbrace{c_{n,1} \cdots c_{n,t}}^{C_n}
$$
$$
\underbrace{\phantom{r_{1,1} \cdots r_{1,s} r_{2,1} \cdots , r_{2,s} \cdots r_{n,1} \cdots r_{n,s}}}_{R} \qquad \underbrace{\phantom{c_{1,1} \cdots c_{1,t} c_{2,1} \cdots c_{2,t} \cdots c_{n,1} \cdots c_{n,t}}}_{C}
$$

Because each R_i under labeling L forms a single interval, for each column j,

$$
N(L, \Gamma(C_j)) = N(L, \bigcup_{M[i,j]=1} R_i) = N(\pi, \{i|M[i,j]=1\})
$$
$$
= \text{the number of consecutive } 1\text{'s blocks in the } j\text{'th column of } M
$$
$$
\leqslant k.
$$

For a similar reason and because of the symmetry of M, for each row i, we have

$$
N(L, \Gamma(R_i)) = N(L, \bigcup_{M[i,j]=1} C_j) = N(\pi, \{j|M[i,j]=1\})
$$
$$
= \text{the number of consecutive } 1\text{'s blocks in the } i\text{'th row of } M
$$
$$
\leqslant k.
$$

Concerning the arc labeling I for all-shortest-path IRS, on each arc $e = (u, v)$, there are only two alternatives, setting $I(u, v)$ to $S(u, v)$ or to $S(u, v) + u$. We make the choice for different arcs that would guarantee each arc receiving no more than $k + 1$ intervals.

At vertex a: Let $I(a, r_{i,l}) = S(a, r_{i,l}) = \{r_{i,l}\}$ and $I(a, c_{j,h}) = S(a, c_{j,h}) = \{c_{j,h}\}$. Obviously they receive one interval. At other vertices, we will use the inner case.

At vertex $r_{i,l}$: On arc $(r_{i,l}, a)$, let $I(r_{i,l}, a) = S(r_{i,l}, a) + \{r_{i,l}\} = \{a\} + R + C - \Gamma(R_i) = V - \Gamma(R_i)$; then $N(L, I(r_{i,l}, a)) \leqslant N(\Gamma(R_i)) + 1 \leqslant k + 1$. On arc $(r_{i,l}, c_{j,h})$, let $I(r_{i,l}, c_{j,h}) = S(r_{i,l}, c_{j,h}) + \{r_{i,l}\} = \{c_{j,h}\} + \Gamma(C_j)$; then $N(L, I(r_{i,l}, c_{j,h})) \leqslant 1 + N(\Gamma(C_j)) \leqslant k + 1$.

At vertex $c_{j,h}$: Similarly for the last case, let $I(c_{j,h}, a) = S(c_{j,h}, a) + \{c_{j,h}\} = V - \Gamma(C_j)$ and let $I(c_{j,h}, r_{i,l}) = S(c_{j,h}, r_{i,l}) + \{c_{j,h}\} = \{r_{i,l}\} + \Gamma(R_i)$; then both kinds of arcs receive at most $k + 1$ intervals.

The *only if* part of the lemma is proved.

Suppose conversely that G admits an all-shortest-path *(k+1)-IRS*, $\langle L, I \rangle$. Consider a permutation π on $\{1, 2, \ldots, n\}$ induced from L in such a way that

$$\pi(i) < \pi(j) \iff \min\{L(v) | v \in R_i\} < \min\{L(v) | v \in R_j\}.$$

We need only to justify that, for every $1 \leqslant j \leqslant n$, $N(\pi, \{i | M[i, j] = 1\})$, the number of blocks in the j'th column of M when its rows are permuted using π, is no more than k.

For $1 \leqslant i \leqslant n$, let $\min R_i$ denote the minimum vertex in R_i, i.e., $L(\min R_i) = \min\{L(r_{i,l}) | r_{i,l} \in R_i\}$. If $M[i_1, j] = 1$ and $M[i_2, j] = 1$, and in $\bigcup_{M[i,j]=1} R_i$ under L, the two vertices $\min R_{i_1}$ and $\min R_{i_2}$ belong to the same interval, then in the matrix M permuted according to π, the two rows i_1 and i_2 must belong to the same block of consecutive 1's. Thus we have $N(\pi, \{i | M[i, j] = 1\}) \leqslant N(L, \bigcup_{M[i,j]=1} R_i) = N(L, \Gamma(C_j))$. So we need only to prove that for every $1 \leqslant j \leqslant n$, $N(L, \Gamma(C_j)) \leqslant k$.

Let us first show that for any column j, $N(L, \Gamma(C_j)) \leqslant k + 3$. Otherwise, there would be some j such that $N(L, \Gamma(C_j)) > k + 3$. Let us consider the interval number $N(L, I(r_{i,l}, c_{j,h}))$ on an arc $(r_{i,l}, c_{j,h})$ at a vertex $r_{i,l} \in \Gamma(C_j)$. $I(r_{i,l}, c_{j,h})$ is either $S(r_{i,l}, c_{j,h})$ or $S(r_{i,l}, c_{j,h}) + \{r_{i,l}\}$. If it is $S(r_{i,l}, c_{j,h})$, then

$$N(L, I(r_{i,l}, c_{j,h})) = N(L, \{c_{j,h}\} + \Gamma(C_j) - \{r_{i,l}\}) \geqslant N(L, \Gamma(C_j)) - 2 > k + 1.$$

If it is $S(r_{i,l}, c_{j,h}) + \{r_{i,l}\}$, then

$$N(L, I(r_{i,l}, c_{j,h})) = N(L, \{c_{j,h}\} + \Gamma(C_j)) \geqslant N(L, \Gamma(C_j)) - 1 > k + 2.$$

Both cases imply that under the labeling L the edge $(r_{i,l}, c_{j,h})$ must receive more than $k + 1$ intervals, contradicting that $\langle L, I \rangle$ is a $(k + 1)$-IRS for G.

Now, we prove that for any column j, $N(L, \Gamma(C_j)) \leqslant k$. Supposing that it is not the case, then there must be a j such that $N(L, \Gamma(C_j)) \geqslant k + 1$. In this case, we can show that there exists an $r_{i,l} \in \Gamma(C_j)$ and a $c_{j,h} \in C_j$ such that the arc $(r_{i,l}, c_{j,h})$ at vertex $r_{i,l}$ receives at least $k + 2$ intervals, contradicting that L is a vertex labeling of the *(k+1)-IRS*.

Since $|C_j| = 2n(k + 4) + 1 = 2|R| + 1 \geqslant 2|\Gamma(C_j)| + 1$, and each $r_{i,l} \in \Gamma(C_j)$ is adjacent to no more than two vertices of C_j (with respect to the order defined by the labeling L), there must be a vertex $c_{j,h} \in C_j$ such that $L(c_{j,h})$ is adjacent to no labels of the vertices in $\Gamma(C_j)$. In the set $S(r_{i,l}, c_{j,h}) = \{c_{j,h}\} + \Gamma(C_j) - \{r_{i,l}\}$, under the labeling L, $c_{j,h}$ itself has to form a single interval. If $I(r_{i,l}, c_{j,h})$ is $S(r_{i,l}, c_{j,h}) + \{r_{i,l}\}$, then the interval number of arc $(r_{i,l}, c_{j,h})$ is

$$N(L, I(r_{i,l}, c_{j,h})) = N(L, \{c_{j,h}\} + \Gamma(C_j)) = 1 + N(L, \Gamma(C_j)) \geqslant k + 2,$$

which is a contradiction. If $I(r_{i,l}, c_{j,h})$ is $S(r_{i,l}, c_{j,h})$, by further selecting an appropriate $r_{i,l}$ from $\Gamma(C_j)$ we can arrive at another contradiction. Because there are at least $k + 4$ vertices in $\Gamma(C_j)$ and all of the vertices of $\Gamma(C_j)$ are distributed among the $N(L, \Gamma(C_j)) \leqslant k + 3$ intervals, there must be an interval $[L(r_{i,l}), L(r_{i',l'})]$ of $\Gamma(C_j)$ under L having at least two vertices. Selecting the boundary vertex $r_{i,l}$ in this interval we have $N(L, \Gamma(C_j) - \{r_{i,l}\}) = N(L, \Gamma(C_j))$. Hence the number of the intervals assigned on edge $(r_{i,l}, c_{j,h})$ is

$$
\begin{aligned}
N(L, I(r_{i,l}, c_{j,h})) &= N(L, \{c_{j,h}\} + \Gamma(C_j) - \{r_{i,l}\}) \\
&= 1 + N(L, \Gamma(C_j) - \{r_{i,l}\}) = 1 + N(L, \Gamma(C_j)) \\
&\geqslant k + 2,
\end{aligned}
$$

which again is a contradiction. The *if* part of the lemma is proved. □

It is easy to see that the above transformation can be performed in polynomial time. By the NP-completeness of k-CBS for every fixed $k \geqslant 2$, the next theorem follows.

Theorem 1. *Given a network G, the problem of deciding if there exists an all-shortest-path k-IRS for G is NP-complete for every fixed $k \geqslant 3$.*

Note that, in the *only if* part of the proof of Lemma 1, the $(k + 1)$-IRS for G derived from the permutation of the rows of M is linear. Thus, the following holds.

Theorem 2. *Given a network G, the problem of deciding if there exists an all-shortest-path k-LIRS for G is NP-complete for every fixed $k \geqslant 3$.*

In the transformation, if we let each R_i contain $2k + 7$ vertices and each C_j contain $2n(2k + 7) + 1$ vertices, we will be able to find one of the intervals of $\Gamma(C_j)$ under L containing at least three vertices. Selecting an intermediate vertex from such an interval, a vertex $r_{i,l} \in \Gamma(C_j)$ can be found satisfying $N(L, \Gamma(C_j) - \{r_{i,l}\}) = N(\Gamma(C_j)) + 1$. Note that in all-shortest-path SIRS, the arc labeling L is identical to S. By a similar argument one can show that there exists an all-shortest-path strict $(k + 2)$-IRS for G if and only if there exists a permutation for M leading to at most k consecutive 1's blocks per column.

Theorem 3. *Given a network G, the problem of deciding if there exists an all-shortest-path k-SIRS (k-SLIRS) for G is NP-complete for every fixed $k \geqslant 4$.*

4   Discussions

We have proved that to recognize networks that admit all-shortest paths k-IRS (k-SIRS) for every $k \geqslant 3$ ($k \geqslant 4$) is NP-complete for unweighted graphs (and of course also for weighted graphs). Our transformation takes advantage of the symmetry of the matrix in the instance of the NP-complete problem k-CBS. For general binary matrices, Booth and Lueker [1] gave a linear algorithm for $k = 1$; Goldberg et al. [8] proved NP-completeness for every fixed $k \geqslant 2$; Flammini et al. [4] showed the same for general k even if the matrices are restricted to each row having no more than k blocks of consecutive 1's (although that is not stated explicitly in their paper).

Note that if we apply our transformation to an arbitrary binary m by n matrix, a graph will be constructed such that the matrix has no more than k blocks in each column and no more than l blocks in each row if and only if the constructed graph supports an all-shortest-path $(\max\{k, l\}+1)$-IRS $((\max\{k, l\}+2)$-SIRS). Thus, the transformation can start from an instance of Flammini's NP-complete problem in [4] to prove the NP-completeness of all-shortest-path k-IRS (and its variants) for general (but not constant) integer k.

The results of this paper clearly imply that the optimization problem of determining the minimal k such that a given network supports an all-shortest-path k-IRS (or its variants) is NP-hard. They also imply that we cannot in polynomial time approximate the compactness of IRS (SIRS) within a ratio of less than $4/3$ $(5/4)$, unless P=NP.

The stretch factor of a k-IRS R on a graph G is defined as the smallest real s such that for every pair of vertices, x, y, the length of the routing path induced by R from x to y is at most s times the distance in G between x and y. Since the construction in our transformation is a graph of diameter 2, it follows that if P$\neq$NP then for every $k \geqslant 3$ $(k \geqslant 4)$ we cannot in polynomial time determine if a given graph supports an all-path k-IRS $(k$-SIRS) of stretch factor less than $3/2$.

Some remaining open problems: What is the time complexity when $k = 1, 2$ $(k = 2, 3)$ for all-shortest-path k-IRS $(k$-SIRS)?

Acknowledgment

We thank the referees for their useful comments that helped improve the presentation.

References

1. K.S. Booth and S. Lueker, Linear algorithm to recognize interval graphs and test for consecutive ones property, Proc. 7th Ann. ACM Symp. on Theory of Computing, New York, pp. 255–265, 1975.
2. T. Eilam, S. Moran, and S. Zaks, The complexity of the characterization of networks supporting shortest-path interval routing, Theoretical Computer Science, 289, pp. 85–104, 2002. (Also in 4th Internat. Coll. on Structural Information & Communication Complexity (SIROCCO '97), Carleton Scientific, July 1997, pp. 99–111).
3. M. Flammini, On the hardness of devising interval routing schemes, Parallel Processing Lett. 7, pp. 39–47, 1997.
4. M. Flammini, G. Gambosi, and S. Salomone. Interval routing schemes, Algorithmica, 16, pp. 549–568, 1996. (Also in 12th Ann. Symp. on Theoretical Aspects of Computer Science (STACS), Lecture Notes in Computer Science, Vol. 900, Springer, Berlin, March 1995, pp. 279–290).

5. M. Flammini, U. Nanni, and R.B. Tan, Characterization results of all-shortest paths interval routing schemes, Networks, Vol. 37, no. 4, pp. 225-232, 2001. (Also in 5th Internat. Coll. on Structural Information & Communication Complexity (SIROCCO '98), 1998, pp. 201–213).
6. M.R. Garey and D.S. Jonson, Computers and intractability: A guide to the theory of NP-completeness. Freeman, San Francisco, CA, 1979.
7. C. Gavoille, A survey on interval routing, Theoretical Computer Science, 245(2), pp. 217–253, 2000.
8. P.W. Goldberg, M.C. Glumbic, H. Kaplan, and R. Shamir, Four strikes against physical mapping of DNA, Journal of Computational Biology, 2(1), pp. 139–152, 1995.
9. N. Santoro and R. Khatib, Labeling and implicit routing in networks, Comput. J., 28, pp. 5–8, 1985.
10. J. van Leeuwen, R.B. Tan, Computer networks with compact routing tables, The Book of L, Springer, Berlin, pp. 298–307, 1985.
11. R. Wang and F.C.M. Lau, Consecutive ones block for symmetric matrices, Technical Report TR-2003-09, Department of CSIS, The University of Hong Kong, 2003. (www.csis.hku.hk/research/techreps/document/TR-2003-09.pdf)

On-Line Scheduling of Parallel Jobs

Deshi Ye[1,2,*] and Guochuan Zhang[3,**]

[1] Department of Mathematics, Zhejiang University, Hangzhou 310027, China
[2] Institut für Informatik und Praktische Mathematik, Universität Kiel
Olshausenstr. 40, 24098 Kiel, Germany
dye@informatik.uni-kiel.de
[3] Institut für Informatik, Universität Freiburg
Georges-Köhler-Allee 79, 79110 Freiburg, Germany
gzhang@informatik.uni-freiburg.de

Abstract. We study an on-line parallel job scheduling problem, where jobs arrive over list. A parallel job may require a number of machines for its processing at the same time. Upon arrival of a job, its processing time and the number of requested machines become known, and it must be scheduled immediately without any knowledge of future jobs. We present a 8-competitive on-line algorithm, which improves the previous upper bound of 12 by Johannes [8]. Furthermore, we investigate two special cases in which jobs arrive in non-increasing order of processing times or jobs arrive in non-increasing order of sizes. Better bounds are shown.

Keywords: On-line scheduling, parallel computing

1 Introduction

In this paper we study an on-line problem of scheduling parallel jobs. Parallel jobs may require more than one machine simultaneously. They are characterized by two parameters, processing time and size (the number of requested machines). Jobs $\{J_1, J_2, \cdots, J_n\}$ arrive over list, which are not known in advance. Upon arrival of job J_i, its processing time and the number of machines required become known, and it must immediately and irrevocably be scheduled. The next job J_{i+1} appears only after job J_i has been assigned. The goal is to minimize the *makespan*, i.e., the largest completion time over all the jobs.

The model of scheduling parallel jobs has recently been studied extensively, see, e.g., [2]-[4], [10] and [11]. In many applications, a network topology is considered, in which jobs can only be executed on particular machines. Only those machines connected may execute a job together simultaneously. Parallel machines with a specific network topology can be viewed as a graph where each node represents a machine and each edge represents the communication link between the two nodes. In this paper we focus on the parallel system, called *PRAM*, whose underlying network topology is a complete graph.

----

* Research supported by DAAD Sandwich Program and the EU-Project APPOL II.
** Research supported by the DFG Project AL 464/4-1 and NSFC (10231060).

R. Královič and O. Sýkora (Eds.): SIROCCO 2004, LNCS 3104, pp. 279–290, 2004.
© Springer-Verlag Berlin Heidelberg 2004

On-Line Models. We can distinguish on-line scheduling problems by the manner that jobs arrive. In general, there are three different versions.

1. *Jobs arrive over list.* Jobs arrive one by one. The next job appears only after the current one has been scheduled. Upon arrival of a job its size is known.
2. *Jobs arrive over time.* Each job J_i has a release date r_i, which is not known before J_i appears. Upon arrival of job J_i its size s_i becomes known.
3. *Jobs arrive on dependencies.* Each job appears only after all its predecessors are completed, i.e., some special order in which jobs have to be started (they can be expressed in term of chains, trees, series parallel orders, interval order and so on). As soon as a job appears its size is known. However, we are not aware of the dependencies among jobs in advance.

For each of the above three versions, upon arrival of a job, we may also assume that the job processing time becomes known (called *known processing time*) or unknown until they are completed (called *unknown processing times*). In this paper we investigate the model that jobs arrive over list with known processing times.

Known Results. For the model where jobs arrive over time, Naroska et al. [11] presented a list scheduling heuristic with competitive ratio of $2 - 1/m$ for unknown processing times, and pointed out it is optimal. The model where jobs arrive on dependencies was first studied by Feldmann et al. [4]. With unknown processing times, it is shown that the best competitive ratio is m, the number of machines, even if the precedence is known in advance. Bischof et al. [1] considered the problem that the job processing times are equal and present a 2.7-competitive algorithm and a lower bound of 2.691.

Regarding the model where jobs arrive over list, to our best known, it was only studied by Johannes [8]. A 12-competitive algorithm was presented and a lower bound 2.25 was proved with an enumerating program.

Our Contribution. In this paper, we concentrate on the model where jobs arrive over list and improve the upper bound given by Johannes [8]. We present an on-line algorithm with competitive ratio of at most 8. A lower bound of 6 is proved for this algorithm. Then we study two special cases, in which we know some a priori information about the job sizes or job processing times. The first case assumes that *jobs arrive in non-increasing order of processing times*. We obtain an upper bound of 2 by employing a greedy algorithm, which schedules jobs as early as possible. Note that if all job processing times are equal to one, it becomes the classical bin packing problem, for which the best known lower bound and upper bound are $5/3$ [13] and $7/4$ [12], respectively. Then we deal with the second case that *jobs arrive in non-increasing order of sizes*. We show that the greedy algorithm has a competitive ratio of at most 2.75. If all job sizes are equal to one, it is the classical on-line parallel machine scheduling problem, for which the best known lower bound 1.85358 and upper bound 1.9201 were given by Gormley et al. [7] and, Fleischer and Wahl [6], respectively.

Organization of the Paper. The remainder of this paper is organized as follows. Section 2 gives preliminaries. Section 3 presents the on-line algorithm with competitive ratio of at most 8. The two special cases are dealt with in Sections 4 and 5.

2   Preliminaries

An instance I of our problem consists of a list of parallel jobs $\{J_1, J_2, \cdots, J_n\}$ and m identical machines. Each job J_j is characterized by size s_j, the number of required machines, and processing time p_j. Once job J_j appears, its size s_j and processing time p_j become known. We use (s_j, p_j) to denote job J_j. The *efficiency* of a schedule at any time t is defined to be the number of busy machines at time t divided by m. The average efficiency in disjoint intervals (a_i, b_i), $i = 1, \ldots, l$, is thus defined by

$$\sum_{i=1}^{l} \frac{1}{m(b_i - a_i)} \cdot \int_{a_i}^{b_i} (\text{ total number of busy machines at time } t) \, dt.$$

We adopt the standard measure *competitive ratio* to evaluate an on-line algorithm. For any instance I, let $C_A(I)$ and $C^*(I)$ be the makespan given by an on-line algorithm A and the makespan produced by an optimal off-line algorithm, respectively. The *competitive ratio* of algorithm A is defined as

$$R_A = \sup_I \{C_A(I)/C^*(I)\}.$$

Algorithm A is called ρ-competitive if $C_A(I) \leq \rho C^*(I)$ holds for any instance I. A lower bound R of a problem means that any on-line algorithm has competitive ratio of at least R.

When job J_j is coming, one can schedule it in a time interval which has a duration at least p_j and during which the number of idle machines is at least s_j. The simplest on-line algorithm is the following greedy algorithm.

Algorithm *Greedy.* As a job is coming, schedule it as early as possible.

As shown in [8] algorithm *Greedy* has a competitive ratio of m. The bound can be obtained with the following example of $2m$ jobs. Let $\varepsilon > 0$ be a sufficiently small number. The $(2i-1)$-st job is $(1, m+(i-1)\varepsilon)$, while the $2i$-th job is (m, ε), $i = 1, \ldots, m$. Algorithm *Greedy* processes the job one after another and gives a makespan of $m^2 + m(m-1)\varepsilon/2 + m\varepsilon$. However, one can first process the m jobs of size m and then process the m jobs of size 1 each on a machine. It gives a makespan of $m + (m-1)\varepsilon + m\varepsilon$. The competitive ratio goes to m as ε tends to zero.

The example tells us that it is not wise to be so greedy. To improve the bound we may leave some space for the future jobs in case that the current schedule is not in a "tight" manner. Johannes [8] designed an on-line algorithm alone this line, in which the time axis is partitioned into intervals. The length of the first interval I_1 depends on the processing time of the first job, and the next intervals I_i always have a double length of the interval I_{i-1}. The jobs are scheduled in different way in some intervals. It was shown that the competitive ratio of the algorithm is at most 12. Note that in her algorithm the partition of the time interval is almost independent on the current schedule. In the next section we propose an improved algorithm which partitions the time axis into intervals relying on the current schedule. We prove that the proposed algorithm has a competitive ratio in between 6 and 8.

3   An Improved On-Line Algorithm

A job is called *big* if its size is larger than $m/2$, otherwise it is called *small*. No two big jobs can be processed at the same time. To present the algorithm more clearly we show first the structure of the schedule given by the algorithm. The schedule looks like an aligned house, which consists of *rooms* and *walls*. Rooms and walls appear alternatively. The rooms are constructed by small jobs, while the walls are constructed by big jobs. A big job is assigned to an old wall or builds a new wall by itself. A small item is always assigned to a room. The schedule starts from a wall and ends at a wall. We assume that there is a wall at time zero with a length (thickness) of zero. In case that the last room has no wall on the right side after all jobs are scheduled, we add a dummy wall with length of zero starting at the completion of the last completed job. Therefore the schedule can be described as

$$W_1 \cup R_1 \cup W_2 \cup R_2 \cup \cdots \cup W_N \cup R_N \cup W_{N+1},$$

where W_i denotes the i-th wall and R_i denotes the i-th room. The length of wall W_i is the total processing time of the big jobs involved in the wall. The length of room R_i is the distance between two walls, more precisely, the time difference between the starting time of the wall W_{i+1} and the ending time of the wall W_i. The length of the schedule is the total length of the rooms and the walls.

A wall is open if there are no small jobs scheduled after it and a room is called open if there is no wall on the right side.

Algorithm *DW* (Dynamic Waiting)

1. Create a wall (W_1) of length zero at time zero.
2. If the incoming job J_i is big, i.e., $s_i > m/2$, schedule J_i immediately after the open wall if such a wall exists. If there is no open wall, then there must be an open room. Let h be the length of the interval of efficiency less than $1/2$ in the open room and let L be the current length of the schedule. Start job J_i at time of $L + h$.
3. If the incoming job J_i is small, i.e., $s_i \leq m/2$, find the first room which can accommodate J_i and schedule J_i as early as possible. If such a room does not exist, we create a new room for J_i following the present latest wall.
4. If no job comes, add a dummy wall if needed and stop.

An Example: There are 10 machines and 8 jobs, where $J_1 = (4,1)$, $J_2 = (3,3)$, $J_3 = (6,1)$, $J_4 = (1,4)$, $J_5 = (1,6)$, $J_6 = (1,7)$, $J_7 = (7,1)$, $J_8 = (6,1)$. The schedule by algorithm DW is shown in Fig. 1, where $W_1 = [0,0]$, $R_1 = [0,5]$, $W_2 = [5,6]$, $R_2 = [6,20]$ and $W_3 = [20,22]$.

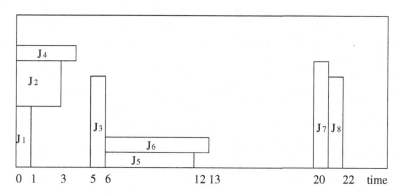

Fig. 1. An illustration of algorithm DW

Lemma 3.1. *Let S be a schedule by an algorithm A. If the length of the schedule S can be divided into two parts, one with average efficiency no less than α, $0 < \alpha \leq 1$, and another with total length of at most β times the optimal makespan, then the competitive ratio of algorithm A is at most $1/\alpha + \beta$.*

Proof. Denote by x the length of the part with average efficiency no less than α and denote by y the length of another part. Thus $C_A = x + y$. Note that $C^* \geq \alpha x$ and $y \leq \beta C^*$. We have $C_A = x + y \leq (1/\alpha + \beta)C^*$. □

Theorem 3.2. *The competitive ratio of algorithm DW is at most 8.*

Proof. Let $\mathcal{L}$ be the final schedule by algorithm DW, i.e., $\mathcal{L} = W_1 \cup R_1 \cup \cdots \cup W_N \cup R_N \cup W_{N+1}$. By the algorithm, all rooms consist of small jobs. We divide room R_i into two time intervals E_i and F_i, where in interval E_i the efficiency of the schedule is at least $1/2$, and $F_i = R_i - E_i$. Since R_i consists of small jobs, it is easy to see that E_i precedes F_i.

The idea to prove the bound of 8 is partitioning the time interval of the schedule $\mathcal{L}$ into two classes. In the first class the average efficiency of the intervals is at least $1/4$ and in the second class the total length of the time intervals is bounded above by 4 times the longest job processing time. The theorem thus follows from Lemma 3.1.

Note that the efficiency of a wall is larger than $1/2$. We only need to consider the rooms. If all $F_i = \emptyset$, the efficiency of the schedule at any time is at least $1/2$, which implies that the competitive ratio is at most 2.

Assume that $F_k \neq \emptyset$ and $F_j = \emptyset$ for all $j < k$. If $k = N$, all intervals but F_N have efficiency at least $1/2$. During the time interval F_N each machine executes at most one job. Otherwise the last job can start earlier since the efficiency of the schedule is less than $1/2$ and all jobs in the room are small. By the algorithm, the length of F_N is at most twice the longest job processing time. It proves that the competitive ratio is at most 4 by Lemma 3.1.

Now we assume that $k < N$. Note that all time intervals before F_k have efficiency of at least $1/2$. We consider the rooms R_j, $j = k, \ldots, N$. For any two adjacent rooms R_i and R_j ($j > i$), the jobs scheduled in R_j must have processing time larger than $|F_i|$, the length of F_i. Otherwise, they could have been assigned to room R_i. Therefore, either $E_j = \emptyset$ or $|E_j| > |F_i|$.

Find those $E_j \neq \emptyset$ for $j \geq k+1$. For each of such E_j, we have $|E_j| > |F_i|$, for $k \leq i < j$. We want to determine some F_i ($i \leq j$) such that the average efficiency of E_j and F_i is at least $1/4$. Then the pair of E_j and F_i is called a *match* and such an F_i is called *matched*. Before we start this procedure, all F_i's are *un-matched* for $i \geq k$. Start from the first non-empty E_j ($j > k$) and continue the following matching procedure until all non-empty E_j's are matched. If $|E_j| \geq |F_j|$, we put E_j and F_j as a match. Otherwise, $|F_j| > |F_i|$ holds for all $k \leq i < j$. We put E_j and the un-matched F_i with the largest index $i < j$ together as a match, and thus F_i is matched. For each match, it is obvious that the average efficiency is at least $1/4$.

We do the matching interval by interval starting from R_k (we do not count E_k) until all nonempty E_j's are matched for $j > k$. We will show that one of the following two assumptions must hold during the matching procedure.

1. There is only one un-matched F_i.
2. Re-index the un-matched F_i's and denote them by $\bar{F}_1, \bar{F}_2, \cdots$. Then $|\bar{F}_{j+1}| > 2|\bar{F}_j|$.

Recall that $F_k \neq \emptyset$. We start from intervals R_k. Clearly, the assumption 1 holds. Now we assume that one of the two assumptions is true when we deal with interval R_p ($p \geq k$). Now we consider interval R_{p+1} with the following two cases.

Case 1. $E_{p+1} = \emptyset$. Consider the last un-matched $F_q \neq \emptyset$ preceding interval R_{p+1}. The small jobs in interval R_{p+1} have processing times greater than $|F_q|$ and the length of interval R_{p+1} is larger than twice of $|F_q|$ when the next wall W_{p+2} is created. Then $|F_{p+1}| > 2|F_q|$, where $q \leq p$.

Case 2. $E_{p+1} \neq \emptyset$. If there is only one un-matched F_i so far, then either F_{p+1} or F_i matches E_{p+1}. Thus there is only one un-matched interval after R_{p+1} is considered. The assumption 1 still holds. If there are more than one un-matched F_i, without loss of generality, re-index them as $\bar{F}_1, \bar{F}_2, \cdots, \bar{F}_q$. By the assumption 2, we have $|\bar{F}_{j+1}| > 2|\bar{F}_j|$, $1 \leq j \leq q-1$. If $|F_{p+1}| > |E_{p+1}|$, then $\bar{F}_q$ and E_{p+1} are matched and F_{p+1} is un-matched. We have $|F_{p+1}| > |\bar{F}_q| > 2|\bar{F}_{q-1}|$ and now F_{p+1} becomes $\bar{F}_q$. If $|F_{p+1}| \leq |E_{p+1}|$, F_{p+1} and E_{p+1} are matched. The un-matched intervals remain unchanged. for interval R_{p+1}.

Finally we get a list of un-matched intervals. Let x be the processing time of the longest job in the last un-matched interval $\bar{F}_t$. Thus $2x \geq |\bar{F}_t|$ by the algorithm. $\sum_{j=1}^{t} |\bar{F}_j| < 2|\bar{F}_t|$. Since $C^* \geq x$, we get that the total length of these intervals is bounded above by $4C^*$. Therefore the theorem follows.   □

In the following, we will give an instance to show that a lower bound of algorithm DW is 6. The instance consists of sequential jobs and block jobs, where a sequential job has a size of one and a block job has a size of m.

Theorem 3.3. *The competitive ratio of the algorithm DW is at least $6m/(m+1)$ for $m \geq 2$.*

Proof. Let k be a sufficiently large integer and let $\varepsilon < 1/2^{k+2}$ be a small positive number. Let $t = \lfloor (m-2)2^{k+1}/(m+1) \rfloor$. In the first t job groups $G_1, \cdots, G_t$, each consists of $\lceil m/2 \rceil$ identical sequential jobs followed by a block job. The processing time of each block job is ε. The processing time of each sequential job in G_i is $1 - (t - i + 1)\varepsilon$.

The last group G_{t+1} of jobs consists of $2k$ jobs $J_1, J_2, \cdots, J_{2k}$, where J_{2i} is a block job of processing time ε and J_{2i-1} is a sequential job of processing time $2^{i-1} + (2^{i-1} - 1)\varepsilon$, for $1 \leq i \leq k$. Clearly, $p_{2i+1} = 2p_{2i-1} + \varepsilon$, for $1 \leq i \leq k - 1$. In the schedule generated by algorithm DW we have $t + k$ rooms and $t + k + 1$ walls (the length of the first wall is zero). The total length of the walls is $(t+k)\varepsilon$. The length of the room i is less than one, for $i = 1, \ldots, t$, while the length of the $t + i$-th room is $2p_{2i-1}$, for $i = 1, \ldots, k$. Therefore, $C_{DW} = t + 2^{k+2} - 2 + O(\varepsilon)$ (see Fig. 2, in which s denotes a sequential job and B denotes a block job).

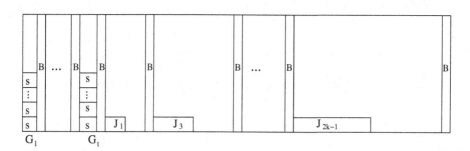

Fig. 2. The schedule by algorithm DW

In an optimal schedule, we can assign the sequential jobs of groups $G_1, \ldots, G_t$ to $m - 2$ machines, the sequential jobs $J_1, J_3, \ldots, J_{2k-3}$ to a machine and the sequential job J_{2k-1} to one machine. The block jobs are scheduled one by one following the sequential jobs (see Fig. 3). It shows that $C^* = 2^k + O(\varepsilon)$.

$$C_{DW}/C^* \geq (2^{k+2} - 3 + (m-2)2^{k+1}/(m+1) + O(\varepsilon))/(2^k + O(\varepsilon)) \to 6m/(m+1)$$

as k tends to infinity.   □

Theorem 3.3 shows that an asymptotic lower bound of algorithm DW is 6.

4   Non-increasing Processing Times

In this section, we study the case that the jobs arrive in non-increasing order of processing times. Note that if all the job processing times are equal to one, the

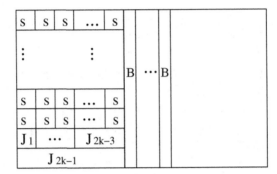

Fig. 3. An optimal schedule

scheduling problem is equivalent to the classical on-line bin packing problem. To see this, we can always assume that an on-line algorithm starts any job at an integral time. As far as we know, the best known lower bound and upper bound of classical on-line bin packing are $5/3$ [13] and $7/4$ [12], respectively. To make the paper self-contained, we give an instance to show the lower bound of $5/3$. Let $k > 0$ be a sufficiently large integer and let $m = 6k$. There are three groups, each of which consists of 6 jobs. The jobs of first group I_1 are of $(k - 2, 1)$, the jobs of the second group I_2 are of $(2k + 1, 1)$, and the jobs of the last group I_3 are of $(3k + 1, 1)$. Clearly, $C^*(I_1) = 1$, $C^*(I_1 \cup I_2) = 3$ and $C^*(I_1 \cup I_2 \cup I_3) = 6$. For any on-line algorithm A, it is easy to verify that

$$\max\{\frac{C_A(I_1)}{C^*(I_1)}, \frac{C_A(I_1 \cup I_2)}{C^*(I_1 \cup I_2)}, \frac{C_A(I_1 \cup I_2 \cup I_3)}{C^*(I_1 \cup I_2 \cup I_3)}\} \geq 5/3.$$

In the following, we analyze algorithm *Greedy* for the special problem. Note that *Greedy* is just FF (*First Fit*) bin packing algorithm if all job processing times are equal. It is known that the (absolute) competitive ratio of FF for bin packing is in between 1.7 [9] and $7/4$ [12].

We distinguish jobs as big ones and small ones as in the previous section.

Theorem 4.1. *The competitive ratio of algorithm Greedy for the on-line parallel jobs scheduling with non-increasing processing times is at most 2.*

Proof. Denote by $\mathcal{L}$ the time interval of the final schedule by *Greedy*. We partition it into parts according to the efficiency. Let $\mathcal{L} = B_1, L_1, B_2, \cdots, L_k, B_{k+1}$, where B_i is the continuous time interval, at anytime of which the efficiency is larger than $1/2$, and L_i is the continuous time interval, at anytime of which the efficiency is at most $1/2$. To avoid trivial cases we assume that $L_i \neq \emptyset$, for $i = 1, \ldots, k$ and $B_i \neq \emptyset$, for $i = 2, \ldots, k$. Note that it is possible that $B_{k+1} = \emptyset$. The jobs processed during an interval $L_i = [a_i, b_i]$ are small. No jobs start at any time $t_i \in (a_i, b_i)$. If such a job exists, it would have been scheduled at time a_i since at that time there are at least $m/2$ machines available. It shows that the length of each L_i is at most the longest processing times among all small jobs. Thus $C^* \geq |L_i|$, for $i = 1, 2, \ldots, k$.

If $B_1 = \emptyset$, any job scheduled after L_1 is big (with a size larger than $m/2$); Otherwise, it could have been assigned to interval L_1. Let T be the total length of the (big) jobs scheduled after L_1. $C^* \geq T$ and $C^* \geq |L_1|$, while $C_{Greedy} = |L_1| + T$. It follows that the competitive ratio of *Greedy* is at most two. In the following, we assume that $B_1 \neq \emptyset$.

Consider L_i, $1 \leq i < k$. B_{i+1} must start with a big job, since, otherwise, a small job can be scheduled in L_i. We denote the big job by J_b with processing time of p_b. The small jobs scheduled in L_{i+1} arrive after J_b; Otherwise it would be scheduled in L_i since at that moment at least $m/2$ machines are idle. The processing time of any small job in L_{i+1} is at most p_b but larger than $|L_i|$. It shows that $p_b > |L_i|$. Then the average efficiency over the intervals L_i and B_{i+1} is at least $1/2$.

Now we consider the interval L_k together with B_1. Denote the longest job in L_k by J_n with processing time p_n. Thus J_n is a small job and $|L_k| \leq p_n$. At the moment that J_n comes, J_n can not be assigned to B_1. Note that the jobs arrive in non-increasing order of processing times. The jobs, which have been assigned to start at time zero before J_n comes, have processing times at least p_n. Then $|B_1| \geq p_n \geq |L_k|$. We have the average efficiency over B_1 and L_k of at least $1/2$.

Combining all the cases considered above, we conclude that the average efficiency of the schedule is at least $1/2$, which implies the competitive ratio of algorithm *Greedy* is at most 2. □

5   Non-increasing Job Sizes

In this section, we deal with the case that jobs arrive in non-increasing order of sizes. The classical on-line scheduling problem which jobs require exactly one machine is a special problem of our case, if we assume that the job sizes are equal to one. To our knowledge, the currently best lower bound and upper bound are 1.85358 [7] and $1 + \sqrt{\frac{1+\ln 2}{2}} < 1.9201$ [6], respectively. In the following, we first analyze algorithm *Greedy*, which is identical to *LS* (List Scheduling) for the classical schedule problem. We show that it is 2.75-competitive and present a lower bound of 2.5.

Theorem 5.1. *Algorithm* Greedy *is 2.75-competitive.*

Proof. Consider the schedule generated by algorithm *Greedy*. Let x be the length of the schedule during which a big job is processed. Then $C^* \geq x$. Denote by t the starting time of the last job J_n with processing time p_n, which determines the makespan. If $t = x$, we get that $C_{Greedy} \leq 2C^*$. Consider the time interval $[x, t)$. The jobs starting in this interval are small. If the efficiency at some time g in $[x, t)$ is less than $2/3$, any job starting at or later than g must have a size larger than $m/3$. Then the size of any earlier job starting in the time interval $[x, g)$ is larger than $m/3$ too. It shows that two such jobs can be scheduled together and the efficiency must be larger than $2/3$. It is a contradiction. Therefore, the efficiency in the interval $[x, t)$ is at least $2/3$. Let $y = t - x$. We have $C^* \geq x/2 + 2y/3$ and

$C^* \geq p_n$. On the other hand, $C_{Greedy} = x + y + p_n$. The theorem is proved with the following cases:

Case 1: $y \leq \frac{3}{4}x$. In this case, $C_{Greedy} \leq \frac{7}{4}x + p_n \leq \frac{11}{4}C^*$.

Case 2: $y > \frac{3}{4}x$. In this case, $\frac{7}{4}C^* \geq \frac{7}{8}x + \frac{7}{6}y = x + y + \frac{1}{6}(y - \frac{3}{4}x) > x + y$. Combining with $C^* \geq p_n$, we get $C_{Greedy} \leq \frac{11}{4}C^*$. □

Theorem 5.2. *The competitive ratio of algorithm Greedy is at least 2.5.*

Proof. Let $k > 0$ be a sufficiently large integer. Let $m = 3 \cdot 2^k$ and $N = 2^{k-2}$. We are given 6 job groups $\{G_1, G_2, \cdots, G_6\}$. Group G_1 consists of N jobs of $(3 \cdot 2^{k-1} + N, 1), (3 \cdot 2^{k-1} + N - 1, 1), \cdots, (3 \cdot 2^{k-1} + 1, 1)$. G_2 consists of two identical jobs of $(3 \cdot 2^{k-1}, 1/N)$. G_3 has N jobs of $(3 \cdot 2^{k-1} - 1, 1/N), (3 \cdot 2^{k-1} - 2, 1/N), \cdots, (3 \cdot 2^{k-1} - N, 1/N)$. G_4 has N identical jobs of $(2^k + 1, 1)$. G_5 consists of three identical jobs of $(2^k, 1)$. Finally, G_6 has only one job of $(1, N + 1/N)$. Clearly the jobs arrive in non-increasing order of sizes.

By algorithm *Greedy*, the jobs in G_1 are processed one by one. Then the two jobs in G_2 start at time N. The i-th job in group G_3 starts at time $N - i$, for $1 \leq i \leq N$. In the time interval $[0, N]$, the longest idle time of the machines is $1 - 1/N$. Then the G_4 jobs are scheduled in the interval $[N + 1/N, 3N/2 + 1/N]$, where two G_4 jobs are processed together at each unit time. The three G_5 jobs start at time $3N/2 + 1/N$. The last job (from G_6) can only start at $3N/2 + 1 + 1/N$. Thus $C_{Greedy} = 5N/2 + 1 + 2/N$ (see Fig. 4).

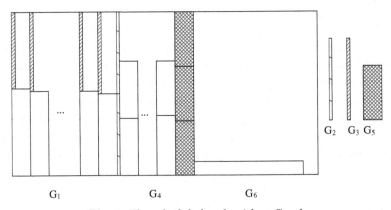

G_2 G_3 G_5

G_1 G_4 G_6

Fig. 4. The schedule by algorithm *Greedy*

However, we can schedule jobs of G_1, G_4 and G_6 by time $N + 1/N$. Each two of the jobs of size $1/N$ (G_2 and G_3 jobs) can be processed at the same time. The three jobs of G_5 are scheduled together at time $N + 2/N + 1$. It implies that $C^* \leq N + 2/N + 2$ (see Fig. 5). Thus,

$$C_{Greedy}/C^* \geq 5/2,$$

as k tends to infinity. □

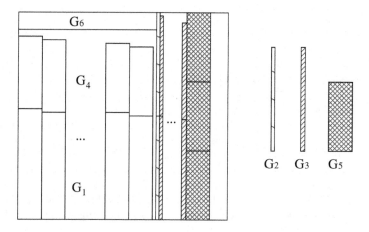

Fig. 5. An optimal schedule

The following *semi* greedy algorithm can be shown to have a competitive ratio of 2.5. Due to the page limit we omit the proof.

- Schedule the big jobs (of size larger than $m/2$) one after another. Let b be the largest completion time among these jobs.
- Schedule the medium jobs (of size at most $m/2$ but larger than $m/3$) from time b with algorithm *Greedy*.
- Schedule the small jobs (of size at most $m/3$) with algorithm *Greedy*.

Note that this algorithm differs algorithm *Greedy* in that it does not schedule medium jobs together with big jobs even there are a sufficient number of available machines.

Final Remarks. There is still a big gap between the upper bound and the lower bound for the general case. Although we feel that the lower bound 2.25 [8] is far away from the tight bound it seems quite difficult to improve it. We can construct a bad example for any on-line algorithm but not able to find a good bound due to the high computation time cost. It might be possible to improve our upper bound by refining the algorithm DW. But the analysis will get much more complicated. The most interesting question is the average case analysis for the problem. In this sense the greedy algorithm would perform much better.

References

1. S. Bischof and E.W. Mayr, On-line scheduling of parallel jobs with runtime restrictions, *Theoretical Computer Science* 268, 67-90, 2001.
2. M. Drozdowski, Scheduling multiprocessor tasks – an overview, *European Journal on Operations Research* 94, 215-230, 1996.
3. J. Du and J. Leung, Complexity of Scheduling Parallel Task System, *SIAM Journal on Discrete Mathematics* 2, 473-487, 1989.

4. A. Feldmann, M.-Y. Kao, J. Sgall, and S.-H. Teng, Optimal online scheduling of parallel jobs with dependencies, *Journal of Combinatorial Optimization* 1, 393-411, 1998.

5. A. Feldmann, J. Sgall, and S.-H. Teng, Dynamic scheduling on parallel machines, *Theoretical Computer Science* 130, 49-72, 1994.

6. R. Fleischer and M. Wahl, Online scheduling revisited, *Proceedings of the 8th Annual European Symposium on Algorithms* (ESA), 202-210, 2000.

7. T. Gormley, N. Reingold, E. Torng, and J. Westbrook, Generating adversaries for request-answer games, *Proceedings of the 11th ACM-SIAM Symposium on Discrete Algorithms* (SODA), 564-565, 2000.

8. B. Johannes, Scheduling parallel jobs to minimize makespan, Technical Report 723-2001, TU Berlin. http://www.math.tu-berlin.de/coga/publications/techreports/

9. D.S. Johnson, A. Demers, J.D. Ullman, M.R. Garey, and R.L. Graham. Worst-case performance bounds for simple one-dimensional packing algorithm, *SIAM Journal on Computing* 3, 299-325, 1974.

10. W. Ludwig and P. Tiwari, Scheduling malleable and nonmalleable parallel tasks, *Proceedings of the 5th ACM-SIAM Symposium on Discrete Algorithms (SODA)*, 167-176, 1994.

11. E. Naroska and U. Schwiegelshohn, On an on-line scheduling problem for parallel jobs, *Information Processing Letters* 81, 297-304, 2002.

12. D. Simchi-Levi, New worst-case results for the bin packing problem, *Naval Research Logistics* 41, 79-585, 1994.

13. G. Zhang, Absolute worst-case analysis for on-line bin packing, manuscript, 2002.

The Range Assignment Problem in Static Ad-Hoc Networks on Metric Spaces

Deshi Ye[1,2,*] and Hu Zhang[2,**]

[1] Department of Mathematics, Zhejiang University, Hangzhou 310027, China
[2] Institute of Computer Science and Applied Mathematics, University of Kiel,
Olshausenstraße 40, D-24098 Kiel, Germany
{dye,hzh}@informatik.uni-kiel.de

Abstract. In this paper we study the range assignment problem in static ad-hoc networks on metric spaces. We consider the h-strong connectivity and h-broadcast problems on trees, high dimensional Euclidean spaces and general finite metric spaces. Both homogeneous and non-homogeneous cases are explored. We show that the h-broadcast problem is polynomial solvable on trees and present an $O(n^2)$-approximation algorithm for the h-strong connectivity problem on trees, where n is the number of stations. Furthermore, we propose a probabilistic $O(\log n \log \log n)$-approximation algorithm for the h-broadcast problem and a probabilistic $O(n^2 \log n \log \log n)$-approximation algorithm for the h-strong connectivity problem on high dimensional Euclidean spaces and general metric spaces. In the case of high dimensional real normed spaces, if the distance-power gradient $\alpha \leq 1 + O(\log \log \log n / \log \log n)$, an $O(\log^\alpha n)$-approximation algorithm and an $O(n^2 \log^\alpha n)$-approximation algorithm are developed for the h-broadcast problem and the h-strong connectivity problem, respectively. They are the first algorithms for the range assignment problem in static ad-hoc networks on general metric spaces. And the approximation ratio of $O(\log n \log \log n)$ for the h-broadcast problem on general metric spaces is close to the known lower bound $O(\log n)$ [19].

1 Introduction

Nowadays the *ad-hoc wireless network* based on *multi-hop* plays an increasingly important role in communication [17]. In *static* ad-hoc networks, the locations of stations are fixed. Typical applications of the static ad-hoc network model include emergencies, battlefield, monitoring remote geographical region and so on. Two of the main benefits of multi-hop ad-hoc networks are the reduction of the power consumption and the ability to vary the power used in the transmission such that the interference problem can be avoided. Minimizing power

* Supported in part by a DAAD Sandwich Project and NSFC(10231060).
** Supported in part by the DFG Graduiertenkolleg 357, Effiziente Algorithmen und Mehrskalenmethoden, by EU Thematic Network APPOL, Approximation and On-line Algorithms for Optimization Problems, IST-2001-32007, and by EU Project CRESCCO, Critical Resource Sharing for Cooperation in Complex Systems, IST-2001-33135.

R. Královič and O. Sýkora (Eds.): SIROCCO 2004, LNCS 3104, pp. 291–302, 2004.
© Springer-Verlag Berlin Heidelberg 2004

consumption is crucial on such networks, since the wireless devices are portable and benefit only of limited power resources. And the range assignment problem is a key issue to minimize the power consumption, which has been extensively studied in wireless network theory.

The problem is defined as follows. We are given a set S of stations (radio transmitter/receivers) on a metric space, which can communicate with each other by sending/receiving radio signals. In general message communication happens via multi-hop transmission, i.e., a message is sent to its destination through some intermediate stations. A range assignment for S is a function $r : S \to \mathbb{R}^+$, which yields a directed communication graph $G_r = (S, E)$, such that for each pair of stations u and v there exists a directed edge $(u, v) \in E$ if and only if v is at a distance at most $r(u)$ from u. The cost (overall power consumption) of a range assignment r is defined as $cost(r) = \sum_{v \in S} c(v)(r(v))^\alpha$, where $c(v)(r(v))^\alpha$ is the cost of station v and $c(v)$ is an coefficient depending on the station. When $c(v)$ is a constant for all $v \in S$, we call it the homogeneous case. Otherwise it is the non-homogeneous case. The *distance-power gradient* α is usually a real number in $[1, 6]$ in practice. The goal of the problem is to find a range assignment r for S such that G_r satisfies a given property Π and $cost(r)$ is minimized. The detailed description of the problem can be found in [19].

Usually the property Π is one of the follows:

- *h-strong connectivity*: from every station s to any other t, G_r must contain a directed path of length at most h.
- *h-broadcast*: G_r must contain a directed source spanning tree rooted at a given source s with depth at most h.

Previous researches mainly focus on the problems defined on one dimensional spaces, i.e., networks that can be modelled as sets of stations located along a straight line. Polynomial time algorithms by dynamic programming were addressed for both homogeneous and non-homogeneous cases for h-broadcast problems on one dimensional spaces in [16, 10, 1]. For the h-strong connectivity problem an algorithm of dynamic programming for homogeneous case to obtain an optimal solution running in $O(n^4)$ time was presented in [16] when $h = n - 1$ is fixed. Unfortunately, for general $h \in \{1, \ldots, n - 1\}$ it is still open whether the h-strong connectivity problem on one dimensional spaces can be solved in polynomially. Clementi et al. in [7] provided a 2-approximation algorithm for any $h \geq 1$ for homogeneous case, which runs in $O(hn^3)$ time. There are only few results in the case of more general spaces. When $h = n - 1$ is fixed, minimum spanning tree based algorithms can deliver a solution with constant approximation ratio for the broadcast problem [6], and the approximation ratio is 2 for the strong connectivity problem [16] for the case of high dimensional Euclidean spaces. In [9], the h-strong connectivity problem was proved in $\mathcal{Av}$-$\mathcal{APX}$ for any fixed $h > 0$ and the problem is $\mathcal{APX}$-hard on d dimensional Euclidean spaces for $d \geq 3$. The lower bound for the h-broadcast problem on general metric spaces is $O(\log n)$ [19]. Up to our knowledge, there is no approximation algorithm for the range assignment problem on general spaces with a general $h \geq 1$ with a bounded or input size dependent ratio. Some results about restricted version on high dimensions can be found in [8, 19].

However, in real application the problem with $h = n-1$ is not practical as too many hops can reduce the quality of communication and also be hard to control. Thus we need to consider the general case that multi-hop $h \in \{1, \ldots, n-1\}$. In addition, the problem on only one dimensional spaces is far from the reality. Therefore we study the case of high dimensional Euclidean spaces. As far as we know, there was no bounded result for the general multi-hop h on high dimensional Euclidean spaces. Due to the wide applications of ad-hoc networks, it contains also stations in space or in ocean. In this case the model of Euclidean space is not applicable as the transmission distance (depending on environmental condition) is not equivalent to the Euclidean distance. Hence the model of general finite metric space must be introduced.

In this paper, we first study the h-broadcast and h-strong connectivity problems on trees for both homogeneous and non-homogeneous cases. We show that an optimal solution to the h-broadcast problem can be obtained in polynomial time by dynamic programming. For the h-strong connectivity problem, an $O(n^2)$-approximation algorithm is proposed. Based on these algorithms, we study the problems on high dimensional Euclidean spaces and general metric spaces. A probabilistic $O(\log n \log \log n)$-approximation algorithm and a probabilistic $O(n^2 \log n \log \log n)$-approximation algorithm are presented for the h-broadcast problem and the h-strong connectivity problem, respectively. The ratios are further improved for special cases. The main technique we use here is the algorithm to approximate a finite metric space by a collection of tree metrics.

1.1 Approximating Metrics

Given a finite metric space induced by an undirected unweighted or weighted graph, we need to embed it in a simpler metric space such that the distance between each pair of vertices is approximately preserved. With this technique, some hard problems can be solved approximately because an arbitrary metric space may not have enough structure for exploring.

In [2], Bartal proposed the idea of *probabilistic approximation* of metric spaces by a set of simpler metric spaces. Given a graph $G = (V, E)$, let M be a finite metric space defined on V and d_M the distance function. For two metric spaces M_1 and M_2 defined on V, we call M_1 *dominates* M_2 if $d_{M_1}(u,v) \geq d_{M_2}(u,v)$ for any pair $u, v \in V$. Suppose $\mathcal{S}$ is a collection of metric spaces on V. Assuming that each metric space in $\mathcal{S}$ dominates M, $\mathcal{S}$ is defined to ρ-*probabilistically approximate* M, if there is a probability distribution μ over $\mathcal{S}$ such that the expected distance distortion between any vertex pair in M in a space chosen from $\mathcal{S}$ according to μ is at most ρ. A polynomial time algorithm to $O(\log^2 n)$-probabilistically approximate any metric space on $|V| = n$ vertices by a collection of tree metrics was addressed in [2]. The approximation ratio was improved to $O(\log n \log \log n)$ in [3]. However, the numbers of the tree metric spaces are exponentially large in both algorithms. Charikar et al. [5] developed a polynomial time algorithm to construct a probability distribution on a set of $O(n \log n)$ trees metrics for any given metric space induced by a (weighted) graph G on n vertices, such that the expected stretch of each edge is no more than $O(\log n \log \log n)$.

To decide the probability distribution μ, a linear program with exponential number of variables has to be solved. Indeed this linear program is a packing problem, which can be solved approximately by some fast algorithms [5, 15, 23]. We notice that in the algorithms above the triangle inequality is not required. Therefore an instance of the problem on a general metric space can be reduced to the case on a tree with a stretch at most $O(\log n \log \log n)$ (See Section 3 and 4).

1.2   Approximation Algorithms for the Packing Problem

The algorithms for the *packing problem* are employed to compute the probability distribution μ over the collection of trees described in Subsection 1.1. The packing problem is to compute $x^* \in B$ such that $\lambda^* = \lambda(x^*) = \min\{\lambda | f(x) \le \lambda \cdot \mathbf{1}, x \in B\}$, where $f : B \to \mathbb{R}^M_+$ is a vector of M continuous convex functions defined on a nonempty convex compact set $B \subseteq \mathbb{R}^N$, and $\mathbf{1}$ is the vector of all ones. The functions f_m, $1 \le m \le M$, are the packing constraints. $x = (x_1, \dots, x_N) \in B$. Then the approximate packing problem is to compute $x \in B$ such that $f(x) \le c(1 + \varepsilon)\lambda^* \cdot \mathbf{1}$, where $\varepsilon \ge 0$ is the given relative error tolerance and $c \ge 1$ is the approximation ratio of the *block solver*, which is a given oracle to solve the *block problem*: to compute $\hat{x} = \hat{x}(y) \in B$ such that $y^T f(\hat{x}) \le c(1 + t) \min\{y^T f(z) | z \in B\}$, where the *price vector* $y \in P = \{y \in \mathbb{R}^M | \sum_{m=1}^M y_m = 1, y_m \ge 0\}$.

There are many approximation algorithms for the packing problem [12, 13, 15, 18, 20, 22]. However, here the linear program of approximating metrics has only a weak block solver, i.e., $c > 1$. In such a case only the algorithms in [15] and [18] (with generalization in [5]) can be employed. The numbers of iterations are bounded by $O(M(\log M + \varepsilon^{-2} \log \varepsilon^{-1}))$ and $O(\varepsilon^{-2}\rho \log(M\varepsilon^{-1}))$, respectively, where $\rho = \max_{1 \le m \le M} \max_{x \in B} f_m(x)$ is the *width* of B. Here we will use the algorithms for the packing problem in [15, 18] to approximate a general metric space by tree metrics in Section 3 and 4 as [5, 23].

The paper is organized as follows. In Section 2 the (approximation) algorithms for the h-broadcast problem and the h-strong connectivity problem are presented and analyzed. Then the cases of high dimensional Euclidean spaces and general finite metric spaces are studied in Section 3 and 4.

2   Algorithms for Problems on Trees

We will explore the range assignment problem on edge weighted trees in this section. Both homogeneous and non-homogeneous cases are investigated. For the term distance $d(i, j)$ we mean the shortest path distance between station i and station j. We will consider the h-broadcast problem and the h-strong connectivity problem separately. Without loss of generality, we only study the non-homogeneous case and the results can be directly applied in the homogeneous case. The main idea we use is dynamic programming, similar to [16, 1].

2.1   The h-Broadcast Problem

We use similar definitions as in [1] here. Suppose that the set of stations S is connected by a tree T. Let $s \in S = \{1, 2, \ldots, n\}$ be the source station. Given a specified pair i and j, $1 \le i \le j \le n$, three sets $S_L^{i,j}, S_M^{i,j}, S_R^{i,j}$ are defined as follows, for any station $p \in S$,

$$p \in \begin{cases} S_L^{i,j}, \text{ if } d(i, p) \le d(i, j) \text{ and } d(j, p) > d(i, j); \\ S_M^{i,j}, \text{ if } d(j, p) \le d(i, j) \text{ and } d(i, p) \le d(i, j); \\ S_R^{i,j}, \text{ otherwise.} \end{cases}$$

We consider two sets $A, B \in S = S_L^{i,j} \cup S_M^{i,j} \cup S_R^{i,j}$ and $A \cap B = \emptyset$. Let $r_{A,B}$ be a minimum cost range assigned to stations in A such that for any station $b \in B$, there exists a station $a \in A$ with $r_{A,B}(a) \ge d(a, b)$, where $r_{A,B}(a)$ is the range assigned to a and $d(a, b)$ is the distance between a and b in T. Denote by $cost(A, B)$ the cost of $r(A, B)$ and by $cov(A, B)$ the set of stations such that for any $u \in cov(A, B)$, $r_{A,B}(a) \ge d(a, u)$ for some $a \in A$. Finally, we denote $MSB_h^{i,j}(A, B)$ to be any optimal multi-source broadcast where the source set is A and the target set is B, in which every station in B is connected to some station in A with at most h hops, $h \in \{1, 2, \cdots, n - 1\}$. Let $M_h^{i,j}(A)$ be the cost of $MSB_h^{i,j}(A, S - A)$, for any $A \subseteq S$. From above definitions, dynamic programming yields

$$M_h^{i,j}(S_M^{i,j}) = \begin{cases} 0, & S_L^{i,j} = S_R^{i,j} = \emptyset; \\ \min_{u \notin S_M^{i,j}} \{cost(S_M^{i,j}, \{u\}) + M_{h-1}^{i,j}(S_O) \\ \quad |S_O = S_M^{i,j} \cup cov(S_M^{i,j}, \{u\})\}, & \text{otherwise,} \end{cases}$$

for all $S_M^{i,j}$ according to all pairs of i and j, and $h \in \{1, \ldots, n - 1\}$.

Theorem 1. *The h-broadcast problem on trees can be solved in $O(hn^4)$ time.*

Proof. The correctness holds with the same argument as [16, 1]. We need to compute all tables $M_1^{i,j}, \ldots, M_h^{i,j}$ for all pairs $i, j \in S$. Table $M_1^{i,j}$ is directly computed at first. Then by dynamic programming we can compute all $cost(S_M^{i,j}, \{u\})$ in $O(n^2)$ time according to pair of i and j. With the pre-computed cost and $M_{l-1}^{i,j}$, we are able to obtain $M_l^{i,j}$, where $l \in \{1, \ldots, h\}$. This procedure takes $O(n^2)$ time. Overall the algorithm runs in $O(hn^4)$ time and returns the optimal solution $M_h^{s,s}(S_M^{s,s})$.

2.2   The h-Strong Connectivity Problem

For the station set $S = \{1, 2, \cdots, n\}$ connected by a tree T, we define a *path set* $P(i, j)$ as the stations in the path from i to j, for any $1 \le i, j \le n$. Notice that

$|\{P(i,j)|i,j \in S\}| = O(n^2)$. For any path set $P(i,j)$, the stations in it can be regarded as aligned on a line. Therefore for a given h, the algorithm in [7] for the homogeneous h-strong connectivity problem can be applied here and generate a range assignment for $P(i,j)$ with a cost at most twice of the optimum. This algorithm holds also for the non-homogeneous case [1]. For each $l \in \{1, \ldots, h\}$ and each pair $1 \le i, j \le n$, we can compute a range assignment for all $P(i,j)$ by the algorithm in [7] for one dimensional case by dynamic programming. This generates at most $O(n^2)$ different tables, in which every table corresponds to a 2-approximate range assignment $r_{P(i,j)}$ for a specific path set $P(i,j)$.

Now we consider any station $u \in S$. Obviously u is assigned at most $O(n^2)$ ranges because u is in at most $O(n^2)$ path sets. Among them we assign the maximum range in the $O(n^2)$ candidates to u. Denote by r_S the range assignment over S by the above rule. Moreover, we define r_S^* to be the optimal range assignment over S.

Theorem 2. *There is an $O(n^2)$-approximation algorithm for the h-strong connectivity problem on trees with a running time bounded by $O(hn^5)$.*

Proof. Define $r^*(u)$ the range assigned to $u \in S$ in the optimal solution for S. For a fixed pair i and j, it is obvious that $cost(r_{P(i,j)}) \le 2 \sum_{u \in P(i,j)} cost(r^*(u))$ as $r_{P(i,j)}$ is a 2-approximate range assignment for $P(i,j)$. As showed before, there are at most $O(n^2)$ path sets. Hence $cost_{sum} = \sum_i \sum_j cost(r_{P(i,j)}) \le 2 \sum_i \sum_j \sum_{u \in P(i,j)} cost(r^*(u)) \le O(n^2)cost(r_S^*)$. Since in our algorithm, for every station we just take only the maximum range among the solutions over paths containing that station, it holds that $cost(r_S) = \sum_u \max_{u \in P(i,j)} cost(r_{P(i,j)}) \le cost_{sum}$. Thus $cost(r_S) \le O(n^2)cost(r_S^*)$.

As for the running time, there are at most $O(n^2)$ tables. For each table the 2-approximation range assignment can be computed by dynamic programming in at most $O(hn^3)$ time [7]. Therefore the total running time is $O(hn^5)$.

3   Algorithms for Problems on High Dimensional Euclidean Spaces

We now turn to the d-dimensional Euclidean spaces, where $d \ge 2$. We will develop approximation algorithms for the h-broadcast problem and the h-strong connectivity problem based on the (approximation) algorithms on trees in Section 2.

For a given station set S, we also denote by $S \in \mathbb{R}^d$ the space induced by a complete graph defined on station set S with distance function in $\mathbb{R}^d$. The two-phase algorithm $\mathcal{ALG}_\mathcal{E}$ is as follows: In the first phase, a collection $\mathcal{T}$ of N tree metrics spanning S and a probability distribution μ on $\mathcal{T}$ are generated. Then we select a tree metric $T_i \in \mathcal{T}$ according to its probability x_i. In the second phase, the algorithms for the h-broadcast problem and the h-strong connectivity problem in Section 2 are applied for the corresponding problems on the selected tree metric T_i. In this way we are able to obtain an approximate solution for the range assignment problem.

With the analysis in Section 2, for the selected tree metric T_i with a distance function d_{T_i}, an optimal solution for the h-broadcast problem and an $O(n^2)$-approximate solution for the h-strong connectivity problem can be found, where $n = |S|$. Thus we will focus on the first phase of $\mathcal{ALG_E}$, i.e., how to construct the tree metrics in $\mathcal{T}$ and the probability distribution μ.

We first define an edge weighted complete graph M_S on the vertex set S. For each pair of vertices $u, v \in S$, the length of edge (u, v) is $l_{M_S}(u, v) = (d_S(u, v))^\alpha$, where d_S is the distance function in original space S. It is obvious that the shortest path distance between u and v is bounded by the length of edge (u, v) in M_S, i.e., $d_{M_S}(u, v) \leq l_{M_S}(u, v)$. In this case the cost that station u can cover v is $cost(u, v) = d_{M_S}(u, v) = (d_S(u, v))^\alpha$.

Then based on the algorithm in [5] we can find the collection $\mathcal{T}$, in which any tree T_i, $i \in \{1, \ldots, N\}$, induces a metric space that dominates M_S, i.e., for any two stations u and v, $d_{T_i}(u, v) \geq d_{M_S}(u, v)$. The tree metrics in $\mathcal{T}$ will be employed to approximate the graph M_S. To generate a probability distribution over $\mathcal{T}$, we can assign each $T_i \in \mathcal{T}$ a real number $x_i \in [0, 1]$. Denoting by λ the maximum stretch over all edges, simple observation yields the following linear program:

$$\min \lambda$$
$$\sum_{i=1}^{N} d_{T_i}(e)x_i \leq \lambda d_{M_S}(e), \quad \text{for any edge } e;$$
$$\sum_{i=1}^{N} x_i = 1;$$
$$x_i \geq 0. \tag{1}$$

Here the first set of constraints indicates that the expected length of the paths in tree metrics in $\mathcal{T}$ corresponding to edge $e \in M_S$ is no more than the maximum stretch of the length. And the other constraints are directly from the definition of probability distribution. This problem can be formulated as a packing problem described in Subsection 1.2. Theoretically a linear program can be solved efficiently. However, here the number of variables in (1) (i.e., the number of trees in $\mathcal{T}$) is unbounded and can be exponentially large.

We use the algorithms in [15, 18] to solve (1) approximately. Only a polynomial number of tree metrics will be generated (assigned with non-zero probability). In the algorithms an initial solution is set at the beginning. Then the algorithms run the iterative procedure. In each iteration a pair of solutions to the packing problem and its dual problem is computed. There are following three steps in one iteration: First with the known solution, a price vector y related to the dual value is calculated to determine the direction of moving of the iterate. Then a block solver is called to generate an (approximate) block solution for y and the error tolerance ε. Finally a new solution is obtained as a linear combination of the known solution and block solution with an appropriate step length. When certain stopping rules are satisfied (which indicates that the duality gap is small enough to fulfil the accuracy requirement), the iterative procedure terminates and the desired approximate solution is returned by the last iteration. Details of the algorithm can be found in [15, 23].

In order to complete the algorithm for (1), we need to consider the block problem, which is related to the dual problem. As showed in Subsection 1.2, given a price vector $y \in P = \{(y_1, \ldots, y_m) | \sum_{e=1}^{m} y_e = 1, y_e \geq 0\}$, the block problem is to find an $\hat{x}$ such that $y^T f(\hat{x}) = \min_{x \in B} y^T f(x)$. By the definitions, the right hand side can be simplified as follows:

$$\min_{x \in B} y^T f(x) = \min_{x \in B} \sum_{e=1}^{m} \left(y_e \cdot \sum_{i=1}^{N} \frac{d_{T_i}(e)}{d_{M_S}(e)} x_i \right)$$

$$= \min_{x \in B} \sum_{i=1}^{N} \left(x_i \cdot \sum_{e=1}^{m} y_e \frac{d_{T_i}(e)}{d_{M_S}(e)} \right) = \min_{1 \leq i \leq N} \sum_{e=1}^{m} \frac{y_e}{d_{M_S}(e)} d_{T_i}(e).$$

The last equality holds because we can choose one tree T_i with smallest value of $\sum_e y_e d_{T_i}(e)/d_{M_S}(e)$ and set its probability $x_i = 1$ to achieve the minimum. Denote by $w(e) = y_e/d_{M_S}(e)$ the weight on edge e. Therefore the goal of the block problem is to find a tree T connecting all vertices in M_S such that the value $\sum_e w(e) d_T(e)$ is minimized, for all edge e in M_S. This problem is in fact the minimum communication cost spanning tree (**MCCST**) problem.

However, the **MCCST** problem is $\mathcal{NP}$-hard [11, 14, 21]. The best deterministic polynomial time approximation algorithm is addressed in [3] and [4] independently. And the approximation ratio is $O(\log n \log \log n)$. The worst case stretch of any edge is bounded by $O(n)$.

To solve (1), we have to call a solver of the **MCCST** problem as an oracle. Here it is worth noting that the approximation ratio of the block solver is $c = O(\log n \log \log n)$, i.e., a weak block solver. As a result, the approximation ratio of the solution to (1) is also $O(\log n \log \log n)$. Hence many approximation algorithms for the packing problem are not applicable. The approximation algorithm for the packing problem in [18] was generalized to the case of weak block solver in [5]. The number of iterations is bounded by $O(\varepsilon^{-2} \rho \log(m\varepsilon^{-1}))$ for any given error tolerance $\varepsilon \in (0, 1]$, where $\rho = \max_e \max_{x \in B} \sum_i d_{T_i}(e) x_i / d_{M_S}(e)$ as in Subsection 1.2. In the problem of approximating metric spaces by tree metrics, $\rho = O(n)$. Hence totally at most $O(n\varepsilon^{-2} \log(m\varepsilon^{-1}))$ iterations are required to obtain an $O(\log n \log \log n)$-approximate solution to (1) and in each iteration an $O(\log n \log \log n)$-approximate **MCCST** solver is called once and one tree metric in $\mathcal{T}$ is generated and assigned a positive probability. Therefore when algorithm halts there are $N = |\mathcal{T}| = O(n\varepsilon^{-2} \log(m\varepsilon^{-1}))$ tree metrics. If the algorithm in [3] is applied, the number of trees in $\mathcal{T}$ will be exponentially large.

Finally, in the second phase of the algorithm $\mathcal{ALG}_\varepsilon$, approaches on tree metrics in Section 2 are employed. Notice that the distance of tree metrics in $\mathcal{T}$ in fact corresponds to the cost of original Euclidean space S, and recall that the approximation ratio of the first phase is $O(\log n \log \log n)$. With the property of the solution in the second phase, denoting by β the running time of the **MCCST** solver, we have the following theorems:

Theorem 3. *There exists a probabilistic $O(\log n \log \log n)$-approximation algorithm for the h-broadcast range assignment problem in static ad-hoc networks on d dimensional Euclidean spaces for $d \geq 2$, with a running time $O(hn^4 + n\beta \log n)$.*

Theorem 4. *There exists a probabilistic $O(n^2 \log n \log \log n)$-approximation algorithm for the h-strong connectivity range assignment problem in static ad-hoc networks on d dimensional Euclidean spaces for $d \geq 2$, with a running time $O(hn^5 + n\beta \log n)$.*

The algorithm in [15, 23] can be also applied here with the same approximation ratio but a different bound on running time.

3.1 Improved Algorithms for the Case of Small α

We notice that in [5] a probabilistic $O(\sqrt{d} \log n)$-approximation algorithm for the problem of approximating metrics by tree metrics on d dimensional Euclidean spaces is addressed. In fact here we can apply this algorithm in the case of small α. Since in real applications, the problems are mostly defined on two or three dimensional Euclidean spaces. Therefore in this subsection we consider d as a constant.

If the distance-power gradient $\alpha \leq 1 + O(\log \log \log n / \log \log n)$, then we do not define the complete graph M_S and in the linear program (1), $d_{M_S}(e)$ is replaced by $d_S(e)$. Afterwards we generate the collection $\mathcal{T}$ of trees to approximate S directly. In the first phase we apply the algorithm for real normed spaces in [5] as a weak block solver. And the the stretch of any edge in S is at most $O(\log n)$. Thus in the first phase of the algorithm the cost increases by a factor at most $O(\log^\alpha n)$. We have

$$\log^\alpha n \leq (\log n)^{1+O(\log \log \log n / \log \log n)} = \log n (\log n)^{O(\log \log \log n / \log \log n)}$$
$$= \log n (\log n)^{O(\log_{\log n} \log \log n)} = O(\log n \log \log n).$$

Thus the approximation ratio of the algorithm is $O(\log^\alpha n)$, which is not more than $O(\log n \log \log n)$. For this special case we have the following corollaries:

Corollary 1. *There exists a probabilistic $O(\log^\alpha n)$-approximation algorithm for the h-broadcast range assignment problem in static ad-hoc networks on d dimensional Euclidean spaces for $d \geq 2$, with a running time $O(hn^4 + n\beta \log n)$, for $\alpha \leq 1 + O(\log \log \log n / \log \log n)$.*

Corollary 2. *There exists a probabilistic $O(n^2 \log^\alpha n)$-approximation algorithm for the h-strong connectivity range assignment problem in static ad-hoc networks on d dimensional Euclidean spaces for $d \geq 2$, with a running time $O(hn^5 + n\beta \log n)$, for $\alpha \leq 1 + O(\log \log \log n / \log \log n)$.*

In fact the case $\alpha \leq 1 + O(\log \log \log n / \log \log n)$ does exist. If $n = 256$, then $\log \log \log n / \log \log n = \log 3/3 \approx 0.53$. Then $1 + \log \log \log n / \log \log n \approx 1.53$ We believe that this algorithm can find many real applications.

4   Algorithms for Problems on Finite Metric Spaces

We study the approximation algorithms for the range assignment problem in static ad-hoc networks on any finite metric spaces in this section. Assume that

$G = (S, E)$ is a metric space defined on the set of stations S and E is the set of edges. The length function is defined by $l_G : E \rightarrow R^+ \cup \{0\}$. The idea to solve this problem is essentially the same as Section 3 with a different distance function in $T_i \in \mathcal{T}$, $i = 1, \ldots, N$. Thus we develop the algorithm $\mathcal{ALG}_\mathcal{G}$ for the general metric spaces.

Similarly, we focus on the first phase. We also define a complete graph M_G in advance, which is induced by S and the length of edge (u, v) is defined as $l_{M_G}(u, v) = (d_G(u, v))^\alpha$ (here $d_G(u, v)$ is the metric distance in G). And the shortest path distance between the pair u and v is $d_{M_G}(u, v) \leq l_{M_G}(u, v)$. Hence the cost that the station u can cover v is $c(u, v) = d_{M_G}(u, v) \leq (d_G(u, v))^\alpha$. Thus we obtain a similar linear program to (1) for the probability distribution. The only difference is that the first set of constraints are replaced by

$$\sum_{i=1}^{N} d_{T_i}(e)x_i \leq \lambda d_{M_G}(e), \quad \text{for any edge } e. \tag{2}$$

Then we can also use the approximation algorithm in [5] to solve this linear program. Denoting by weight function $w(e) = y_e/d_{M_G}(e)$, the block problem is also the **MCCST** problem but with a different weight function. The algorithms in [3, 4] are applied as an $O(\log n \log \log n)$-approximate block solver. With the similar arguments in Section 3 we have the following theorems:

Theorem 5. *There exists a probabilistic $O(\log n \log \log n)$-approximation algorithm for the h-broadcast range assignment problem in static ad-hoc networks on finite metric spaces with a running time $O(hn^4 + n\beta \log n)$.*

Theorem 6. *There exists a probabilistic $O(n^2 \log n \log \log n)$-approximation algorithm for the h-strong connectivity range assignment problem in static ad-hoc networks on finite metric spaces with a running time $O(hn^5 + n\beta \log n)$.*

The algorithm in [23] can be also applied here with a number of iterations bounded by $O(n^2(\log n + \varepsilon^{-2} \log \varepsilon^{-1}))$. We notice that in [19] a lower bound $O(\log n)$ for the h-broadcast problem on general metric spaces is proved, which only differs from our result by a factor of $O(\log \log n)$.

4.1   Better Algorithms for the Case of Small α and Real Normed Spaces

We can still use the $O(\log n)$-approximation algorithm in [5] for real normed spaces here to generate the collection $\mathcal{T}$ as the tree metrics are used to approximate the original space G. In fact the algorithm is employed here for the case of $\alpha = 1 + O(\log \log \log n/ \log \log n)$, similar to Subsection 3.1. With the same arguments the following corollaries hold:

Corollary 3. *If $\alpha \leq 1 + O(\log \log \log n/ \log \log n)$, then there exists a probabilistic $O(\log^\alpha n)$-approximation algorithm for the h-broadcast range assignment problem in static ad-hoc networks on real normed spaces with a running time $O(hn^4 + n\beta \log n)$.*

Corollary 4. *If $\alpha \leq 1+O(\log\log\log n/\log\log n)$, then there exists a probabilistic $O(n^2 \log^{\alpha} n)$-approximation algorithm for the h-strong connectivity range assignment problem in static ad-hoc networks on real normed spaces with a running time $O(hn^5 + n\beta \log n)$.*

5 Concluding Remarks

In this paper we have studied the range assignment problem in static multi-hop ad-hoc networks on trees, high dimensional Euclidean spaces and general finite metric spaces. We proposed (approximation) algorithms for these problem in both the homogeneous and the non-homogeneous cases. And the approximation ratio $O(\log n \log \log n)$ for the h-broadcast problem on general metric spaces is close to the known lower bound $O(\log n)$ [19]. We also presented improved approximation algorithms for both problems when α (the distance-power gradient) is small. There are some interesting open problems:

1. Is the h-strong connectivity problem on one dimensional spaces polynomial time solvable?
2. Can we obtain an $O(\log n)$-approximation algorithm for the h-broadcast problem on general metric spaces to reach the lower bound?
3. Can we improve the result of the h-strong connectivity problem such that the factor $O(n^2)$ in the approximation ratio is eliminated?

References

1. C. Ambuehl, A. E. F. Clementi, M. D. Ianni, A. Monti, G. Rossi and R. Silvestri The Range Assignment Problem in Non-Homogeneous Static Ad-Hoc Networks, Proceedings of 4th International Workshop on Algorithms for Wireless, Mobile, Ad Hoc and Sensor Networks, WMAN 2004.
2. Y. Bartal, Probabilistic approximation of metric spaces and its algorithmic applications, Proceedings of the 37th IEEE Annual Symposium on Foundations of Computer Science, FOCS 1996, 184-193.
3. Y. Bartal, On approximating arbitrary metrics by tree metrics, Proceedings of the 30th Annual ACM Symposium on Theory of Computing, STOC 1998.
4. M. Charikar, C. Chekuri, A. Goel and S. Guha, Rounding via trees: deterministic approximation algorithms for group steiner trees and k-median, Proceedings of the 30th Annual ACM Symposium on Theory of Computing, STOC 1998.
5. M. Charikar, C. Chekuri, A. Goel, S. Guha and S. Plotkin, Approximating a finite metric by a small number of tree metrics, Proceedings of the 39th Annual IEEE Symposium on Foundations of Computer Science, FOCS 1998, 379-388.
6. A. E. F. Clementi, P. Crescenzi, P. Penna, G. Rossi and P. Vocca, On the Complexity of Computing Minimum Energy Consumption Broadcast Subgraph. In Proceedings of 18th Annual Symposium on Theoretical Aspect of Computer Science, STACS 2001, LNCS 2010, 121-131.
7. A. E. F. Clementi, A. Ferreira, P. Penna, S. Perennes, R. Silvestri, The Minimum Range Assignment Problem on Linear Radio Networks, Algorithmica, 35(2) 2003, 95-110.

8. A. E. F. Clementi, G. Huiban, P. Penna, G. Rossi and Y. C. Verhoeven, Some Recent Theoretical Advances and Open Questions on Energy Consumption in Ad-Hoc Wireless Networks. ARACNE 2002, 23-38.

9. A. E. F. Clementi, P. Penna, and R. Silvestri, On the power assignment problem in radio networks, Technical Report TR00-054, Electronic Colloquium on Computational Complexity(ECCC), 2000.

10. A. E. F. Clementi, M. D. Ianni and R. Silvestri, The minimum broadcast range assignment problem on linear multi-hop wireless networks, Theoretical Computer Science, 1-3 (299) 2003, 751-761.

11. M. Garey and D. Johnson, Computer and Intractability: A Guide to the Theory of NP-Completeness, W. H. Freeman and Company, NY, 1979.

12. N. Garg and J. Könemann, Fast and simpler algorithms for multicommodity flow and other fractional packing problems, Proceedings of the 39th IEEE Annual Symposium on Foundations of Computer Science, FOCS 1998, 300-309.

13. M. D. Grigoriadis and L. G. Khachiyan, Coordination complexity of parallel price-directive decomposition, Mathematics of Operations Research, 2 (1996), 321-340.

14. T. C. Hu, Optimum communication spanning trees, SIAM Journal on Computing, 3 (1974), 188-195.

15. K. Jansen and H. Zhang, Approximation algorithms for general packing problems with modified logarithmic potential function, Proceedings of 2nd IFIP International Conference on Theoretical Computer Science, TCS 2002.

16. L. M. Kirousis, E. Kranakis, D. Krizanc and A. Pelc, Power Consumption in Packet Radio Networks, Theoretical Computer Science, (243) 2000, 289-305.

17. G. S. Lauer, Packet Radio Routing, (chap. 11), Printice-Hall, Englewood Cliffs, NJ, 1995.

18. S. A. Plotkin, D. B. Shmoys and E. Tardos, Fast Approximation algorithms for fractional packing and covering problems, Mathematics of Operations Research, 2 (1995), 257-301.

19. G. Rossi, The Range Assignment Problem in Static Ad-Hoc Wireless Networks. Ph.D. Thesis, 2003.

20. J. Villavicencio and M. D. Grigoriadis, Approximate Lagrangian decomposition with a modified Karmarkar logarithmic potential, Network Optimization, P. Pardalos, D. W. Hearn and W. W. Hager, Eds, Lecture Notes in Economics and Mathematical Systems 450, Springer-Verlag, Berlin, (1997), 471-485.

21. B. Y. Wu, G. Lancia, V. Bafna, K. Chao, R. Ravi and C. Y. Tang, A polynomial time approximation scheme for minimum routine cost spanning trees, Proceedings of the 9th Annual ACM-SIAM Symposium on Discrete Algorithms, SODA 1998.

22. N. E. Young, Randomized rounding without solving the linear program, Proceedings of the 6th ACM-SIAM Symposium on Discrete Algorithms, SODA 1995, 170–178.

23. H. Zhang, Packing: Scheduling, Embedding and Approximating Metrics, Proceedings of the the 2004 International Conference on Computational Science and its Applications, ICCSA 2004, LNCS.

Author Index